RESONANCE AND BIFURCATION TO CHAOS IN PENDULUM

RESONANCE AND BIFURCATION TO CHAOS IN PENDULUM

Albert C. J. Luo

Higher
Education
Press

World Scientific

Published by

Higher Education Press Limited Company
4 Dewai Dajie, Beijing 100120, P. R. China
and
World Scientific Publishing Co Pte Ltd
5 Toh Tuck Link, Singapore 596224

British Library Cataloguing-in-Publication Data
A catalogue record for this book is available from the British Library.

ISBN 978-981-3231-67-2

水木同而背井，地悠悠而思乡，长河漫漫而渴，博古通今何用。——《游子吟》，朝俊

Preface

This book discusses Hamiltonian chaos and periodic motions to chaos in pendulums. A periodically forced mathematical pendulum is one of the typical and popular nonlinear oscillators that possess complicated and rich dynamical behaviors. It seems that the periodically forced pendulum is one of the simplest nonlinear oscillators. However, until now, a systematical study of periodic motions to chaos cannot be completed. To know periodic motions and chaos in the periodically forced pendulum, the perturbation method has been adopted. Because the sinusoidal function cannot be easy to be handled in the current mathematical tools, one used the Taylor series to expend the sinusoidal function to the polynomial nonlinear terms, and then the traditional perturbation methods were used to obtain the periodic motions of the approximated differential system. One always emphasized that the periodic solutions in the original pendulum are suitable for the small variation of equilibrium. In fact, once vector fields are modified in nonlinear dynamical systems, the new dynamical systems cannot represent the original systems. For example, one used the softening Duffing oscillator to approximate a pendulum oscillator. Such an investigation should not be acceptable.

For a better understanding of complex motions in nonlinear dynamical systems, one considered the periodically forced pendulum. Herein, a few examples are listed to explain how to use the pendulum oscillator for a better understanding of chaos. Since 1960, the nonlinear dynamics of a particle in a traveling electric field was investigated by a nonlinear pendulum. In 1972, Zaslavsky and Chirikov discussed the stochastic (chaotic) instability of nonlinear oscillation through a periodically forced pendulum, and the resonance overlap was discussed. In 1982, Ben-Jacob et al. used the pendulum to investigate intermittent chaos in Josephson junctions. In 1985, Kadanoff investigated the route from a periodic motion to unbounded chaos by investigation of the simple pendulum, and the Chirikov-Taylor model (or the standard mapping model) was obtained. The scaling analysis for the onset of chaos was completed. Gwinn and Westervelt discussed the intermittent chaos and low frequency noise in the driven damped pendulum through the fourth-order

Runge-Kutta method, and the attraction basin was presented. In 2006, Paula et al. established an experimental nonlinear pendulum to investigate chaotic motions.

It is not easy to find periodic motions to chaos in a pendulum system even though the periodically forced pendulum is one of the simplest nonlinear systems. However, the inherent complex dynamics of the periodically forced pendulum is much beyond our imaginations through the traditional thought of the linear dynamical systems. Until now, we have not known complex motions of pendulum yet. What are the mechanism and mathematics of such complex motions in the pendulum? The results presented in this book will give an alternative view of complex motions in the pendulum. In this book, the resonance-based Hamiltonian chaos and bifurcation trees to chaos in pendulums will be analytically and/or semi-analytically predicted for a better understanding of resonant chaos and bifurcation trees to chaos in nonlinear dynamical systems. Hamiltonian chaos in pendulum will be discussed through periodically and parametrically forced pendulums. The bifurcation trees of travelable and non-travelable periodic motions to chaos will be presented through the periodically forced pendulum.

This book includes six chapters. In Chapter 1, the mechanism and criteria of Hamiltonian chaos in stochastic and resonant layers are discussed. Chapter 2 presents Hamiltonian chaos in stochastic and resonant layers of a periodically excited pendulum. In Chapter 3, parametric chaos in the stochastic and resonant layers of the parametrically excited pendulum are presented. The resonant conditions cannot satisfy the traditional Mathieu equation analysis. In Chapter 4, stability and bifurcation theory of nonlinear discrete system are reviewed. In Chapter 5, the methodology for solutions of periodic motions in continuous dynamical systems is presented through the mapping dynamics of discrete implicit mappings under specific truncated errors. The discrete Fourier series of periodic motions are discussed from discrete nodes of periodic motions, and the corresponding, approximate analytical expression can be obtained. Harmonic amplitude quantity levels can be analyzed for periodic motions in continuous nonlinear systems. Chapter 6 discusses the bifurcation trees of periodic motions to chaos in the periodically forced pendulum. Through the aforementioned materials, one will better understand resonant chaos and complicated bifurcation trees of periodic motions to chaos in pendulums, and such materials may help one to understand resonance and bifurcations to chaos in other nonlinear dynamical systems. I hope this book can throw out a different point of view to look into Hamiltonian chaos and periodic motions to chaos in nonlinear dynamics community.

Finally, I would like to appreciate my former student, Dr. Yu Guo, for completing numerical computations in the last chapter. Herein, I thank my wife (Sherry X. Huang) and my children (Yanyi Luo, Robin Ruo-Bing Luo, and Robert Zong-Yuan Luo) for their understanding and support. This book is also dedicated to my parents for their many years expectation to their son.

This gift is too light to them but everything leaving in the book may last very long.

Albert C.J. Luo
Edwardsville, Illinois

Contents

Chapter 1
Resonance and Hamiltonian Chaos

In this Chapter, nonlinear Hamiltonian chaos including stochastic and resonant layers in 2-dimensional nonlinear Hamiltonian systems will be presented. Chaos and resonance mechanism in the stochastic layer of generic separatrix will be discussed that is formed by the primary resonance interaction in nonlinear Hamiltonians systems. However, chaos in the resonant layer of the resonant separatrix will be presented that is formed by the sub-resonance interaction.

1.1 Stochastic layers

In this section, stochastic layers in 2-dimensional nonlinear Hamiltonian systems will be described, and the approximate criterions for onset and destruction of the stochastic layers will be presented.

1.1.1 Definitions

Consider a 2-dimensional Hamiltonian system with a time periodically perturbed vector field, i.e.,

$$\dot{\mathbf{x}} = \mathbf{f}(\mathbf{x}, \boldsymbol{\mu}) + \mu \mathbf{g}(\mathbf{x}, t, \boldsymbol{\pi}); \quad \mathbf{x} = (x, y)^{\mathrm{T}} \in \mathscr{R}^2, \tag{1.1}$$

where $\mathbf{f}(\mathbf{x}, \boldsymbol{\mu})$ is an unperturbed Hamiltonian vector field on $\mathscr{R}^2$ and $\mathbf{g}(\mathbf{x}, t, \boldsymbol{\pi})$ is a periodically perturbed vector field with period $T = 2\pi/\Omega$, and

$$\mathbf{f}(\mathbf{x}, \boldsymbol{\mu}) = (f_1(\mathbf{x}, \boldsymbol{\mu}), f_2(\mathbf{x}, \boldsymbol{\mu})^{\mathrm{T}} \text{ and } \mathbf{g}(\mathbf{x}, t, \boldsymbol{\pi}) = (g_1(\mathbf{x}, t, \boldsymbol{\pi}), g_2(\mathbf{x}, t, \boldsymbol{\pi})^{\mathrm{T}} \tag{1.2}$$

are sufficiently smooth $(C^r, r \geqslant 2)$ and bounded on a bounded set $D \subset \mathscr{R}^2$ in phase space. $f_1 = \partial H_0(x, y, \boldsymbol{\mu})/\partial y$, $f_2 = -\partial H_0(x, y, \boldsymbol{\mu})/\partial x$; $g_1 =$

$\partial H_1(x, y, \Omega t, \boldsymbol{\pi})/\partial y$, $g_2 = -\partial H_1(x, y, \Omega t, \boldsymbol{\pi})/\partial x$. If the perturbation (or forcing term) $\mathbf{g}(\mathbf{x}, t, \boldsymbol{\pi})$ vanishes, equation (1.1) is a complete nonlinear Hamiltonian system $\dot{\mathbf{x}}^{(0)} = \mathbf{f}(\mathbf{x}^{(0)}, \boldsymbol{\mu})$. Thus, the total Hamiltonian of Eq. (1.1) can be expressed by

$$H(x, y, t, \mathbf{p}) = H_0(x, y, \boldsymbol{\mu}) + \mu H_1(x, y, \Omega t, \boldsymbol{\pi}), \qquad (1.3)$$

with excitation frequency Ω and strength μ of the perturbed Hamiltonian $H_1(x, y, t, \boldsymbol{\pi})$ as well. To compare with the other approximate analysis, such a perturbation parameter is introduced herein. The Hamiltonian of the integrable system in Eq.(1.1) is $H_0(x, y, \boldsymbol{\mu})$. Once the initial condition is given, the Hamiltonian $H_0(x, y, \boldsymbol{\mu})$ is invariant (i.e., $H_0(x, y, \boldsymbol{\mu}) = E$), which is the first integral manifold.

To restrict this investigation to the 2-dimensional stochastic layer, four assumptions for Eq.(1.1) are introduced as follows:

(H1.1) The unperturbed system of Eq.(1.1) possesses a bounded, closed separatrix $q_0(t)$ with at least one hyperbolic point $p_0 : (x_h, y_h)$.

(H1.2) The neighborhood of $q_0(t)$ for the point $p_0 : (x_h, y_h)$ is filled with at least three families of periodic orbits $q_\sigma(t)(\sigma = \alpha, \beta, \gamma)$ with $\alpha, \beta, \gamma \in (0, 1]$.

(H1.3) For the Hamiltonian energy E_σ of $q_\sigma(t)$, its period T_σ is a differentiable function of E_σ.

(H1.4) The perturbed system of Eq.(1.1) possesses a *perturbed* orbit $q(t)$ in the neighborhood of the *unperturbed* separatrix $q_0(t)$.

From the aforementioned hypothesis, the phase portrait of the unperturbed Hamiltonian system in the vicinity of the separatrix is sketched in Fig.1.1. The following point sets and the corresponding Hamiltonian energy are introduced, i.e.,

$$\Gamma_0 \equiv \{(x, y)|(x, y) \in q_0(t), t \in \mathscr{R}\} \cup \{p_0\} \text{ and } E_0 = H_0\left(q_0(t)\right) \qquad (1.4)$$

for the separatrix,

$$\Gamma_\sigma \equiv \{(x, y)|(x, y) \in q_\sigma(t), t \in \mathscr{R}\} \text{ and } E_\sigma = H_0\left(q_\sigma(t)\right) \qquad (1.5)$$

for the unperturbed, σ-periodic orbit and

$$\Gamma = \{(x, y)|(x, y) \in q(t), t \in \mathscr{R}\} \text{ and } E = H_0\left(q(t)\right) \qquad (1.6)$$

for the perturbed orbit $q(t)$.

The Hamiltonian energies in Eqs.(1.4) and (1.5) are constant for any periodic orbit of the unperturbed system but the Hamiltonian energy in Eq.(1.6) varies with $(x, y) \in q(t)$ of the perturbed system. The unperturbed Hamiltonian $H_0(q_\sigma(t))(\sigma = \alpha, \beta, \gamma)$ and $H_0(q_0(t))$ are $C^r(r \geqslant 2)$ smooth (also see, Luo and Han, 2001). The hypotheses (H1.2)–(H1.3) imply that $T_\sigma \to \infty$ monotonically as $\sigma \to 0$ (i.e., the periodic orbit $q_\sigma(t)$ approaches to $q_0(t)$ as $\sigma \to 0$).

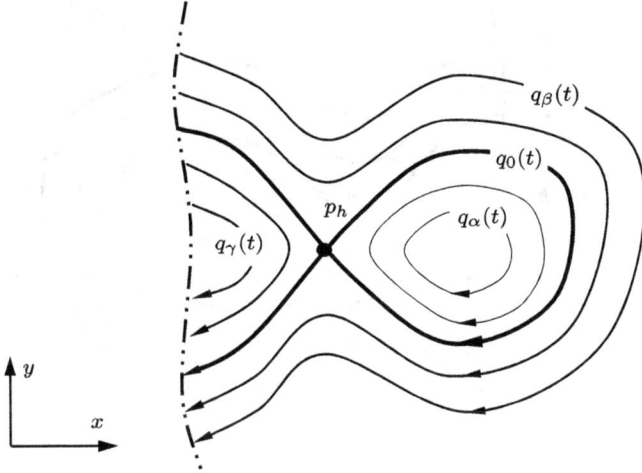

Fig. 1.1 The phase portrait of the unperturbed system of Eq.(1.1) near a hyperbolic point p_h and $q_0(t)$ is a separatrix going through the hyperbolic point and splitting the phase into three parts near the hyperbolic point, and the corresponding orbits $q_\sigma(t)$ are termed the σ-orbit ($\sigma = \{\alpha, \beta, \gamma\}$).

The δ-sets of the first integral quantity (or the energy) of the unperturbed Eq. (1.1) in $\Gamma_\sigma(\sigma = \alpha, \beta, \gamma)$, are defined as

$$N_\sigma^\delta(E_0) = \{E_\sigma \big| |E_\sigma - E_0| < \delta_\sigma, \text{ for small } \delta_\sigma > 0\} \tag{1.7}$$

and the union of the three δ-sets with E_0 is

$$N^\delta(E_0) = \bigcup_\sigma N_\sigma^\delta(E_0) \cup \{E_0\}. \tag{1.8}$$

For some time t, there is a point $\mathbf{x}_\sigma = (x_\sigma(t), y_\sigma(t))^{\mathrm{T}}$ on the orbit $q_\sigma(t)$ and this point is also on the normal $\mathbf{f}^\perp(\mathbf{x}_0) = (-f_2(\mathbf{x}_0), f_1(\mathbf{x}_0))^{\mathrm{T}}$ of the tangential vector of the separatrix $q_0(t)$ at a point $\mathbf{x}_0 = (x_0(t), y_0(t))^{\mathrm{T}}$, as shown in Fig.1.2. Therefore, the distance is defined as

$$\|q_\sigma(t) - q_0(t)\| = \max_{t \in \mathscr{R}} \|\mathbf{x}_\sigma(t) - \mathbf{x}_0(t)\|$$

$$= \max_{t \in \mathscr{R}} \sqrt{[x_\sigma(t) - x_0(t)]^2 + [y_\sigma(t) - y_0(t)]^2}. \tag{1.9}$$

Lemma 1.1 *For* a perturbed Hamiltonian systems in Eq.(1.1) *with* (H5.1)— (H5.4), *for any positive* $\varepsilon > 0$, *there is a positive* $\delta_\sigma > 0 (\sigma = \alpha, \beta, \gamma)$ *so that* $\|q_\sigma(t) - q_0(t)\| < \varepsilon$ *for* $E_\sigma \in N^\delta(E_0)$ *at a specific time* t.

Proof. For any positive $\varepsilon > 0$ let $\delta_\sigma = \varepsilon \|H_0\| > 0$ satisfying

$$|E_\sigma - E_0| = |H_0(q_\sigma) - H_0(q_0)| \leqslant \|H_0\| \cdot \|q_\sigma - q_0\| < \delta_\sigma,$$

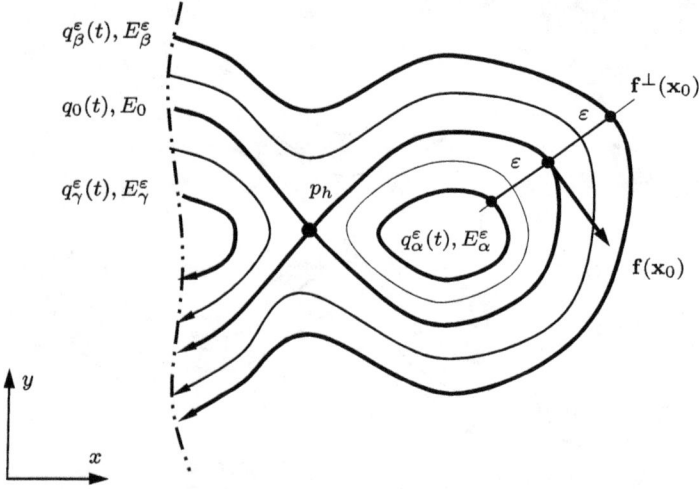

Fig. 1.2 The ε-neighborhood of orbit $q_0(t)$. The bold solid curves represent the separatrix $q_0(t)$ and the ε-neighborhood boundaries $q_\sigma^\varepsilon(t)$ are determined by $\max\limits_{t \in [0,\infty)} \|q_\sigma^\varepsilon(t) - q_0(t)\| = \varepsilon(\sigma = \alpha, \beta, \gamma)$. The solid curves depict all orbits $q_\sigma(t)$ in the ε-neighborhood. The energies on the boundary orbits are given through $E_\sigma^\varepsilon = H_0(q_\sigma^\varepsilon(t))$.

where

$$\|H_0\| \equiv \sup_{\sigma \neq 0} \left[|H_0(q_\sigma) - H_0(q_0)| / \|q_\sigma - q_0\| \right].$$

Since the unperturbed Hamiltonians H_0 of orbits q_σ and q_0 are C^r-smooth ($r \geqslant 2$) and $0 < \|H_0\| < \infty$ for bounded and closed orbits. Therefore, one obtains

$$\|q_\sigma(t) - q_0(t)\| < \delta_\sigma / \|H_0\| = \varepsilon.$$

This lemma is proved.   ∎

The ε-neighborhood of orbit $q_0(t)$ is formed by the three ε-sets of Γ_0 for the σ-orbits ($\sigma = \alpha, \beta, \gamma$), as shown in Fig.1.2. The bold solid curves denote the separatrix $q_0(t)$ and the ε-neighborhood boundaries, $q_\sigma^\varepsilon(t)(\sigma = \alpha, \beta, \gamma)$, are determined through $\max\limits_{t \in [0,\infty)} \|q_\sigma^\varepsilon(t) - q_0(t)\| = \varepsilon$ as $E_\sigma^\varepsilon = H_0(q_\sigma^\varepsilon(t))$. The solid curves represent all the σ-orbits $q_\sigma(t)$ in the ε-neighborhood.

The three ε-sets of Γ_0 for the σ-orbits ($\sigma = \alpha, \beta, \gamma$) are defined by

$$\Gamma_\sigma^\varepsilon = \{(x,y) | (x,y) \in q_\sigma(t), \|q_\sigma(t) - q_0(t)\| < \varepsilon, t \in \mathscr{R}\}. \tag{1.10}$$

Furthermore, from Eq.(1.8), the unions of the ε-sets with Γ_0 are

$$\Gamma_0^\varepsilon = \cup_\sigma \Gamma_\sigma^\varepsilon \cup \Gamma_0, \quad \Gamma_{\sigma 0}^\varepsilon = \Gamma_\sigma^\varepsilon \cup \Gamma_0. \tag{1.11}$$

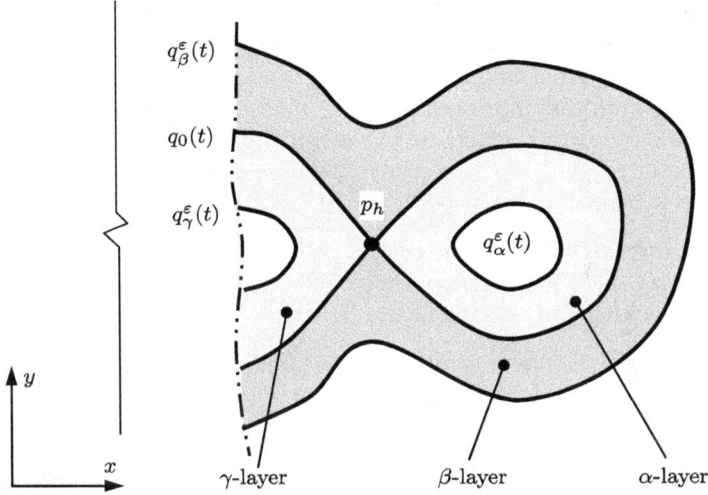

Fig. 1.3 A stochastic layer of Eq.(1.1) formed by the Poincaré mapping set of $q(t)$ in the ε-neighborhood of $q_0(t)$ for $t \in [0, \infty)$. The separatrix separates the stochastic layer into three sub-stochastic layers (i.e., α-layer and β-layer and γ-layer).

The Poincaré map $P : \Gamma^P \to \Gamma^P$, where the Poinaré mapping set in phase space is

$$\Gamma^P \equiv \{(x_N, y_N) | (x_N, y_N) \in q(t), t_N = 2\pi N/\Omega + t_0, N = 0, 1, \cdots\} \subset \Gamma,$$
(1.12)

where t_0 is the initial time. Using the above notations, a stochastic layer is defined through the Poincaré mapping set with nonzero measure as follows:

Definition 1.1 The Poincaré mapping set Γ^P is termed the stochastic layer in the ε-sense if the compact dense set Γ^P belongs to Γ_0^ε (or $\Gamma^P \subset \Gamma_0^\varepsilon$) for $t_N = 2N\pi/\Omega + t_0 (N = 0, 1, \cdots)$. Similarly, the Poincaré mapping subset $U_\sigma \subset \Gamma^P$ is the σ-stochastic layer if $U_\sigma \subset \Gamma_{\sigma0}^\varepsilon$ for $t_N = 2N\pi/\Omega + t_0$.

A stochastic layer of system in Eq.(1.1) is formed through the Poincaré mapping set of $q(t)$ in the ε-neighborhood for time $t \in [0, \infty)$, as shown in Fig.1.3. The separation of the stochastic layer by the separatrix gives three sub-stochastic layers shaded. The sub-layers relative to the σ-orbits ($\sigma = \alpha, \beta, \gamma$) are termed the σ-stochastic layer. The more detail description can be referred to Luo (2008).

1.1.2 Approximate criteria

The predictions of resonance in the stochastic layer of a 2-dimensional nonlinear Hamiltonian system will be presented. The incremental energy technique will be presented first from the approximate first integral quantity increment (or approximate energy increment). The whisker mapping will be obtained, and the corresponding criterion will be presented. The linearization of the whicker mapping, the improved standard mapping will be presented and the approximate prediction of chaos onset will be given. From the exact first integral quantity, the energy spectrum technique will be developed for a numerical prediction.

(A) *An Incremental Energy Method.* As in Luo and Han (2001), the incremental energy method will be presented for a understanding of the resonant mechanism of chaos in the stochastic layer.

Lemma 1.2 *For the dynamical system in Eq.(1.1) with (H1.1)—(H1.4), if a point $(x, y) \in \Gamma \cap \Gamma_\sigma$ for some $\sigma = \{\alpha, \beta, \gamma\}$, then $H_0(q(t)) = H_0(q_\sigma(t))$ for some time t.*

Proof. If the perturbed orbit $q(t)$ in the set Γ is intersected with an unperturbed orbit $q_\sigma(t)$ in the set Γ_σ for some $\sigma \in \{\alpha \in [-1,0), \beta \in (0,1], \gamma \in [-1,0)\}$ at time t, there is a single point $(x, y) \in \Gamma \cap \Gamma_\sigma$. Therefore, for $(x, y) \in \Gamma \cap \Gamma_\sigma$, we have $(x, y) \in q_\sigma(t)$ and $(x, y) \in q(t)$. Thus, $H_0(q(t)) = H_0(x, y) = H_0(q_\sigma(t))$, which implies that the conservative energy is equal for the same point in phase space. This lemma is proved.   ∎

The detailed discussion is given as follows. Because the conservative energy H_0 is the first integral quantity, for the σ-layer, the map describing the changes of both energy H_0 and phase φ for time transition from t_i to $t_i + T_\sigma$ in Eq.(1.1) is obtained, i.e.,

$$E_{i+1} = E_i + \Delta H^\sigma(\varphi_i) \text{ and } \varphi_{i+1} = \varphi_i + \Delta\varphi^\sigma(E_{i+1}), \qquad (1.13)$$

where $E_i = H(q(t_i))$ and $\varphi_i = \varphi(q(t_i))$. For a specific external frequency Ω, the initial phase is defined by $\varphi_i = \Omega t_i$. Notice that the energy relationship in the foregoing can be expressed through the action variable. As in Chirikov (1979) and Lichtenberg and Lieberman (1992), the phase and energy changes, $\Delta\varphi^\sigma(E_{i+1})$ and $\Delta H^\sigma(\varphi_i)$, are approximately computed by

$$\Delta\varphi^\sigma(E_{i+1}) \approx \Omega T_\sigma(E_{i+1}), \text{ and}$$

$$\Delta H^\sigma(\varphi_i) \approx \mu \int_{t_i}^{T_\alpha(E_i)+t_i} [H_0, H_1] dt$$

$$= \mu \int_{t_i}^{T_\sigma(E_i)+t_i} (f_1 g_2 - f_2 g_1) dt, \qquad (1.14)$$

where $[\cdot, \cdot]$ represents the Poisson bracket. The energy and phase changes in Eq.(1.14) for the system in Eq.(1.1) over one period T_σ of the σ-orbit are sketched in Fig.1.4. If $E_i = E_0$ expresses the energy of the separatrix, equation (1.13) becomes a generalized separatrix map (or a generalized whisker map). When the σ-orbit ($\sigma = \alpha, \beta, \gamma$) is close to the separatrix (i.e., $T_\sigma \to \infty$), the energy increment reduces to the one along the separatrix in Luo and Han (2001):

$$\Delta H^h(\varphi_i) \approx \lim_{T_\sigma \to \infty} \mu \int_{t_i}^{T_\sigma(E_i)+t_i} [H_0, H_1]dt$$

$$= \lim_{T_\sigma \to \infty} \mu \int_{t_i}^{T_\sigma(E_i)+t_i} (f_1 g_2 - f_2 g_1)\, dt \qquad (1.15)$$

which can also be obtained through the Melnikov function with a small parameter μ, i.e., $\Delta H^h(\varphi_i) = \mu M(t_i)$ (e.g., Rom-Kedar, 1990, 1994, 1995; Zaslavsky and Abdullaev, 1995; Abdullaev and Zaslavsky, 1995, 1996; Ahn et al., 1996; Iomin and Fishman, 1996).

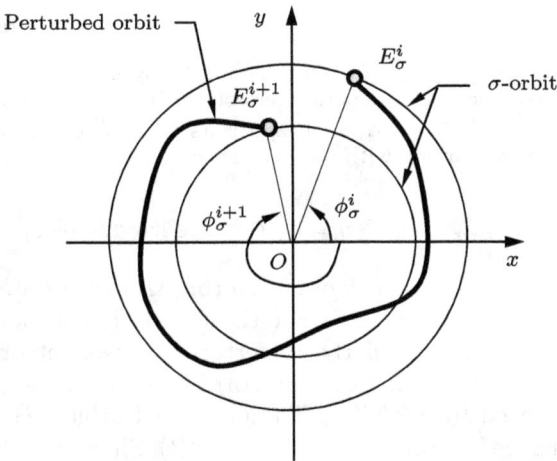

Fig. 1.4 The energy and phase changes of a perturbed orbit over one period T_α based on E_σ^i.

After the KAM torus in the vicinity of separatrix is destroyed, the stochastic layer (or the instability zone) will form in such vicinity, and the resonant-separatrix webs appear as in Luo et al. (1999). Such resonance-separatrix webs are generated by the interaction of resonances between the unperturbed orbits and the periodic forcing in the stochastic layer. In this section, the prediction of the onset of the primary resonance in the stochastic layer will be of great interest. Such resonant interactions in the ε-neighborhood of the separatrix are qualitatively illustrated in Fig.1.5. The hollow circles depict intersection points of the perturbed and unperturbed orbits. When the perturbed orbit arrived to the $(m_\sigma : n_\sigma)$ primary resonant unperturbed orbit of Eq.(1.1), the resonance between the unperturbed Hamiltonian and perturbation appears, and the resonant condition for a periodically forced system with one degree of freedom is obtained from Eq.(1.3), i.e.,

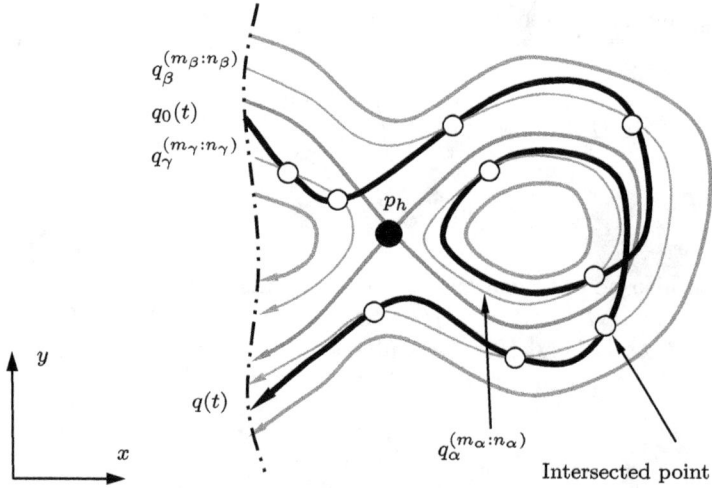

Fig. 1.5 A perturbed orbit $q(t)$ for Eq.(1.1) and the resonant interactions with the unperturbed orbits. The circles denote intersections between the perturbed and unperturbed orbits. $q_\sigma^{(m_\sigma:n_\sigma)}$ ($\sigma = \alpha, \beta, \gamma$) represents an unperturbed orbit having the $(m_\sigma : n_\sigma)$-resonant interaction with the perturbed orbit.

$$m_\sigma \omega_\sigma = n_\sigma \Omega, m_\sigma, n_\sigma \in \mathbb{N} \text{ are irreducible}, \tag{1.16}$$

where $\omega_\sigma = 2\pi/T_\sigma$ is a frequency of the σ-orbit, Ω is the excitation frequency and $\mathbb{N}$ is the natural number set. For the $(i + 1)$th iteration of Eq.(1.13), if $q(t_{i+1}) = (x_{i+1}, y_{i+1})$ is on the unperturbed, resonant orbit, the phase change in the second equation of Eq.(1.13) becomes $\varphi_{i+1} - \varphi_i = 2\pi m_\sigma/n_\sigma$ and E_{i+1} is replaced by $E^{(m_\sigma:n_\sigma)}$. The perturbed orbit $q(t)$ relative to the $(m_\sigma : n_\sigma)$-resonance is represented by $q^{(m_\sigma:n_\sigma)}(t)$. Since $\omega_\sigma = 2\pi/T_\sigma$ depends on the corresponding energy E_σ, the resonant conditions can be directly expressed through E_σ and Ω, as shown in Fig.1.6. In Fig.1.6(a), $E_\alpha^{(m_\alpha:n_\alpha)}$ is an unperturbed Hamiltonian energy relative to the $(m_\alpha : n_\alpha)$-resonant orbit and $E_\beta^{(m_\beta:n_\beta)}$ is the Hamiltonian energy pertaining to the $(m_\beta : n_\beta)$-resonant orbit. The resonant number sets in the stochastic layer for $\sigma = \alpha, \beta, \gamma$ are introduced by

$$R_\sigma^\varepsilon = \left\{ (m_\sigma : n_\sigma) \left| \begin{array}{l} m_\sigma \omega_\sigma = n_\sigma \Omega, m_\sigma, n_\sigma \in \mathbb{N} \text{ are irreducible,} \\ \text{and } \|q_\sigma(t) - q_0(t)\| < \varepsilon \end{array} \right. \right\}. \tag{1.17}$$

For a given external frequency Ω with constant $(m_\sigma : n_\sigma)$, frequency ω_σ is relative to E_σ. In Fig.1.6(b), the zoomed view of the resonant relations in the neighborhood of the separatrix with energy $E = E_0$ at a specific excitation frequency Ω is illustrated through the α- and β-layers. Once the resonance energy is close to the energy of the separatrix, the density of the resonance separatrix increases in the stochastic layer because of the rapid change of elliptic modulus function. Namely, as $E_\sigma \to E_0$, we have $\omega_\sigma \to 0$. Therefore,

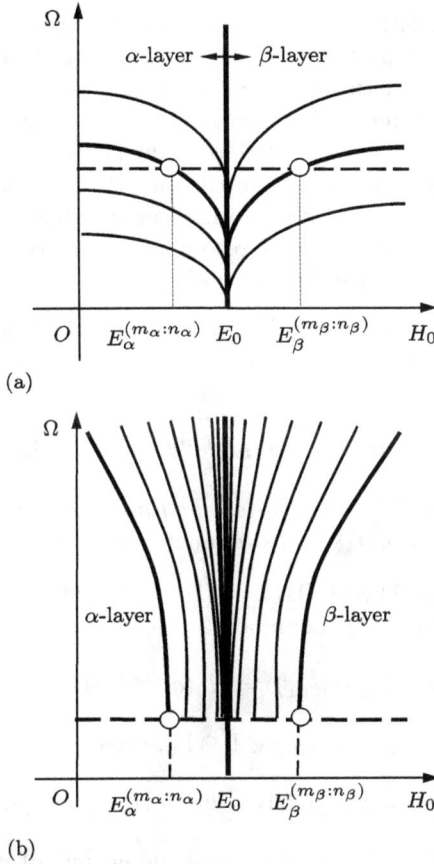

Fig. 1.6 (a) Resonant conditions of the perturbed system in the stochastic layer and (b) a zoomed view of resonant conditions in neighborhood of separatrix.

from Eq.(1.17), one obtains $m_\sigma \to \infty$. A resonance with large m_σ will be included in the stochastic layer.

For each m_σ, the $(m_\sigma : n_\sigma)$-resonant separatrix possesses m_σ-center points and m_σ-hyperbolic points under the n_σ-external periods. In the stochastic layer, for $E_\sigma \to E_0$, the number for center points and hyperbolic points will approach the infinity. When the energy E_σ arbitrarily approaches E_0, there are infinite centers and infinite hyperbolic points on the resonant separatrix. Such an issue in resonant layers will be discussed in detail. As popularly accepted (e.g., Lu, 2007), the chaos is formed by infinite stable periodic solutions and infinite unstable periodic solutions. The total resonant separatrix number is the summation of the resonant orbit relative to the $(m_\sigma : n_\sigma)$-condition with $m_\sigma \in [\min\{m_\sigma\}, \infty) \subset \mathbb{N}$. This implies that the infinite resonances exist in the σ-layer ($\sigma = \alpha, \beta, \gamma$). The stochastic layer complexity will be formed by the resonant web, which can be referred to Luo et al. (1999).

When the perturbation becomes stronger, the perturbed orbit will cross over more and more unperturbed, periodic orbits, as shown in Fig.1.5. Therefore, the resonant interaction between the Hamiltonian system and its perturbation increases with increasing perturbation strength. The resonant interaction leads to a new resonant overlap, and such resonant overlap generates a new stochastic layer different from the old stochastic layer because a new, specific resonant-separatrix web is formed in the stochastic layer. Thus, the prediction of the onset of a specific resonant-separatrix web in the stochastic layer is given by the following theorems.

Theorem 1.1 *For a perturbed Hamiltonian Eq.(1.1) with (H1.1)—(H1.4), for an arbitrarily small $\varepsilon_\sigma > 0$, there is a positive $\delta_\sigma > 0$, so that if for $(m_\sigma : n_\sigma) \in R_\sigma^\varepsilon (\sigma = \alpha, \beta, \gamma)$,*

$$|\Delta H^\sigma(\varphi_i)| = |E_\sigma^{(m_\sigma:n_\sigma)} - E_0| \leqslant \delta_\sigma \qquad (1.18)$$

then $\|q^{(m_\sigma:n_\sigma)}(t_i) - q_0(t_i)\| < \varepsilon_\sigma$ for all the time t_i, i.e., the Poincaré mapping set of $q^{(m_\sigma:n_\sigma)}(t)$ is in a σ-stochastic layer as $t \in [0, \infty)$.

Proof. Consider Eq.(1.1) with (H1.1)—(H1.4). If $q^{(m_\sigma:n_\sigma)}(t_{i+1}) = (x_{i+1}, y_{i+1}) \in \Gamma \cap \Gamma_{\sigma 0}^\varepsilon$, for $\sigma \in \{\alpha, \beta, \gamma\}$, Lemma 1.2 gives

$$E_{i+1} = H_0(q^{(m_\sigma:n_\sigma)}(t_{i+1})) = H_0(q_\sigma^{(m_\sigma:n_\sigma)}(t_{i+1})) = E_\sigma^{(m_\sigma:n_\sigma)}.$$

Similarly, if $q^{(m_\sigma:n_\sigma)}(t_i) = (x_i, y_i) \in \Gamma \cap \Gamma_0$, then

$$E_i = H_0(q_0(t_i)) = E_0 = H_0(q_0(t_{i+1})),$$

where $q_0(t_{i+1}) = (x_{i+1}^0, y_{i+1}^0) \in q_0(t)$ and the normal of its tangential vector intersects with $q_\sigma^{(m_\sigma:n_\sigma)}(t)$ at (x_{i+1}, y_{i+1}). From Eq. (1.13),

$$|\Delta H^\sigma(\varphi_0)| = |E_{i+1} - E_i| = |E_\sigma^{(m_\sigma:n_\sigma)} - E_0|.$$

For an arbitrary $\varepsilon > 0$, let $\delta_\sigma = \varepsilon \|H_0\| > 0$ satisfying

$$|E_\sigma^{(m_\sigma:n_\sigma)} - E_0| \leqslant \|H_0\| \cdot \|q_\sigma^{(m_\sigma:n_\sigma)}(t_{i+1}) - q_0(t_{i+1})\| \leqslant \varepsilon \|H_0\| = \delta_\sigma.$$

Since the unperturbed Hamiltonian H_0 of orbits q_σ and q_0 is $C^r(r \geqslant 2)$ smooth,

$$0 < \|H_0\| \equiv \sup_{\sigma \neq 0} \left[|H_0(q(t)) - H_0(q_0(t))| / \|q(t) - q_0(t)\|\right] < \infty,$$

for bounded and closed orbits. Therefore, one obtains

$$\|q^{(m_\sigma:n_\sigma)}(t_{i+1}) - q_0(t_{i+1})\| < \delta_\sigma / \|H_0\| = \varepsilon$$

for all t_{i+1}, that is,

$$\|q^{(m_\sigma : n_\sigma)}(t_{i+1}) - q_0(t_{i+1})\| < \varepsilon.$$

This theorem is proved.   ∎

In Eq.(1.18), the incremental energy $\Delta H^\sigma(\varphi_0)$ is a function of the amplitude and frequency of perturbation. The conservative energy $E_\sigma^{(m_\sigma : n_\sigma)}$ relative to the unperturbed resonant orbit is determined by the resonant condition given in Eq.(1.16). With $\max\limits_{t \in [0,\infty)} |H(q^{(m_\sigma^{(1)} : n_\sigma^{(1)})}(t)) - E_0| = \delta_\sigma^{(1)}$, the foregoing theorem gives the following corollary.

Corollary 1.1 *For a perturbed Hamiltonian in Eq.(1.1) with (H1.1)—(H1.4), for an arbitrarily small $\varepsilon_\sigma > 0$, there is a positive $\delta_\sigma^{(1)} > 0$, so that if for $(m_\sigma^{(1)} : n_\sigma^{(1)}) \in R_\sigma^\varepsilon (\sigma = \alpha, \beta, \gamma)$,*

$$\max_{\sigma \in \{\alpha,\beta,\gamma\}} |\Delta H^\sigma(\varphi_i)| = \max_{R_\sigma^\varepsilon} |E_\sigma^{(m_\sigma : n_\sigma)} - E_0| = |E_\sigma^{(m_\sigma^{(1)} : n_\sigma^{(1)})} - E_0| \leqslant \delta_\sigma^{(1)}, \quad (1.19)$$

then $\|q^{(m_\sigma^1 : n_\sigma^1)}(t) - q_0(t)\| < \varepsilon$, i.e., the Poincaré mapping sets of $q^{(m_\sigma^{(1)} : n_\sigma^{(1)})}(t) \in \Gamma$ is the last one absorbed in the ε-stochastic layer.

Proof. Consider Eq.(1.1) with (H1.1)—(H1.4). For a σ-resonant orbit $q^{(m_\sigma : n_\sigma)}(t) \in \Gamma^\varepsilon$, for arbitrary $\varepsilon > 0$, choosing $\delta_\sigma > 0$, find $(m_\sigma^{(1)} : n_\sigma^{(1)}) \in R_\sigma^\varepsilon$ satisfying

$$\max_{R_\sigma^\varepsilon} |E_\sigma^{(m_\sigma : n_\sigma)} - E_0| \leqslant |E_\sigma^{(m_\sigma^{(1)} : n_\sigma^{(1)})} - E_0| = \delta_\sigma.$$

Let $\delta = \max\limits_{\sigma \in \{\alpha,\beta,\gamma\}} \{\delta_\sigma\} > 0$, we have

$$\max_{\sigma \in \{\alpha,\beta,\gamma\}} \max_{R_\sigma^\varepsilon} |E_\sigma^{(m_\sigma : n_\sigma)} - E_0| \leqslant \max_{\sigma \in \{\alpha,\beta,\gamma\}} |E_\sigma^{(m_\sigma^{(1)} : n_\sigma^{(1)})} - E_0| = \delta.$$

Therefore, from Theorem 1.1,

$$\|q^{(m_\sigma^{(1)} : n_\sigma^{(1)})}(t) - q_0(t)\| < \varepsilon.$$

This corollary is proved.   ∎

After a new stochastic layer is formed, the stochastic layer becomes thicker and thicker with increasing the perturbation strength until its destruction occurs. The increase of perturbation strength also leads to the energy increment not satisfying Eq.(1.18), and then, the enlarged stochastic motion domain is termed *the global stochastic layer* in the ε-sense. For an approximate prediction of resonant-separatrix webs in the global stochastic layer, we have the following theorem:

Theorem 1.2 *For a perturbed Hamiltonian in Eq.(1.1) with (H1.1)—(H1.4), for an arbitrarily small $\varepsilon_\sigma > 0$, there exists $\delta_\sigma > 0$ so that if for $(m_\sigma : n_\sigma) \notin R_\sigma^\varepsilon (\sigma = \alpha, \beta, \gamma)$,*

$$|\Delta H^\sigma(\varphi_0)| = |E_\alpha^{(m_\sigma:n_\sigma)} - E_0| \geq \delta_\sigma, \tag{1.20}$$

then $\|q^{(m_\alpha:n_\alpha)}(t_i) - q_0(t_i)\| > \varepsilon_\sigma$.

Proof. The proof of contradiction is used herein. This theorem can be proved in a same manner as for Theorem 1.1.   ∎

The criterion in Eq.(1.18) can be expressed through the action and natural frequency of unperturbed Hamiltonian system. The onset of a specified resonance in the stochastic layer is predicted through the incremental energy approach. This approach is also applicable for strong excitations when the energy increments still maintain in good accuracy. For the more accurate predictions of the resonance interaction in the stochastic layer, a new computational method for the energy increment should be developed because the sub-resonance is not considered. If the exact energy increment is given, the above theorems give the exact prediction of the resonance in the stochastic layer.

(B) *Accurate Standard Mapping Technique.* In Luo (2001), the accurate standard mapping technique was developed to determine the resonant mechanism of the stochastic layer. For linearization of the second equation in Eq.(1.13) at the period-1 fixed point on the $(m_\sigma : n_\sigma)$-resonance, for $E_{i+1} = E_i = E_\sigma^{(m_\sigma:n_\sigma)}$ and $\varphi_{i+1} = \varphi_i + 2\pi m_\sigma/n_\sigma = \varphi^{m_\sigma:n_\sigma} + 2\pi m_\sigma/n_\sigma$, equation (1.13) gives

$$\Delta H_0^\sigma(\varphi_\sigma^{(m_\sigma:n_\sigma)}) = 0 \quad \text{and} \quad 2\pi m_\sigma/n_\sigma = \Delta\varphi^\sigma(E_\sigma^{(m_\sigma:n_\sigma)}). \tag{1.21}$$

From Eq.(1.21), $\varphi_\sigma^{(m_\sigma:n_\sigma)}$ and $E_\sigma^{(m_\sigma:n_\sigma)}$ can be obtained. A near energy is

$$E_i = E_\sigma^{(m_\sigma:n_\sigma)} + \Delta E_i \quad \text{and} \quad I_i = G_\sigma^{(m_\sigma:n_\sigma)}\Delta E_i, \tag{1.22}$$

where $G_\sigma^{(m_\sigma:n_\sigma)} = \partial(\Delta\varphi(E_{i+1}))/\partial E_{i+1}|_{E_{i+1}=E_\sigma^{(m_\sigma:n_\sigma)}}$. With Eq.(1.22), linearization of the second equation of Eq.(1.13) leads to

$$I_{i+1} = I_i + G_\sigma^{(m_\sigma:n_\sigma)}\Delta H_0^\sigma(\varphi_i) \quad \text{and} \quad \varphi_{i+1} \approx \varphi_i + I_{i+1}, \tag{1.23}$$

which gives a generalized standard map. In the above derivation, no approximations of the period are required. The criteria for the $(m_\sigma : n_\sigma)$-resonance in the stochastic layer can be obtained through the transition from the local stochasticity to global stochasticity in Eq.(1.23). Setting $G_\sigma^{(m_\sigma:n_\sigma)}\Delta H_0^\sigma(\varphi_i) = K\sin\varphi_i$, such an equation presents a standard map (or the Chirikov-Taylor map), i.e.,

$$I_{i+1} = I_i + K\sin\varphi_i \quad \text{and} \quad \varphi_{i+1} \approx \varphi_i + I_{i+1}. \tag{1.24}$$

In Greene (1968, 1979), a method was developed to numerically determine the strength of the stochasticity parameter when the transition to global stochasticity for Eq. (1.23) occurs. Therefore, for the special case, the perturbation strength of Eq.(1.1) is estimated from

$$|G_\sigma^{(m_\sigma:n_\sigma)}\Delta H^\sigma(\varphi_{i_i})| \approx 0.9716\cdots. \tag{1.25}$$

Other estimates for the strength of stochasticity parameter can be referred to the references (e.g., Chirikov,1979; Lichtenberg and Lieberman,1992). Luo (2001) developed an accurate standard map approach based on the accurate whisker map for such a prediction of the resonance in the stochastic layer. Luo et al. (1999) developed an energy spectrum approach (numerical method) for prediction of the onset of resonance in the stochastic layer. A comparison of analytical and numerical predictions is presented in Luo and Han (2000). From numerical results, the previous approaches presented in Luo and Han (2001) are not very accurate for strong excitations. When the excitation strength is very weak, the accurate and approximate standard-map methods are in good agreement, and the accurate one is applicable to nonlinear systems with strong excitations. However, the further improvement should be completed for a more accurate, analytical prediction of the onset of a new resonant-overlap in stochastic layers of nonlinear Hamiltonian systems with periodical excitations. The energy spectrum is based on the exact energy increments. Thus, the energy spectrum will be discussed in next section and the corresponding layer width can be estimated.

(C) *Energy Spectrum and Layer Width.* For the numerical prediction of resonances in the stochastic layer, Luo et al. (1999) developed an energy spectrum technique, and the resonant characteristics in stochastic layers are investigated through the energy spectra. This technique computes the maximum and minimum energies of the Poincaré mapping points as in Eq.(1.3). From Lemma 1.2, the perturbed energy can be measured by the unperturbed Hamiltonian. Thus, the unperturbed Hamiltonian for each Poincare mapping point of Eq.(1.1) is

$$H_0^{(N)} = H_0(\mathbf{x}_N, \boldsymbol{\mu}) \equiv H_0(x_N, y_N, \boldsymbol{\mu}), \tag{1.26}$$

and its minimum and maximum energies are determined by

$$E_{\max} = \max_{N\to\infty}\{H_0^{(N)}\} \quad \text{and} \quad E_{\min} = \min_{N\to\infty}\{H_0^{(N)}\}. \tag{1.27}$$

The minimum layer width defined in Luo et al. (1999) is

$$w \equiv \min_{t\in[0,\infty)}\|\mathbf{x}(E_{\max},t) - \mathbf{x}(E_{\min},t)\| \equiv \|\mathbf{x}^{\max} - \mathbf{x}^{\min}\|, \tag{1.28}$$

where $\|\cdot\|$ is a norm. Two points $\mathbf{x}^{\max}$ and $\mathbf{x}^{\min}$ on the normal vector $\mathbf{f}^\perp(\mathbf{x}_0) = (-f_2(\mathbf{x}_0), f_1(\mathbf{x}_0))^{\mathrm{T}}$ of the tangential vector of separatrix at point $\mathbf{x}_0$ are the closest between the maximum and minimum energy orbits $\mathbf{x}(E_{\max},t)$ and $\mathbf{x}(E_{\min},t)$ which can be obtained by Eq.(1.27) with $E_{\max}$ and $E_{\min}$, as shown in Fig.1.2.

1.2 Resonant separatrix layers

Consider a 2-dimensional Hamiltonian system as

$$\dot{\mathbf{x}} = \mathbf{f}(\mathbf{x}, \boldsymbol{\mu}) + \mathbf{g}(\mathbf{x}, t, \boldsymbol{\pi}); \quad \mathbf{x} = (x, y)^{\mathrm{T}} \in \mathscr{R}^2 \qquad (1.29)$$

where $\mathbf{f}(\mathbf{x}, \boldsymbol{\mu})$ is a Hamiltonian vector field defined on $\mathscr{R}^2$ and $\mathbf{g}(\mathbf{x}, t, \boldsymbol{\pi})$ is a $T = 2\pi/\Omega$-periodic (fixed period) Hamiltonian vector field in time t, and Ω denotes excitation frequency. Specifically, they are in the form of

$$\mathbf{f}(\mathbf{x}, \boldsymbol{\mu}) = (f_1(\mathbf{x}, \boldsymbol{\mu}), f_2(\mathbf{x}, \boldsymbol{\mu}))^{\mathrm{T}} \quad \text{and} \quad \mathbf{g}(\mathbf{x}, t, \boldsymbol{\pi}) = (g_1(\mathbf{x}, t, \boldsymbol{\pi}), g_2(\mathbf{x}, t, \boldsymbol{\pi}))^{\mathrm{T}} \qquad (1.30)$$

and are assumed to be sufficiently smooth $(C^r, r \geqslant 2)$ and bounded on bounded sets $D \subset \mathscr{R}^2$ in phase space. The total energy of such a system is

$$H(x, y, t, \mathbf{p}) = H_0(x, y, \boldsymbol{\mu}) + H_1(x, y, t, \boldsymbol{\pi}), \qquad (1.31)$$

where $H_0(x, y, \boldsymbol{\mu})$ and $H_1(x, y, t, \boldsymbol{\pi})$ are energy functions of the conservative and perturbed Hamiltonians, respectively.

To restrict the discussion on the 2-dimensional resonant layer (or band) in perturbed nonlinear Hamiltonian system, the following hypothesis will be used.

(A1.1a) There is a bounded open domain $D \subset \mathscr{R}^2$ and in such a domain, there is only one center equilibrium $p_c : (x_c, y_c)$ around which a family of periodic flows $q_\alpha(t) = (x_\alpha(t), y_\alpha(t))(\alpha \in [1, \infty))$ of the unperturbed Hamiltonian exists.

(A1.1b) There is an open domain $D \subset \mathscr{R}^2$ bounded by a separatrix (i.e., $q_0(t) = (x_0(t), y_0(t)) : \mathbf{x}_0 \in q_0(t))$ with hyperbolic points, and then in such a domain there is a center equilibrium $p_c : (x_c, y_c)$ around which a family of periodic flows (i.e., $\mathbf{x}_\alpha \in q_\alpha(t)$ for $\alpha \in (0, 1]$) of unperturbed Hamiltonian with $\lim\limits_{\alpha \to 0} \sup\limits_{t \in \mathscr{R}} \inf\limits_{\mathbf{x}_0 \in q_0(t), \mathbf{x} \in q_\alpha(t)} \|\mathbf{x}_\alpha(t) - \mathbf{x}_0(t)\| = 0$ exists.

(A1.1c) There is an open domain $D \subset \mathscr{R}^2$ bounded by an internal boundary formed by a separatrix (i.e., $\mathbf{x}_0 \in q_0(t)$) with hyperbolic points. On the outside of the separatrix, a family of periodic flows (i.e., $\mathbf{x}_\alpha \in q_\alpha(t)$ for $\alpha \in (0, \infty))$ of the unperturbed Hamiltonian with $\lim\limits_{\alpha \to 0} \sup\limits_{t \in \mathscr{R}} \inf\limits_{\mathbf{x}_0 \in q_0(t), \mathbf{x} \in q(t)} \|\mathbf{x}_\alpha - \mathbf{x}_0\| = 0$ exists.

(A1.2) $H_0(q(t)) = E_\alpha$ and T_α is the period of $q_\alpha(t)$ and its frequency ω_α is greater than zero (i.e. $\omega_\alpha > 0$). The frequency is a differentiable function of E_α (i.e., $d\omega_\alpha/dE_\alpha \neq 0$). Namely, $d\omega_\alpha/dE_\alpha > 0$, $d\omega_\alpha/dE_\alpha < 0$ and $d\omega_\alpha/dE_\alpha > 0$ are for cases in (A1.1a), (A1.1b) and (A1.1c), respectively.

Without loss of generality, consider the second type of domain is bounded by the separatrix, as shown in Fig.1.7(a). In this domain, there is a center point. The separatrix is sketched by a dashed curve. All the periodic flows in this domain will be formed around the center point in Fig.1.7(b). The

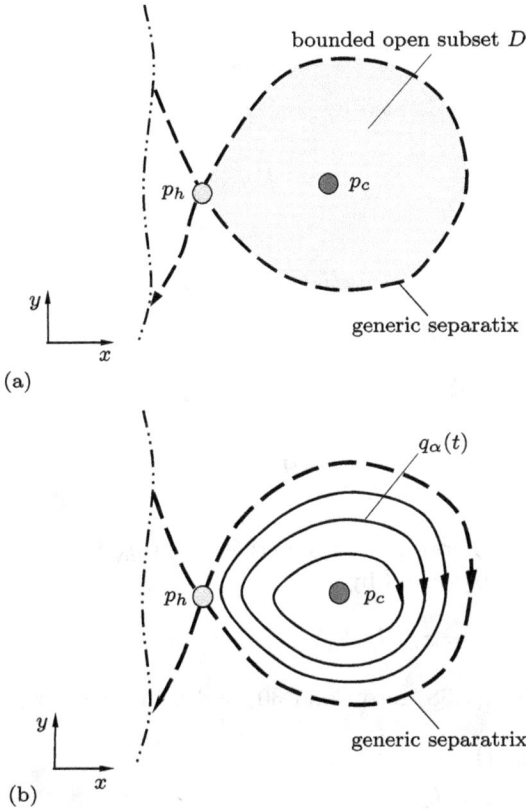

Fig. 1.7 (a) An open domain D bounded by the separatrix and (b) periodic flow of the unperturbed system of Eq.(1.29). $q_\alpha(t)$ is the periodic flow of the unperturbed system in domain $D \subset \mathscr{R}^2$.

natural frequency at the center point is maximum. With increasing energy, the frequency will decrease (i.e., $d\omega_\alpha/dE_\alpha < 0$) or the flow period will increase (i.e., $dT_\alpha/dE_\alpha > 0$). When a periodic flow in the family of periodic flows in such a domain approaches the separatrix, the natural frequency will approach zero (i.e., $\lim_{E_\alpha \to E_0} \omega_\alpha = 0$) or the corresponding period approaches infinity (i.e., $\lim_{E_\alpha \to E_0} T_\alpha = \infty$).

1.2.1 Layer dynamics

To investigate chaotic motions in a resonant separatrix layer, for a given energy E_α, the Hamiltonian is

$$H_0(x_\alpha, y_\alpha) = E_\alpha \tag{1.32}$$

from which

$$y_\alpha = y_\alpha(x_\alpha, E_\alpha). \tag{1.33}$$

The action variable is defined by

$$J_\alpha = \oint y_\alpha dx_\alpha. \tag{1.34}$$

So we have

$$H_0(J_\alpha) = E_\alpha. \tag{1.35}$$

The angle variable is defined by

$$\theta_\alpha = \omega_\alpha t + \theta_{\alpha 0}, \tag{1.36}$$

with

$$\dot\theta_\alpha = \frac{\partial H_0(J_\alpha)}{\partial J_\alpha} = \omega_\alpha. \tag{1.37}$$

From the foregoing hypotheses, the periodic flow $q_\alpha(t) = (x_\alpha(t), y_\alpha(t))$ in the domain D can be expressed by

$$x_\alpha = x_\alpha(J_\alpha, \theta_\alpha) \quad \text{and} \quad y_\alpha = y_\alpha(J_\alpha, \theta_\alpha). \tag{1.38}$$

Substitution of Eq.(1.38) into Eq.(1.30) and using Fourier expansion gives

$$
\begin{aligned}
&H(J_\alpha, \theta_\alpha, t) \\
&= H_0(J_\alpha) + H_1(J_\alpha, \theta_\alpha, \Omega t) \\
&= H_0(J_\alpha) + \sum_n \sum_m \left\{ H^{(-)}_{1(m:n)}(J_\alpha) \cos(m\omega_\alpha t - n\Omega t + \psi^{(-)}_{\alpha n}) \right. \\
&\left. \qquad\qquad + H^{(+)}_{1(m:n)}(J_\alpha) \cos(m\omega_\alpha t + n\Omega t + \psi^{(+)}_{\alpha n}) \right\}. \tag{1.39}
\end{aligned}
$$

Letting $\varphi_\alpha = (m\omega_\alpha - n\Omega)t$, we have

$$
\begin{aligned}
&H(J_\alpha, \theta_\alpha, t) \\
&= H_0(J_\alpha) + H_1(J_\alpha, \theta_\alpha, \Omega t) \\
&= H_0(J_\alpha) + \sum_n \sum_{m_1} \left\{ H^{(-)}_{1(m_1:n)}(J_\alpha) \cos\left[\frac{m_1}{m}\varphi_\alpha + \frac{n(m_1 - m)}{m}\Omega t + \psi^{(-)}_{\alpha n}\right] \right. \\
&\left. \qquad\qquad + H^{(+)}_{1(m_1:n)}(J_\alpha) \cos\left[\frac{m_1}{m}\varphi_\alpha + \frac{n(m_1 + m)}{m}\Omega t + \psi^{(+)}_{\alpha n}\right] \right\}. \tag{1.40}
\end{aligned}
$$

This periodic flow existing on the inside of separatrix can be called the librational (or local) periodic flows. If the following resonance condition holds as

$$m\omega_\alpha = n\Omega, \tag{1.41}$$

then we have

$$J_\alpha = J_\alpha^{(m:n)} \quad \text{and} \quad E_\alpha = E_\alpha^{(m:n)}. \tag{1.42}$$

To define a generating function, we have

$$H_0(J_\alpha, \theta_\alpha, t) + \frac{\partial G}{\partial t}$$

$$= H_0(J_\alpha^{(m:n)}) + \left.\frac{\partial H_0(J_\alpha)}{\partial J_\alpha}\right|_{J_\alpha = J_\alpha^{(m:n)}} (J_\alpha - J_\alpha^{(m:n)}) + \frac{\partial G}{\partial t}$$

$$+ \frac{1}{2} \left.\frac{\partial^2 H_0(J_\alpha)}{\partial J_\alpha^2}\right|_{J_\alpha = J_\alpha^{(m:n)}} (J_\alpha - J_\alpha^{(m:n)})^2 + \cdots + h.o.t. \tag{1.43}$$

$$H_1(J_\alpha, \theta_\alpha, t)$$

$$= \sum_n \sum_{m_1} \left\{ H_{1(m_1:n)}^{(-)}(J_\alpha) \cos\left[\frac{m_1}{m}\varphi_\alpha + \frac{n(m_1 - m)}{m}\Omega t + \psi_{\alpha n}^{(-)}\right]\right.$$

$$\left. + H_{1(m_1:n)}^{(+)}(J_\alpha) \cos\left[\frac{m_1}{m}\varphi_\alpha + \frac{n(m_1 + m)}{m}\Omega t + \psi_{\alpha n}^{(+)}\right]\right\}. \tag{1.44}$$

Letting

$$\frac{\partial H_0(J_\alpha)}{\partial J_\alpha}(J_\alpha - J_\alpha^{(m:n)}) = -\frac{\partial G}{\partial t}, \tag{1.45}$$

the generating function can be defined as

$$G = -\omega_\alpha t (J_\alpha - J_\alpha^{(m:n)}) = -\frac{(\varphi_\alpha + n\Omega t)}{m}(J_\alpha - J_\alpha^{(m:n)}). \tag{1.46}$$

Furthermore, we have a new coordinate $(\bar{p}_\alpha, \varphi_\alpha)$

$$\bar{p}_\alpha = -\frac{\partial G}{\partial \varphi_\alpha} = \frac{1}{m}(J_\alpha - J_\alpha^{(m:n)}) \quad \text{and} \quad \theta_\alpha = \omega_\alpha t = -\frac{\partial G}{\partial J_\alpha} = \frac{1}{m}(\varphi_\alpha + n\Omega t). \tag{1.47}$$

If $\bar{p}_\alpha = 0$, we have $J_\alpha = J_\alpha^{(m:n)}$. The variable $\bar{p}_\alpha$ gives the difference between the instant J_α and the resonance $J_\alpha^{(m:n)}$, which defines the gap of the resonant separatrix layer. Because the natural frequency ω_α is a function of energy E_α or the action variable J_α, the relation between the resonance frequency and energy is illustrated in Fig.1.8 from the resonance relation in Eq.(1.41). The resonant frequency distributions along the conservative energy are distinguishing for four resonant layers. Further, their resonant structures are different from each other. The specific $(m : n)$-resonant frequency and natural frequency are expressed by $\Omega^{(m:n)}$ and $\omega_\alpha^{(m:n)}$, respectively. The corresponding resonant condition in Eq.(1.41) becomes

$$m\omega_\alpha^{(m:n)} = n\Omega^{(m:n)}. \tag{1.48}$$

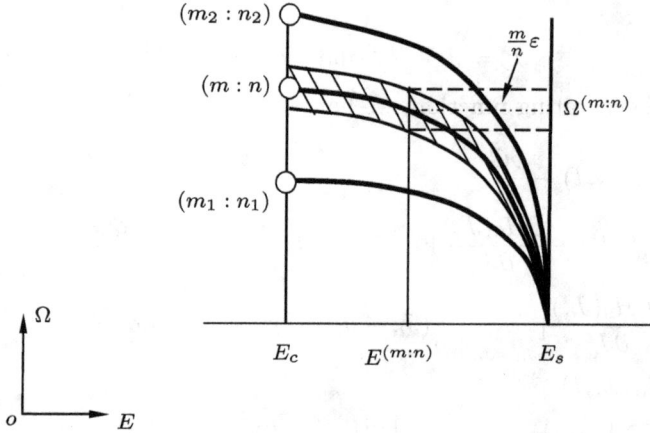

Fig. 1.8 The neighborhood of the $(m:n)$-resonant frequency for the inside separatrix.

To guarantee the resonant layers in a certain gap, consider a neighborhood of a natural frequency ω_α close to $\omega_\alpha^{(m:n)}$, i.e.,

$$\left|\omega_\alpha - \omega_\alpha^{(m:n)}\right| = \left|\omega_\alpha - \frac{n}{m}\Omega^{(m:n)}\right| \leqslant \varepsilon \text{ and}$$

$$T(E_\alpha^{(m:n)}) = \frac{2\pi}{\omega_\alpha^{(m:n)}}, \quad T(E_\alpha) = \frac{2\pi}{\omega_\alpha}, \tag{1.49}$$

where $\varepsilon \ll 1$ and $T(E)$ is the nonlinear period. From the foregoing condition, with Eqs.(1.43) and (1.44), one obtains

$$H(J_\alpha, \theta_\alpha, t) + \frac{\partial G}{\partial t}$$
$$\approx E_\alpha^{(m:n)} + \frac{1}{2}\frac{\partial^2 H_0(J_\alpha)}{\partial J_\alpha^2}\bigg|_{J_\alpha = J_\alpha^{(m:n)}} m^2 \overline{p}_\alpha^2 \tag{1.50}$$
$$+ \sum_n \sum_{m_1} \left\{ H_{1(m_1:n)}^{(-)}(J_\alpha^{(m:n)}, \overline{p}_\alpha) \cos\left[\frac{m_1}{m}\varphi_\alpha + \frac{n(m_1 - m)}{m}\Omega t + \psi_{\alpha n}^{(-)}\right] \right.$$
$$\left. + H_{1(m_1:n)}^{(+)}(J_\alpha^{(m:n)}, \overline{p}_\alpha) \cos\left[\frac{m_1}{m}\varphi_\alpha + \frac{n(m_1 + m)}{m}\Omega t + \psi_{\alpha n}^{(+)}\right] \right\}.$$

Rescaling gives the following variables as

$$\overline{H}(J_\alpha, \theta_\alpha, t) = H(J_\alpha, \theta_\alpha, t) + \frac{\partial G}{\partial t} - E_\alpha, \quad B = \frac{\partial^2 H_0(J_\alpha)}{\partial J_\alpha^2}\bigg|_{J_\alpha = J_\alpha^{(m:n)}},$$
$$mB\overline{p}_\alpha = p_\alpha, H(p_\alpha, \varphi_\alpha, t) = B\overline{H}(J_\alpha, \theta_\alpha, t), \quad \Omega_1 = 2n\Omega, \tag{1.51}$$
$$U_{(m:n)}^{(+)} = BH_{1(m:n)}^{(+)}(J_\alpha^{(m:n)}, p_\alpha), \quad U_{(m:n)}^{(-)} = BH_{1(m:n)}^{(-)}(J_\alpha^{(m:n)}, p_\alpha).$$

The new Hamiltonian becomes

$$H(p_\alpha, \varphi_\alpha, t) \approx \frac{1}{2} p_\alpha^2 + U_{1(m:n)}^{(-)} \cos(\varphi_\alpha + \psi_{\alpha n}^{(-)}) \tag{1.52}$$

$$+ U_{1(m:n)}^{(+)} \cos(\varphi_\alpha + 2n\Omega t + \psi_{\alpha n}^{(+)})$$

$$+ \sum_{n_1} \sum_{m_1} \left\{ U_{1(m_1:n_1)}^{(-)} \cos\left[\frac{m_1}{m} \varphi_\alpha + \frac{n_1(m_1 - m)}{m} \Omega t + \psi_{\alpha n_1}^{(-)} \right] \right.$$

$$\left. + U_{1(m_1:n_1)}^{(+)} \cos\left[\frac{m_1}{m} \varphi_\alpha + \frac{n_1(m_1 + m)}{m} \Omega t + \psi_{\alpha n_1}^{(+)} \right] \right\}.$$

Because the primary resonance is relatively isolated, herein the other resonance terms in Eq.(1.52) are ignored except for the $(m : n)$-resonance, the approximate Hamiltonian is expressed by

$$H(p_\alpha, \varphi_\alpha, t)$$

$$\approx \frac{1}{2} p_\alpha^2 + U_{1(m:n)}^{(-)} \cos(\varphi_\alpha + \psi_{\alpha n}^{(-)}) + U_{1(m:n)}^{(+)} \cos(\varphi_\alpha + \Omega_1 t + \psi_{\alpha n}^{(+)}). \tag{1.53}$$

It is assumed that the two parameters $U_{1(m:n)}^{(-)}$ and $U_{1(m:n)}^{(+)}$ are independent of p_α, and the corresponding dynamical system is

$$\dot{\varphi}_\alpha = p_\alpha,$$

$$\dot{p}_\alpha = -U_{1(m:n)}^{(-)} \sin(\varphi_\alpha + \psi_{\alpha n}^{(-)}) - U_{1(m:n)}^{(+)} \sin(\varphi_\alpha + \Omega_1 t + \psi_{\alpha n}^{(+)}). \tag{1.54}$$

This equation represents a kind of parametrically excited pendulum. The dynamics in the neighborhood of the $(m : n)$-resonant separatrix can be investigated through Eq.(1.54). The resonance effects of Eq.(1.52) give the sub-resonance for the $(m : n)$-resonance. Because $U_{1(m:n)}^{(-)}$ is relative to external excitation and energy orbit, the resonant separatrix will be changed with frequency and amplitude. The sub-resonance can be obtained from the perturbed system in Eq.(1.53). On the other hand, from the differential equation, the self-similar structure may not exist. The sub-resonant structure is strongly dependent on the energy analysis of Eq.(1.54). Based on a certain sub-resonance in the $(m : n)$-resonant separatrix layer, the sub-sub-resonance can be obtained by repeating the same procedure to obtain a new equation similar to Eq.(1.54). This renormalization procedure cannot lead to a self-similar sub-resonance structure. Indeed, the sub-resonant structure in the resonant layer is relative to the corresponding parent-resonance separatrix of Eq.(1.54), but such a sub-resonant structure cannot be generated by simply copying from its parent primary-resonance. From the foregoing discussion, the dynamics of the parametric pendulum is a key to understand the mechanism of stochasticity in a neighborhood of the $(m : n)$-resonant separatrix.

1.2.2 Approximate criteria

To develop the criteria for appearance, growth and destruction of primary resonance layers, consider an $(m_1 : n_1)$-primary resonance closest to the $(m : n)$-resonant separatrix. In the Chirikov overlap criterion and renormalization technique, from Eq.(1.52), the following new energy form is considered.

$$H(p_\alpha, \varphi_\alpha, t) \approx \frac{1}{2} p_\alpha^2 + U_{1(m:n)}^{(-)} \cos(\varphi_\alpha + \psi_{\alpha n}^{(-)}) \tag{1.55}$$

$$+ U_{1(m_1:n_1)}^{(-)} \cos\left[\frac{m_1}{m}\varphi_\alpha + \frac{n_1(m_1 - m)}{m}\Omega t + \psi_{\alpha n_1}^{(-)}\right].$$

As in Chirikov (1979) and Reichl (1992), the corresponding Chirikov resonance overlap criterion becomes

$$\sqrt{U_{1(m:n)}^{(-)}} + \sqrt{U_{1(m_1:n_1)}^{(-)}} = 1. \tag{1.56}$$

From renormalization, the criterion becomes

$$\sqrt{U_{1(m:n)}^{(-)}} + \sqrt{U_{1(m_1:n_1)}^{(-)}} \approx 0.7. \tag{1.57}$$

From the author's points of view, once the $(m : n)$-resonant separatrix is formed, with increasing excitation strength, the sub-resonance in its neighborhood will be developed first from Eq.(1.54). Of course, from Eq.(1.52), the other primary resonance may have a certain effect as external excitations. Until such a resonant layer approaches an unperturbed orbit from which the $(m_1 : n_1)$-primary resonance can be formed, the $(m : n)$-resonant layer almost cannot be destroyed, and the effects of the $(m_1 : n_1)$-primary resonance to the $(m : n)$-resonant layer are very small compared to the $U_{1(m:n)}^{(+)}$-term. In other words, before such a resonant layer is destroyed, the $(m_1 : n_1)$-primary resonance will not be strongly involved in the $(m : n)$-resonant layer. Therefore, it is very doubtable that the criterions in Eqs.(1.56) and (1.57), given by the Chirikov overlap criterion and the renormalization technique, can provide a reasonable prediction of the global stochasticity of the resonant layer. In addition, it is not clear that the two existing criteria can be used for the appearance or disappearance of the resonant layer.

(A) *Layer Onset Conditions.* To discuss the appearance, growth and destruction of the $(m : n)$-resonant layer, the geometrical intuitions of both just after onset and just before destruction of the resonant layer are sketched. The corresponding resonant layer appearance with a thin layer is sketched in Fig.1.9(a). The $(m : n)$-resonant layer is formed in the neighborhood of its primary resonant separatrix. With increasing excitation strength, the width of the resonant layer will increase.

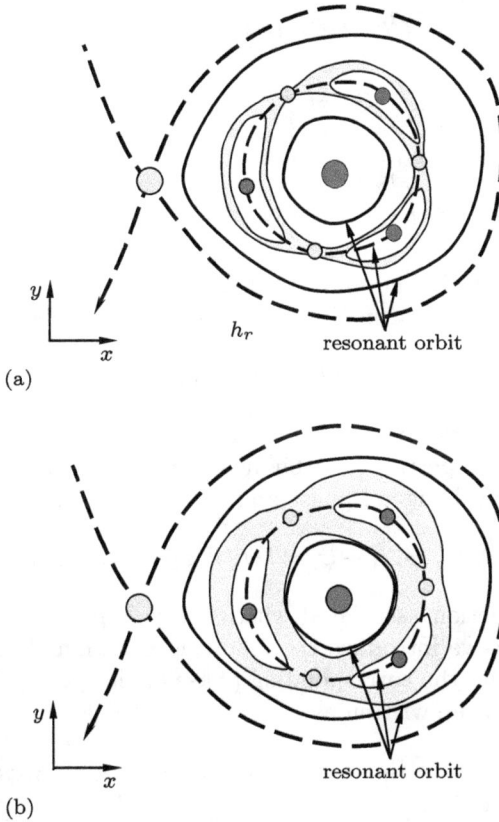

Fig. 1.9 Resonant layer in nonlinear Hamiltonian systems with separatrix: (a) appearance and (b) disappearance. The solid and hollow circles represent the center and hyperbolic points. The irregular small curcles are the sub-resonances in the neighborhood of the primary resoance.

For an approximate estimate of the resonant layer width, the theorem given in Luo (2008) is adopted. In other words, if the farthest energy boundary of the primary $(m : n)$-resonant layer is the energy of the $(m_1 : n_1)$-primary resonant orbit, the primary $(m : n)$-resonant layer will be destroyed. However, the onset of the primary resonant layer will be estimated through the standard mapping technique. To understand the formation mechanism of the primary resonant layer, the onset of the primary $(m : n)$-resonant layer will be discussed first as follows.

The energy increment along the $(m : n)$-resonant orbit of the perturbed conservative system is approximated by

$$\Delta H_0 = \int_{t_i}^{T_\alpha(E_i)+t_i} \frac{dH(x,y,t)}{dt} dt$$

$$= \int_{t_i}^{T_\alpha(E_i)+t_i} \{H_0, H_1\}_{\text{Possion}} dt$$

$$= \int_{t_i}^{T_\alpha(E_i)+t_i} (f_1 g_2 - f_2 g_1) dt = U_0 f(\varphi_i), \qquad (1.58)$$

where t_i is the initial time and $f(\varphi_i)$ is a bounded and periodic function. Without loss of generality, in Eq.(1.58), consider the following form,

$$\Delta H_0 = U_0 \sin \varphi_i, \qquad (1.59)$$

in which U_0 is a system parameter function excluding the initial phase angle $\varphi_i = \Omega t_i$. For a perturbed orbit in the neighborhood of the prescribed resonant orbit, the change of phase angle over one period is

$$\Delta \varphi = \varphi_{i+1} - \varphi_i = \Omega T(E_{i+1}) = V_0(E_{i+1}), \qquad (1.60)$$

where V_0 is a function associated with energy E_{i+1}. To calculate this new energy iteratively, we introduce the following notation: $E_{i+1} = w_{i+1}$ at the $(i+1)$th period and the corresponding phase angle is φ_{i+1}. Equations (1.59) and (1.60) can now be written as

$$w_{i+1} = w_i + U_0 \sin \varphi_i \quad \text{and} \quad \varphi_{i+1} = \varphi_i + V_0(w_{i+1}). \qquad (1.61)$$

The resonant separatrix layer can be investigated by iteration of the mapping in Eq.(1.61). Although this mapping is not based on the accurate energy increment, it is good enough as an approximate, analytical expression because the exact computation of the energy increment only can be done numerically. In the neighborhood of the resonant separatrix layer, equation (1.61) can be linearized about a fixed point and the standard mapping can be obtained. Consider a period-1 iteration of the iterative map for a specific resonance of $(m : n)$. Its fixed point can be easily determined by $w_{i+1} = w_i = w_0$ and $\varphi_{i+1} = \varphi_i + 2m\pi/n = \varphi_0 + 2m\pi/n$. This implies

$$U_0 \sin \varphi_0 = 0, \quad V_0(w_0) = \frac{2\pi m}{n}. \qquad (1.62)$$

Defining a new dimensionless energy

$$I_i = \frac{\partial V_0(w_{i+1})}{\partial w_{i+1}}\bigg|_{w_{i+1}=w_0} (w_i - w_0), \qquad (1.63)$$

and linearization of Eq.(1.61) about the fixed point yields

$$I_{i+1} = I_i + \Xi \sin \varphi_i \quad \text{and} \quad \varphi_{i+1} = \varphi_i + I_{i+1}, \qquad (1.64)$$

where $\Xi = U_0 \partial V_0 / \partial w_{i+1}|_{w_{i+1}=w_0}$. From Eq.(1.64), the mechanism involved in the transition to the global stochasticity in a nonlinear Hamiltonian system is very clear. The coefficient Ξ is the only control parameter for the characterization of the KAM tori. For the standard map, a critical value of Ξ is attained when $\Xi = \Xi^* = 0.9716354 \cdots$. At this value, the last remaining KAM torus is broken. As this occurs, we have

$$U_0 \frac{\partial V_0}{\partial w_0} = \Xi^*. \tag{1.65}$$

The transition from the local stochasticity to global stochasticity implies chaos appears in such a standard mapping. This appearance condition is as a condition for the appearance of resonant layers in the neighborhood of the $(m : n)$-resonant separatrix. For a generalized case in Eq.(1.61), it can be discussed in a similar fashion. The corresponding criteria can be developed for the global stochasticity of motion in the primary resonant layer.

(B) *Layer Vanishing Conditions.* Once the resonant separatrix layer is formed, with increasing the excitation, the other sub-resonant separatrix layers will merge in the resonant layer until they come into contact the closest resonant orbit. When this case occurs, the resonant layer will be destroyed, and a new stochastic motion near that resonant orbit will be involved in, and suddenly, the width of the resonant layer will become large. The two primary resonant layers will be overlapped each other. Based on this reason, the Chirikov resonant overlap criterion and the renormalized criterion may be used as a condition for the destruction of a certain, primary resonant layer. Such a mechanism is qualitatively sketched in Fig.1.9(b). Again it is postulated that when the resonant layer is destroyed, the energy increment in Eq.(1.28) is given by the energy difference between the two closest resonant orbits, one of which is associated with the destroyed resonant layer. From Luo (2008), we have

$$\min \left(|E_\alpha^{(m_2:n_2)} - E_\alpha^{(m:n)}|, |E_\alpha^{(m_1:n_1)} - E_\alpha^{(m:n)}| \right) = |\Delta H_0^{(m:n)}| \approx U_0|f(\varphi_0)|. \tag{1.66}$$

The foregoing equation gives the critical condition for the disappearance of the $(m : n)$-resonant separatrix layer. From the foregoing equation, the excitation strength for *disappearance* of the resonant layer can be computed. To determine the excitation strength for *appearance* of the resonant layer, equation (1.65) will be used. For a better prediction of resonant layers, the effects of the secondary resonances should be considered in the vicinity of the primary resonant layer. Because the energy increment is computed by an approximate expression, with increasing excitation strength, such a prediction is not accurate. To verify the approximate prediction, the numerical prediction should be completed through the energy increment for nonlinear Hamiltonian system.

(C) *Energy Increment Spectrum.* The exact energy increment can be computed numerically. Luo et al. (1999) developed the energy spectrum approach, which was used for the numerical prediction of the onset of resonance in the stochastic layer. In the energy spectrum, the maximum and minimum conservative energies are computed through the Poincaré mapping section. As discussed in Section 1.1.2(c), the energy spectrum for resonant layer can be determined by the energy increment spectrum. Using the Poincaré map section of Eq.(1.29), the Poincaré map is defined by $P : \Sigma \to \Sigma$. Such a technique computes the maximum and minimum energies of the Poincaré mapping points as in Eq.(1.31). The unperturbed Hamiltonian energy for each Poincaré mapping point of Eq.(1.29) is

$$H_0^{(N)} = H_0(\mathbf{x}_N, \boldsymbol{\mu}) \equiv H_0(x_N, y_N, \boldsymbol{\mu}). \tag{1.67}$$

However, in computation of conservative energy for a specified resonant layer, the energy changes in energy spectrums cannot be observed clearly. To observe the energy changes caused by the sub-resonance for the specified resonant layer, the minimum and maximum energy increments with respect to the unperturbed resonant orbit are introduced herein. On the other hand, the initial condition is chosen from the specific resonant orbit. The initial energy is $H_0(\mathbf{x}_0, \boldsymbol{\mu}) = E_\alpha^{(m:n)}$, so the energy increment (or the first integral quantity increment) should be computed by $\Delta H_0(t_0, kT) = H_0^{(N)}(\mathbf{x}_N, \boldsymbol{\mu}) - H_0(\mathbf{x}_0, \boldsymbol{\mu})$, i.e.,

$$\begin{aligned} \Delta E_{\max} &= \max_{N \to \infty} \left\{ H_0^{(N)} - E_\alpha^{(m:n)} \right\}, \\ \Delta E_{\min} &= \min_{N \to \infty} \left\{ H_0^{(N)} - E_\alpha^{(m:n)} \right\}. \end{aligned} \tag{1.68}$$

For the appearance and disappearance of the resonant layer for a specific resonance, the maximum or minimum energy increment will have a big jump between the two closest resonant separatrices. From the minimum and maximum energy increment spectra, the width of a resonant separatrix layer is computed like the one of a stochastic layer in Luo et al. (1999), i.e.,

$$w \equiv \min_{t \in [0, \infty)} \|\mathbf{x}(E_\alpha^{\max}, t) - \mathbf{x}(E_\alpha^{\min}, t)\| \equiv \|\mathbf{x}_\alpha^{\max} - \mathbf{x}_\alpha^{\min}\| \tag{1.69}$$

where $\| \cdot \|$ is a norm and the minimum and maximum energies are computed by $E_\alpha^{\max} = \Delta E_{\max} + E_\alpha^{(m:n)}, E_\alpha^{\min} = \Delta E_{\min} - E_\alpha^{(m:n)}$. Two points $\mathbf{x}_\alpha^{\max}$ and $\mathbf{x}_\alpha^{\min}$ on the normal vector $\mathbf{f}^\perp(\mathbf{x}_\alpha^{(m:n)}) = (-f_2(\mathbf{x}_\alpha^{(m:n)}), f_1(\mathbf{x}_\alpha^{(m:n)}))^{\mathrm{T}}$ of the tangential vector of unperturbed resonant orbit at point $\mathbf{x}_\alpha^{(m:n)}$ are the closest between the maximum and minimum energy orbits $\mathbf{x}(E_\alpha^{\max}, t)$ and $\mathbf{x}(E_\alpha^{\min}, t)$ which can be obtained from Eq.(1.31) with $E_\alpha^{\max}$ and $E_\alpha^{\min}$. The detailed discussion on the energy increment spectrum can be referred to Luo (2002).

References

Abdullaev, S.S. and Zaslavsky, G.M., 1995, Self-similarity of stochastic magnetic field lines near the X-point, *Physics of Plasmas*, **2**, 4533–4541.

Abdullaev, S.S. and Zaslavsky, G.M., 1996, Application of the separatrix map to study perturbed magnetic field lines near the separatrix, *Physics of Plasmas*, **3**, 516–528.

Ahn, T., Kim, G. and Kim, S., 1996, Analysis of the separatrix map in Hamiltonian systems, *Physica D*, **89**, 315–328.

Chirikov, B.V., 1979, A universal instability of many-dimensional oscillator systems, *Physics Reports*, **52**, 263–379.

Feng K. and Qin M.Z., 1991, Hamiltonian algorithms for Hamiltonian systems and a comparative numerical study, *Computer Physics Communications*, **65**, 173–187.

Greene, J.M., 1968, Two-dimensional measure-preserving mappings, *Journal of Mathematical Physics*, **9**, 760–768.

Greene, J.M., 1979, A method for computing the stochastic transition, *Journal of Mathematical Physics*, **20**, 1183–1201.

Han, R.P.S. and Luo, A.C.J., 1998, Resonant layers in nonlinear dynamics, *ASME Journal of Applied Mechanics*, **65**, 727–736.

Iomin, A. and Fishman, S., 1996, Semiclassical quantization of a separatrix map, *Physical Review E*, **54**, R1–R5.

Lichtenberg, A.J. and Lieberman, M.A., 1992, *Regular and Chaotic Dynamics*, 2nd Edition, Springer, New York.

Lu, C., 2007, Chaos of a parametrically excited undamped pendulum, *Communications in Nonlinear Science and Numerical Simulation*, **12**, 45–57.

Luo, A.C.J., 1995, Analytical modeling of bifurcations, chaos and multifractals in nonlinear dynamics. *Ph.D. Dissertation*, University of Manitoba, Winnipeg, Manitoba, Canada.

Luo, A.C.J., 2001, Resonant-overlap phenomena in stochastic layers of nonlinear Hamiltonian systems with periodical excitations, *Journal of Sound and Vibration*, **240**(5), 821–836.

Luo, A.C.J., 2002, Resonant layers in a parametrically excited pendulum, *International Journal of Bifurcation and Chaos*, **12**(2), 409–419.

Luo, A.C.J., 2008, *Global Transversality, Resonance and Chaotic Dynamics*, World Scientific, Singapore.

Luo, A.C.J., Gu, K. and Han, R.P.S., 1999, Resonant-separatrix webs in stochastic layers of the twin-well Duffing oscillator, *Nonlinear Dynamics*, **19**, 37–48.

Luo, A.C.J. and Han, R.P.S., 2000, Investigations of stochastic layers in nonlinear dynamics, *Journal of Vibration and Acoustics*, **122**, 36–41.

Luo, A.C.J. and Han, R.P.S., 2001, The resonance theory for stochastic layers in nonlinear dynamical systems, *Chaos, Solitons and Fractals*, **12**, 2493–2508.

McLachlan, R. and Atela, P., 1992, The accuracy of symplectic integrators, *Nonlinearity*, **5**, 541–562.

Melnikov, V.K., 1963, On the stability of the center for time periodic perturbations, *Transaction Moscow Mathematical Society*, **12**, 1–57.

Reichl, L.E., 1992, *The Transition to Chaos in Conservative Classic System: Quantum Manifestations*, Springer-Verlag, New York.

Rom-Kedar, V., 1990, Transport rates of a class of two-dimensional maps and flow, *Physica D*, **43**, 229–268.

Rom-Kedar, V., 1994, Homoclinic tangles-classification and applications, *Nonlinearity*, **7**, 441–473.

Rom-Kedar, V., 1995, Secondary homoclinic bifurcation theorems, *Chaos*, **5**, 385–401.

Zaslavsky, G.M. and Abdullaev, S.S., 1995, Scaling properties and anomalous transport of particles inside the stochastic layer, *Physical Review E*, **51**, 3901–3910.

Chapter 2
Hamiltonian Chaos in Pendulum

In this chapter, the mechanism of stochastic layer characterized by resonance in a periodically forced pendulum will be discussed, and the analytical conditions for the onset of a specific resonance in the stochastic layer will be obtained. The stochastic layer in the vicinity of the heteroclinic orbit is based on the generic separatrix (i.e., the heteroclinic orbit), which can be easily achieved. However, the specific resonant layer is based on the specific resonant separatrix. It is very difficult to obtain the resonant separatrix generated by the resonance. Thus, the onset and disappearance of the resonant layers will be analytically predicted. From the analytical predictions of stochastic and resonant layers, numerical simulations will be completed to illustrate the stochastic and resonant layers through the Poincaré mapping sections. Such results can help one better understand chaotic motions in nonlinear Hamiltonian systems.

2.1 Resonance conditions

As in Luo and Han (2000), consider a periodically-driven pendulum as

$$\ddot{x} + \alpha \sin x = Q_0 \cos \Omega t, \qquad (2.1)$$

where Q_0 and Ω are excitation amplitude and frequency, respectively. The Hamiltonian of Eq.(2.1) is

$$H(x, y, t) = \frac{1}{2} y^2 - \alpha \cos x - x Q_0 \cos \Omega t, \qquad (2.2)$$

and its time-independent H_0 and time-dependent H_1 are

$$H_0 = \frac{1}{2} y^2 - \alpha \cos x \quad \text{and} \quad H_1 = -x Q_0 \cos \Omega t \qquad (2.3)$$

where $y = \dot{x}$.

2.1.1 Conservative system

The conservative system in Eq.(2.1) possesses elliptic points $(\pm 2j\pi, 0)$ and hyperbolic points $(\pm(2j+1)\pi, 0)(j = 0, 1, \cdots)$. Two heteroclinic orbits connecting all the hyperbolic points separate the phase space in the conservative system of Eq.(2.1) into the libration and rotation. The solutions for the heteroclinic motion, libration and rotation are as follows.

(A) *Heteroclinic motion:*
The energy of the heteroclinic orbit passing through the hyperbolic point is $H_0 = E_h = \alpha$, where the subscript h denotes the heteroclinic motion, and the solution to the heteroclinic motion is

$$x_h^0 = \pm 2j\pi \pm 2\arcsin\left[\tanh(\sqrt{\alpha}t)\right],$$
$$y_h^0 = \pm 2\sqrt{\alpha}\,\text{sech}(\sqrt{\alpha}t); \tag{2.4}$$

where the superscript 0 denotes the non-forced pendulum (conservative system).

(B) *Libration:*
For $H_0 = E_l$ $(-\alpha \leqslant E_l < \alpha)$ where the subscript l denotes the libration, its solution is

$$x_l^0 = \pm 2j\pi \pm 2\arcsin\left[k_l\text{sn}\left(\frac{2K(k_l)\theta_l}{\pi}, k_l\right)\right],$$
$$y_l^0 = 2k_l\sqrt{\alpha}\,\text{cn}\left(\frac{2K(k_l)\theta_l}{\pi}, k_l\right); \tag{2.5}$$

where sn and cn are the Jacobi elliptic functions, and $k_l = \sqrt{1 + E_l/\alpha}/\sqrt{2}$ is the elliptic modulus. $K(k_l)$ is the elliptic integral of the first kind. The phase angle θ_l and natural frequency ω_l (or period T_l) are

$$\theta_l = \omega_l t + \theta_{l0} \quad \text{and} \quad \omega_l = \frac{2\pi}{T_l} = \frac{\pi\sqrt{\alpha}}{2K(k_l)}; \tag{2.6}$$

where θ_{l0} is the initial phase angle.

(C) *Rotation:*
For $H_0 = E_r > \alpha$ where the subscript r denotes the rotation, its solutions are

$$x_r^0 = \pm 2\text{am}\left(\frac{K(k_r)\theta_r}{\pi}, k_r\right),$$
$$y_r^0 = \pm\frac{2\sqrt{\alpha}}{k_r}\text{dn}\left(\frac{K(k_r)\theta_r}{\pi}, k_r\right) \tag{2.7}$$

where am is the amplitude and dn is a Jacobi elliptic function. $k_r = \sqrt{2\alpha}/\sqrt{E_r + \alpha}$ and $\theta_r = \omega_r t + \theta_{r0}$. The natural frequency ω_r (or period T_r) is

$$\omega_r = \frac{2\pi}{T_r} = \frac{\pi\sqrt{\alpha}}{k_r K(k_r)}. \tag{2.8}$$

2.1.2 Resonance and energy increments

The stochastic layer in the periodically forced pendulum is separated into the librational layer (shaded) and rotational layer (hatched) by the homoclinic orbit (dotted curve), as shown in Fig.2.1. The layers formed in the vicinity of librational and rotational resonant-sparatrices are termed the librational and rotational resonant layers (e.g., Luo,1995; Han and Luo,1998). The time-dependent Hamiltonian H_1 in Eq.(2.3) for the forced pendulum motion related to the libration is approximately computed by

$$H_1^l = -xQ_0 \cos \Omega t \approx -x_l^0 Q_0 \cos \Omega t. \tag{2.9}$$

For the libration in a potential well, the Fourier expansion of elliptic function (cn) in Eq.(2.5) and its integration with respect to time interval $[0, t]$ yields a Fourier expansion of x_l^0, then substitution of the Fourier expansion of x_l^0 into Eq.(2.9) leads to

$$H_1^l \approx -\frac{Q_0}{k_l\sqrt{\alpha}} \sum_{n=1}^{\infty} \frac{1}{(2n-1)} \text{sech} \left[\frac{(2n-1)\pi K(k_l')}{2K(k_l)} \right] \tag{2.10}$$
$$\times \{\sin\left[(2n-1)\omega_l + \Omega\right]t + \sin\left[(2n-1)\omega_l - \Omega\right]t\}$$

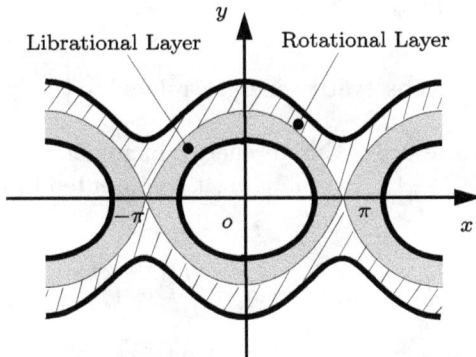

Fig. 2.1 A stochastic layer separated into the librational layer (shaded) and the rotational layer (hatched) by the heteroclinic orbit in a periodically-driven pendulum.

where n is an integer. As in Lichtenberg and Lieberman (1992), in the vicinity of the $((2n-1):1)$ primary resonance, all the other time-dependent terms in H_1^l will average to zero over one period $T = 2\pi/\Omega$ except for the term related to that resonance. Hence, from Eq.(2.10), the librational resonant condition and the corresponding modulus are at $H_0 < \alpha$

$$(2n-1)\omega_l = \Omega \quad \text{and} \quad k_l^{2n-1} = \frac{\sqrt{\alpha + E_l^{2n-1}}}{\sqrt{2\alpha}}, \tag{2.11}$$

where E_l^{2n-1} denotes the conservative energy related to the $(2n-1)^{\text{th}}$-order librational resonance.

Similarly, the time-dependent Hamiltonian H_1 in Eq.(2.3) for the rotation in the forced pendulum is approximated by

$$H_1^r = -xQ_0 \cos \Omega t \approx -x_r^0 Q_0 \cos \Omega t. \tag{2.12}$$

Substitution of Eq.(2.7) into H_1 in Eq.(2.3) and the Fourier expansion of the elliptic function (am) gives

$$H_1^r \approx \frac{\pi \sqrt{\alpha} Q_0}{k_r K(k_r)} t \cos \Omega t - Q_0 \sum_{m=1}^{\infty} \frac{1}{m} \text{sech} \left[\frac{m\pi K(k_r')}{K(k_r)} \right]$$
$$\times [\sin(m\omega_r + \Omega)t + \sin(m\omega_r - \Omega)t], \tag{2.13}$$

where m is another integer. From the foregoing equation, it implies that the energy travels from one potential well to another, but in the vicinity of the $(m:1)$ primary resonance, all other terms in H_1^r will average to zero over a period $T = 2\pi/\Omega$ except for the term pertaining to that resonance. Thus, the rotational resonant condition and the corresponding modulus are at $H_0 > \alpha$

$$m\omega_r = \Omega \quad \text{and} \quad k_r^m = \frac{\sqrt{2\alpha}}{\sqrt{\alpha + E_r^m}}, \tag{2.14}$$

where E_r^m denote the conservative energies related to the mth-order rotational resonance.

Using Eq.(2.5), the approximate energy increment based on the unperturbed, librational $(2n-1)^{\text{th}}$ resonant orbit is computed from Chirikov (1979), i.e.,

$$\Delta H_0^l = \int_{t_0}^{T_l+t_0} [H_0, H_1] dt = \int_{t_0}^{T_l+t_0} (f_1 g_2 - f_2 g_1) dt$$
$$\approx 4Q_0 \pi \text{sech} \left[\frac{(2n-1)\pi K(k_l')}{2K(k_l)} \right] \sin \Omega t_0, \tag{2.15}$$

where $k_l' = \sqrt{1-k_l^2}$, $[\cdot,\cdot]$ represents the Poisson bracket and

$$f_1 = y, \quad g_1 = 0;$$
$$f_2 = \alpha \sin\theta, \quad g_2 = Q_0 \cos\Omega t_0. \tag{2.16}$$

Using Eq.(2.7), the approximate energy increment based on the unperturbed, rotational resonant orbit is

$$\Delta H_0^r = \int_{t_0}^{T_r+t_0} [H_0, H_1] dt = \int_{t_0}^{T_r+t_0} (f_1 g_2 - f_2 g_1) dt$$

$$\approx 2Q_0 \pi \mathrm{sech} \left[m\pi \frac{K(k_r')}{K(k_r)} \right] \sin\Omega t_0. \tag{2.17}$$

In the stochastic layer, two approximate energy increments based on the resonant orbits reduce to the approximate energy increment based on the heteroclinic orbit. That is,

$$\Delta H_0^h = \lim_{T_r \to \infty} 2 \int_{t_0}^{T_r+t_0} [H_0, H_1] dt = \lim_{T_l \to \infty} \int_{t_0}^{T_l+t_0} [H_0, H_1] dt$$

$$\approx 4Q_0 \pi \mathrm{sech} \left(\frac{\pi\Omega}{2\sqrt{\alpha}} \right) \sin\Omega t_0. \tag{2.18}$$

Note that the energy increments can be computed through the sub-harmonic, Melnikov functions in Guckenheimer and Holmes (1983), i.e., $\Delta H = \varepsilon M$, and M is the sub-harmonic Melnikov function and the small parameter $\varepsilon = 1$.

2.2 Stochastic layers

From the energy increments (ΔH_0^l and ΔH_0^r) and the phase change, the whisker maps are

$$E_{i+1} \approx E_i + 4Q_0 \pi \mathrm{sech} \left[\frac{(2n-1)\pi K(k_l')}{2K(k_l)} \right] \sin\phi_i,$$

$$\phi_{i+1} = \phi_i + \frac{4\Omega K(k_l)}{\sqrt{\alpha}} \tag{2.19}$$

for the presence of the $(2n-1)^{\mathrm{th}}$-order librational resonance in the stochastic layer, and

$$E_{i+1} \approx E_i + 4Q_0 \pi \mathrm{sech} \left[\frac{m\pi K(k_r')}{K(k_r)} \right] \sin\phi_i,$$

$$\phi_{i+1} = \phi_i + \frac{2\Omega k_r K(k_r)}{\sqrt{\alpha}} \tag{2.20}$$

for the presence of the mth-order rotational resonance in the stochastic layer, where $\Delta H_0 = E_{i+1} - E_i, \Delta \phi_i = \phi_{i+1} - \phi_i, \phi_i = \Omega t_i$ and E_i is the conservative energy.

In the second equation of Eq.(2.19), $\phi_{i+1} - \phi_i = 2(2n-1)\pi$ gives the resonant energy E_l^{2n-1}. When $E_{i+1} = E_l^{2n-1}$ and $E_i = \alpha$, the excitation strength Q_0 for the onset of the $(2n-1)^{\text{th}}$-order librational resonance in the stochastic layer, as in Luo and Han (1999) or Luo (2008), is estimated by

$$Q_0 = \frac{|E_l^{2n-1} - \alpha|}{4\pi} \cosh \left[\frac{(2n-1)\pi K(k_l'^{2n-1})}{2K(k_l^{2n-1})} \right], \tag{2.21}$$

where

$$\Omega = \frac{(2n-1)\pi\sqrt{\alpha}}{2K(k_l^{2n-1})}, \quad k_l^{2n-1} = \frac{\sqrt{\alpha + E_l^{2n-1}}}{\sqrt{2\alpha}}. \tag{2.22}$$

When $E_{i+1} = E_r^m, E_i = \alpha$ and the second equation of Eq.(2.20) satisfies the resonant condition in Eq.(2.14), the excitation strength Q_0 for the onset of the mth-order rotational resonance in the stochastic layer is

$$Q_0 = \frac{|E_r^m - \alpha|}{4\pi} \cosh \left[\frac{m\pi K(k_r'^m)}{K(k_r^m)} \right] \tag{2.23}$$

where

$$\Omega = \frac{m\pi\sqrt{\alpha}}{k_r^m K(k_r^m)}, \quad k_r^m = \frac{\sqrt{2\alpha}}{\sqrt{\alpha + E_r^m}}. \tag{2.24}$$

The whisker maps based on ΔH_0^h for the librational and rotational stochastic layer are

$$E_{i+1} \approx E_i + 4Q_0 \pi \text{sech} \left(\frac{\pi\Omega}{2\sqrt{\alpha}} \right) \sin \phi_i,$$
$$\phi_{i+1} = \phi_i + \frac{4\Omega K(k_l)}{\sqrt{\alpha}}; \tag{2.25}$$

and

$$E_{i+1} \approx E_i + 4Q_0 \pi \text{sech} \left(\frac{\pi\Omega}{2\sqrt{\alpha}} \right) \sin \phi_i,$$
$$\phi_{i+1} = \phi_i + \frac{2\Omega k_r K(K_r)}{\sqrt{\alpha}}. \tag{2.26}$$

If Eqs.(2.25) and (2.26), the elliptic integral of the first kind is approximated by $\log(4/\sqrt{1-k^2})$, the approximate whisker map will be obtained in Chirikov (1979) (see also Zaslavakii and Filonenko, 1968). Similarly, the excitation strengths based on Eqs.(2.25) and (2.26) are approximately predicted by

$$Q_0 = \frac{|E_l^{2n-1} - \alpha|}{4\pi} \cosh \left(\frac{\pi\Omega}{2\sqrt{\alpha}} \right), \tag{2.27}$$

where

$$\Omega = \frac{(2n-1)\pi\sqrt{\alpha}}{2K(k_l^{2n-1})}, \quad k_l^{2n-1} = \frac{\sqrt{\alpha + E_l^{2n-1}}}{\sqrt{2\alpha}} \tag{2.28}$$

for the onset of the $(2n-1)^{\text{th}}$-order librational resonance in the stochastic layer, and

$$Q_0 = \frac{|E_r^m - \alpha|}{4\pi} \cosh\left(\frac{\pi\Omega}{2\sqrt{\alpha}}\right) \tag{2.29}$$

where

$$\Omega = \frac{m\pi\sqrt{\alpha}}{k_r^m K(k_r^m)}, \quad k_r^m = \frac{\sqrt{2\alpha}}{\sqrt{\alpha + E_r^m}} \tag{2.30}$$

for the onset of the mth-order rotational resonance in the stochastic layer.

The resonant conditions for the librational and rotational stochastic layers are illustrated in Fig.2.2 at $\alpha = 1.0$ through the excitation frequency Ω versus Hamiltonian H_0. When the onset of a new resonance in the stochastic layer occurs, the excitation strength Q_0 predicted by Eqs.(2.21) and (2.23) (solid line) and Eqs.(2.27) and (2.29) (dash line) are shown in Fig.2.3 for the librational stochastic layer (upper) and the rotational stochastic layer (lower).

For a given excitation frequency, Figs.2.2 and 2.3 tell us how to choose H_0 determining the initial conditions and Q_0, and under these parameters the resonance in the stochastic layer will be observed. Note that Points A–D in Fig. 2.3 are to be used later to demonstrate the stochastic layers characterized by the resonance.

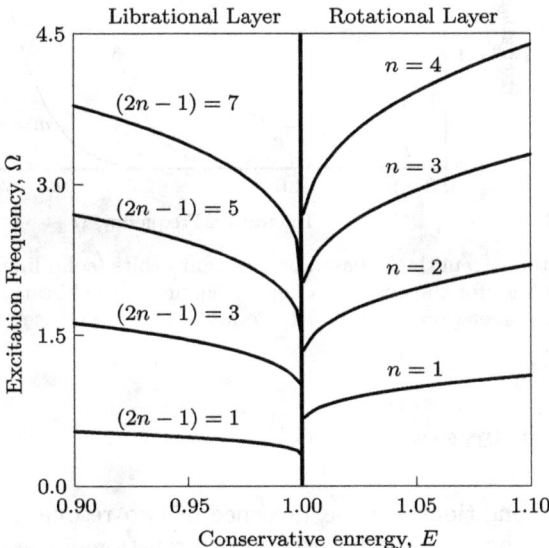

Fig. 2.2 Resonant conditions near the heteroclinic orbit for the stochastic layer ($\alpha = 1$).

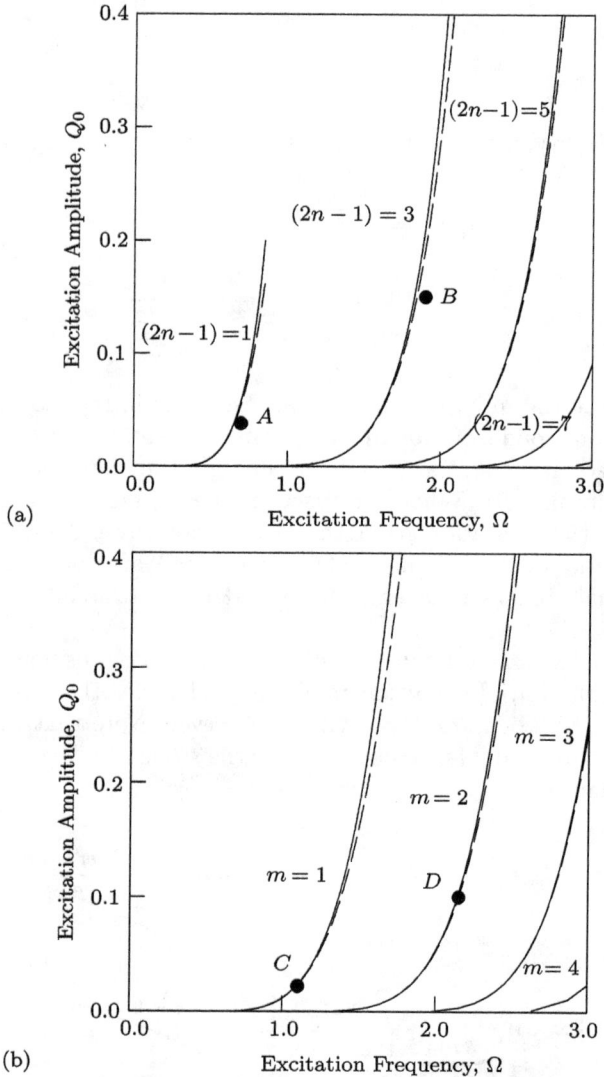

Fig. 2.3 Excitation strength Q_0 based on resonant orbits (solid line) and the heteroclinic orbit (dash line) for the onset of a new resonance in the librational layer (upper) and the rotational layer (lower) at $\alpha = 1$. Points A and B are used for simulation of stochastic layers.

2.3 Resonant layers

The analytical conditions for the presence of two resonant layers are presented separately because the librational and rotational, resonant conditions in Eqs.(2.11) and (2.14) are distinguishing themselves.

2.3.1 Librational resonant layers

The whisker map in Eq.(2.19) is used for investigation of the librational resonant layer. Following the procedure of Han and Luo (1998), equation (2.19) at $E_{i+1} = E_i = E_l^{2n-1}$ and $\phi_{i+1} - \phi_i = 2(2n-1)\pi$ becomes

$$\phi_i^{2n-1} = \{0, \pi, \cdots, (2n-1)\pi\} \quad \text{and} \quad (2n-1)\pi = \frac{2\Omega K(k_l^{2n-1})}{\sqrt{\alpha}}. \quad (2.31)$$

The modulus k_l^{2n-1} is determined from the foregoing equation, and the energy E_l^{2n-1} is obtained from Eq.(2.11). An energy near E_l^{2n-1} is expressed by

$$E_i = E_l^{2n-1} + \Delta E_i, \quad w_i = G_l^{2n-1} \Delta E_i \quad \text{and} \quad \phi_i = \phi_l^{2n-1} + \varphi_i; \quad (2.32)$$

where $G_l^{2n-1} = \partial(\Delta\phi(E_{i+1}))/\partial E_{i+1}\big|_{E_{i+1}=E_l^{2n-1}}$. Equation (2.19) becomes

$$w_{i+1} \approx w_i \pm \Xi \sin\varphi_i \quad \text{and} \quad \varphi_{i+1} \approx \varphi_i + w_{i+1}; \quad (2.33)$$

where

$$\Xi = 4Q_0\pi \operatorname{sech}\left[\frac{(2n-1)\pi K(k_l'^{2n-1})}{2K(k_l^{2n-1})}\right]|G_l^{2n-1}|, \quad (2.34)$$

where

$$G_l^{2n-1} = \frac{\Omega}{(\sqrt{\alpha})^3 (k_l^{2n-1})^2 \left[1 - (k_l^{2n-1})^2\right]}$$
$$\cdot \left\{E(k_l^{2n-1}) - [1 - (k_l^{2n-1})^2]K(k_l^{2n-1})\right\}, \quad (2.35)$$

and $E(k_l^{2n-1})$ is the elliptic complete integral of the second kind. From the analysis of the standard map, we have the critical value $\Xi^* \approx 0.9716354\cdots$ when the KAM torus is destroyed in Lichtenberg and Lieberman (1992). Application of the critical value ($\Xi^* = 0.9716354\cdots$) to Eq.(2.34) yields an critical condition for the approximate prediction of the onset of the librational resonant layer, i.e.,

$$Q_0 \approx \frac{0.24290885}{\pi|G_l^{2n-1}|}\cosh\left[\frac{(2n-1)\pi K(k_l'^{2n-1})}{2K(k_l^{2n-1})}\right]. \quad (2.36)$$

When the librational resonant layer of the $(2n-1)^{\text{th}}$-order interacts with the resonance of the $(2n+1)^{\text{th}}$-order, this layer will be destroyed. Therefore, as in Luo (2008), the excitation strength for the disappearance of the $(2n-1)^{\text{th}}$-order librational resonant layer is approximately estimated through

$$Q_0 \approx \frac{|E_l^{2n+1} - E_l^{2n-1}|}{4\pi}\cosh\left[\frac{(2n-1)\pi K(k_l'^{2n-1})}{2K(k_l^{2n-1})}\right]. \quad (2.37)$$

In Fig.2.4, the resonant conditions (upper) and excitation strengths (lower) for the appearance (solid line) and disappearance (dash-dot line) of the resonant layer are illustrated at $\alpha = 1$. The values at points E and F are used later for illustration of the librational resonant layers.

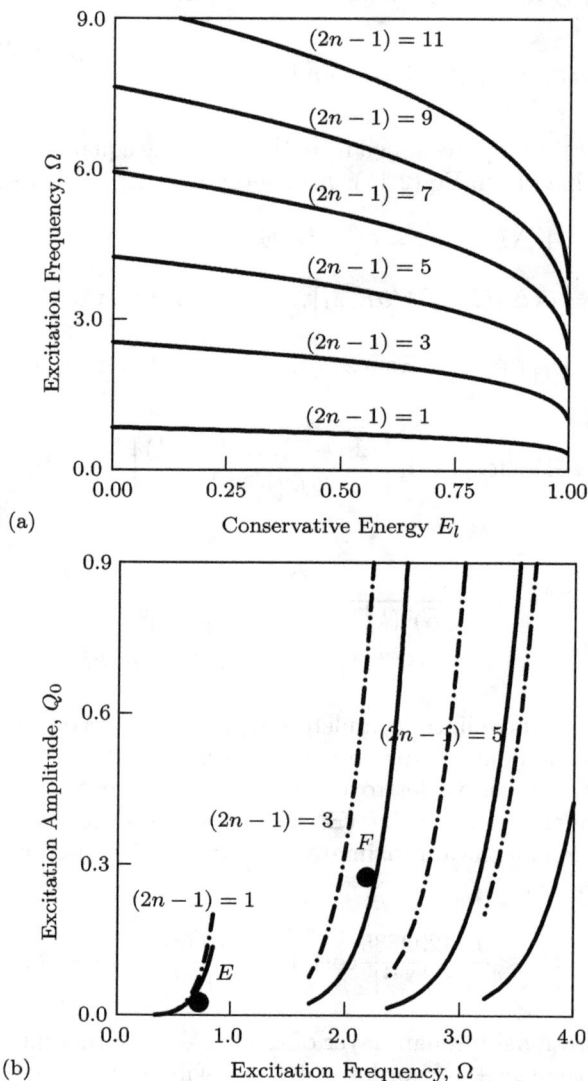

Fig. 2.4 The resonant conditions (upper) and the excitation strengths (lower) for the appearance (solid line) and disappearance (dash-dot line) of the resonant layer at $\alpha = 1$. Points E and F are used for simulation of the librational resonant layers.

2.3.2 *Rotational resonant layers*

In a similar manner, the linearization of Eq.(2.20) near $E_{i+1} = E_i = E_r^m$ and $\phi_{i+1} - \phi_i = 2m\pi$ gives a standard map as in Eq.(2.33), but

$$\Xi = 2Q_0\pi\text{sech}\left[\frac{m\pi K(k_r')}{K(k_r)}\right]|G_r^m| \qquad (2.38)$$

where

$$G_r^m = -\frac{\Omega(k_r^m)^3}{2(\sqrt{\alpha})^3[1-(k_r^m)^2]}E(k_r^m). \qquad (2.39)$$

Application of the critical value of the critical value ($\Xi^* = 0.9716354\cdots$) to Eq.(2.38) produces an analytical condition for the onset of the rotational resonant layer of the mth-order, i.e.,

$$Q_0 \approx \frac{0.4858177}{\pi|G_r^m|}\text{sech}\left[\frac{m\pi K(k_r')}{K(k_r)}\right]. \qquad (2.40)$$

When the overlap of two closest primary resonance occurs, the rotational resonant layer of the mth-order will disappear. Therefore, an approximate, analytical condition for the disappearance of the rotational resonant layer of the mth-order obtained from Luo (2008) is

$$Q_0 \approx \frac{|E_r^{m+1} - E_r^m|}{2\pi}\text{sech}\left[\frac{m\pi K(k_r')}{K(k_r)}\right]. \qquad (2.41)$$

The resonant conditions (upper) computed by Eq.(2.16) and the excitation strengths (lower) for the appearance (solid line) and disappearance (dash-dot line) of the rotational resonant layer by use of Eqs.(2.40) and (2.41) are shown in Fig.2.5. Also, points G and H will be used later to demonstrate the rotational resonant layer.

2.4 Numerical simulations

The motion of the periodically driven pendulum may move from a potential well into another and thus, the Poincaré mapping section is defined as

$$\Sigma = \{(x(t), y(t))|x(t) \bmod 2\pi, t = NT, N = 1, 2, 3, \cdots\}. \qquad (2.42)$$

To observe numerically the stochastic layers characterized by the librational and rotational resonance and resonant layers of the driven pendulum, the input parameters must be appropriately selected using the theoretical results given herein.

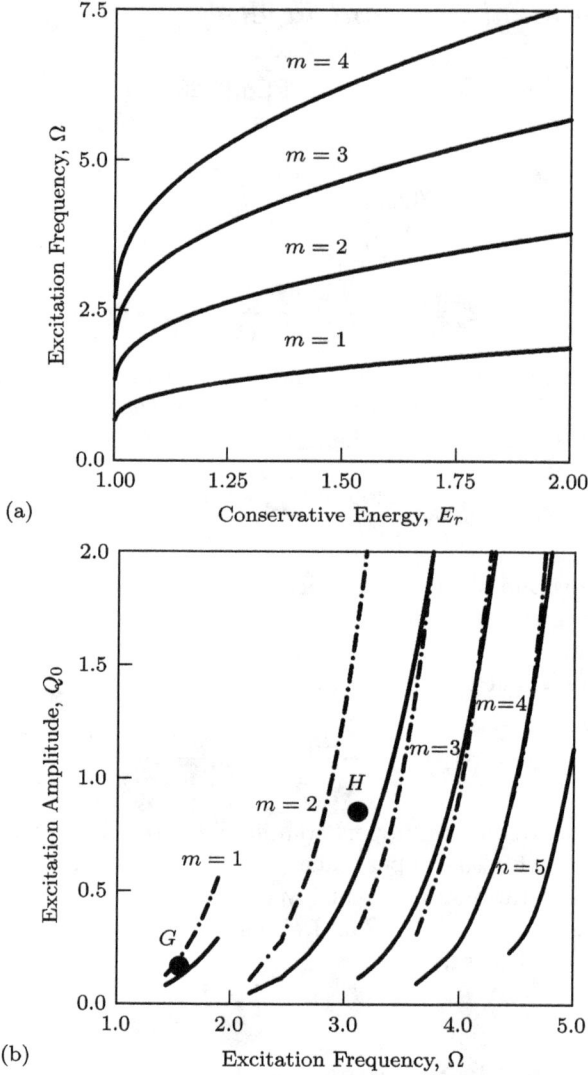

Fig. 2.5 The resonant conditions (upper) and the excitation strengths (lower) for the appearance (solid line) and disappearance (dash-dot line) of the resonant layer at $\alpha = 1$. Points E and F are used for simulation of the librational resonant layers.

Numerical simulations of the stochastic and resonant layers illustrated via Poincare mapping sections are performed by an adaptive Runge-Kutta integration technique with the computational precision 10^{-9} and the maximum and minimum time steps are respectively less than $10^{-5}T$ and $10^{-9}T$, where T is the excitation period. Also as an additional check of the accuracy, a symplectic integration scheme (see Feng and Qin (1991), Robert and Pau (1992)) is used. The results from two approaches are found to be the same.

(A) *Stochastic layers*

For the stochastic layer, analytical conditions plotted in Fig.2.3 and the initial condition at the saddle point $(\pi, 0)$ are employed.

Point $A(\Omega = 0.70, Q_0 = 0.0375, (2n - 1) = 1)$ is relative to the librational resonance of the first-order, and the stochastic layer is characterized by that librational resonance, as shown in Fig.2.6(a). The Poincaré mapping section at point $B(\Omega = 1.9, Q_0 = 0.15, (2n - 1) = 3)$ pertaining to the librational resonance of the third-order is shown in Fig.2.6(b). For the two cases, the stochastic layers are characterized by the librational resonance. For the stochastic layers characterized by the rotational resonance, the Poincaré mapping section of the stochastic layer at point $C(\Omega = 1.15, Q_0 = 0.022, m = 1)$

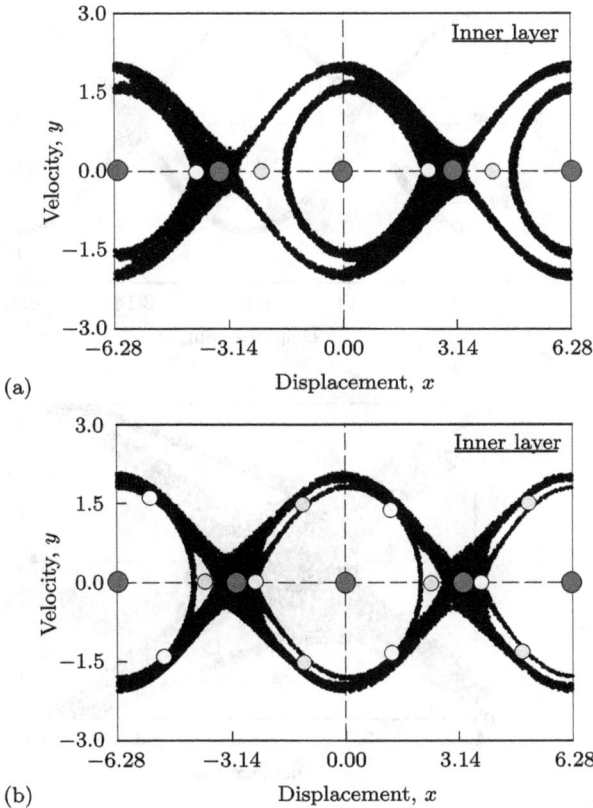

(a)

(b)

Fig. 2.6 (a,b) The stochastic layers characterized by the librational resonance for point $A(\Omega = 0.70, Q_0 = 0.0375, (2n - 1) = 1)$ and for point $B(\Omega = 1.9, Q_0 = 0.15, (2n - 1) = 3)$. (c,d) The stochastic layers characterized by the rotational resonance for point $C(\Omega = 1.15, Q_0 = 0.022, m = 1)$ and (d) for point $D(\Omega = 2.15, Q_0 = 0.1, m = 2)$. (e,f) Zoomed views for point $D(\alpha = 1)$.

(c)

(d)

(e)

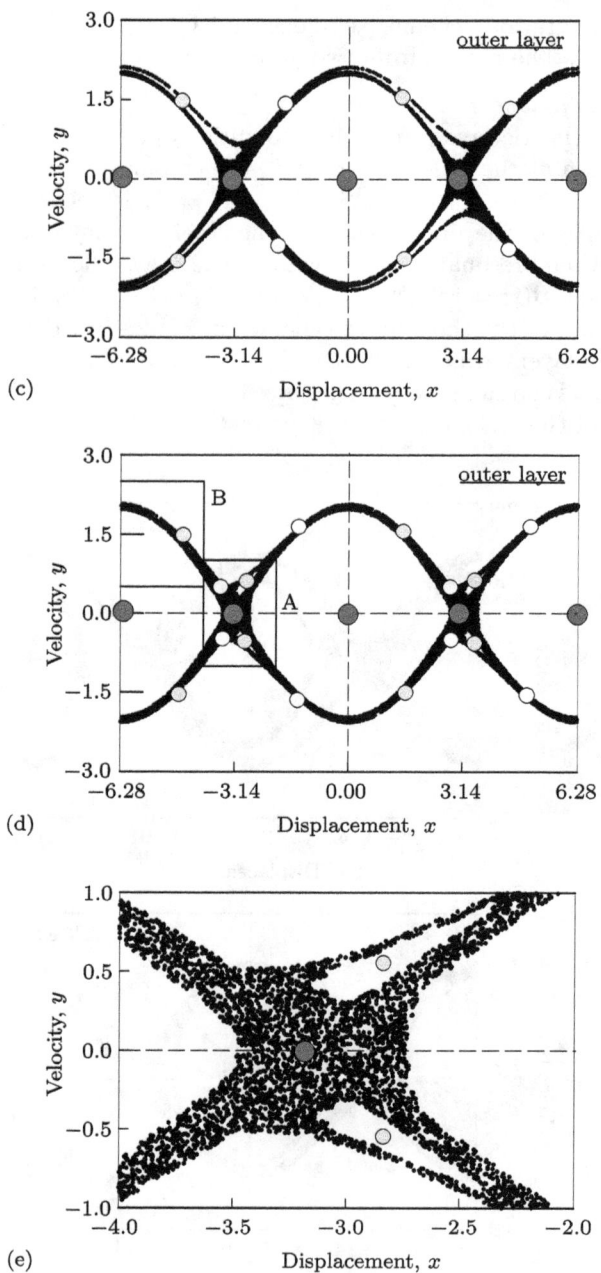

Fig. 2.6 Continued.

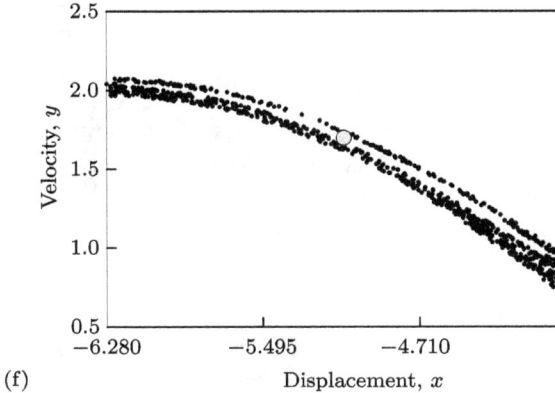

(f)   Displacement, x

Fig. 2.6 Continued.

related to the rotational resonance of the first-order is plotted in Fig.2.6(c). Similarly, at point $D(\Omega = 2.15, Q_0 = 0.1, m = 2)$, the second-order resonance embedded in the stochastic layer is observed in Fig.2.6(d). Zoomed views of the two windows in Fig.2.6(d) are plotted in Fig.2.6(e) and (f), respectively. It is observed that the resonant islands are generated by the rotational resonance of the second-order.

(B) *Resonant layers*

To demonstrate the resonant layers of the periodically forced pendulum, the excitation strength and frequency are chosen at points E to H. Based on the excitation frequency and resonant conditions, the initial conditions are at the saddles generated by the resonant orbits. For point $E(\Omega = 0.7284, Q_0 = 0.025, (2n-1) = 1)$, the resonant condition gives $E_0 = 0.5$ for the librational resonance of the first-order and the initial condition $(x_0 = 2.094395, y_0 = 0.0)$ at a saddle. Using such initial condition, the numerical simulation gives the librational resonant layer of the first-order in Fig.2.7(a). For point $F(\Omega = 2.18519, Q_0 = 0.275, (2n-1) = 3)$ and $E_0 = 0.5$, the initial condition for the librational resonance of the third-order is at a saddle $x_0 = 1.1476, y_0 = 1.3496$. Poincaré map sections in Fig.2.7(b) give the librational resonant layer of the third-order. For point $G(\Omega = 1.5561, Q_0 = 0.165, m = 1)$, the resonant condition gives $E_0 = 1.5$ and the initial conditions $(x_0 = 3.1416, y_0 = \pm 1.0)$ are two saddle points for the upper and lower branches of the rotational resonance of the first-order. The rotational resonant layer of the first-order is shown in Fig.2.7(c). Finally, for point $H(\Omega = 3.1122, Q_0 = 0.85, m = 2)$, two saddles $(x_0 = 1.9627, y_0 = \pm 1.4953)$ at $E_0 = 1.5$ are chosen to be the initial conditions. For this case, we demonstrate the rotational resonant layer of the second-order in Fig.2.7(d). The rotational resonant layer possessing an upper and lower branches is distributed in the vicinity of the separatrix generated by the rotational resonance of the second order.

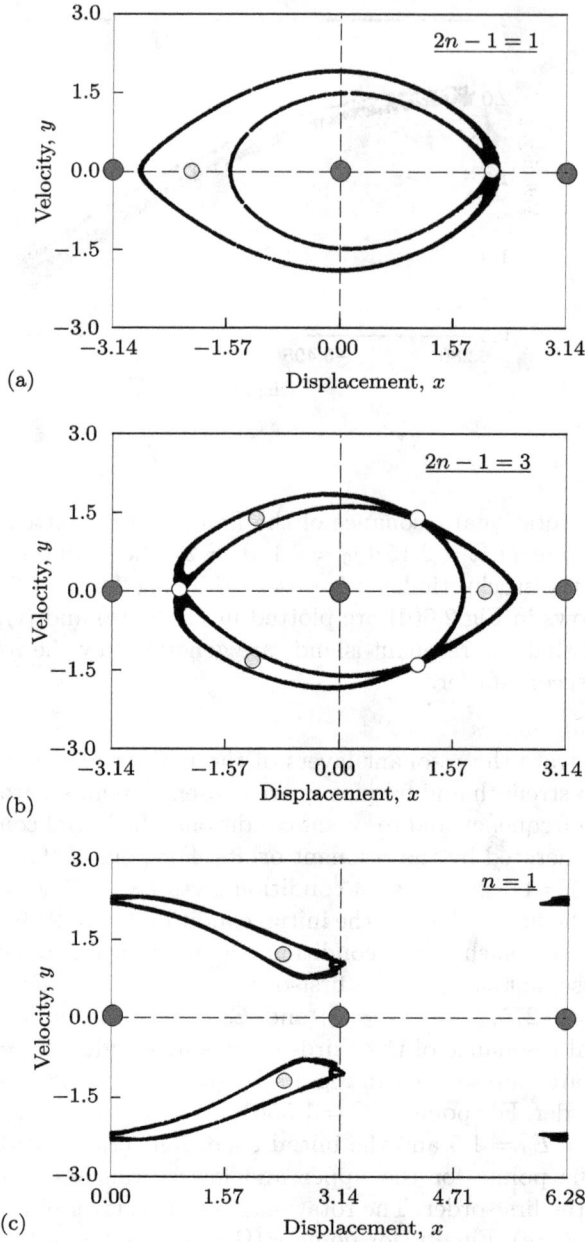

Fig. 2.7 The librational resonant layer: (a) for point $E(\Omega = 0.7284, Q_0 = 0.025, (2n - 1) = 1)$ (b) for point $F(\Omega = 2.18519, Q_0 = 0.275, (2n - 1) = 3)$. The rotational layer: (c) for point $G(\Omega = 1.5561, Q_0 = 0.165, m = 1)$, (d) for point $H(\Omega = 3.1122, Q_0 = 0.85, m = 2)$.

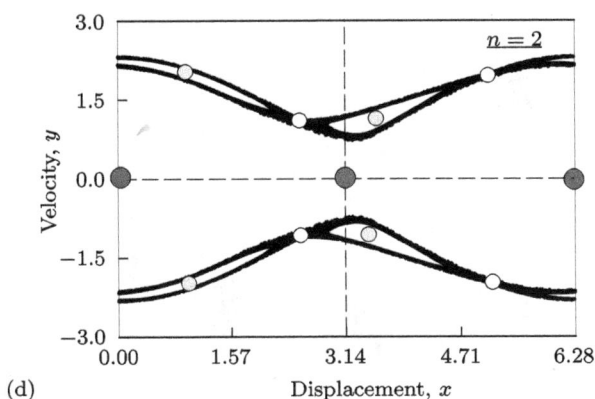

Fig. 2.7 Continued.

References

Chirikov, B.V., 1979, A universal instability of many-dimensional oscillator systems, *Physics Reports*, **52**, 263–379.

Feng K. and Qin M.Z., 1991, Hamiltonian algorithms for Hamiltonian systems and a comparative numerical study, *Computer Physics Communications*, **65**, 173–187.

Han, R.P.S. and Luo, A.C.J., 1998, Resonant layers in nonlinear dynamics, *ASME Journal of Applied Mechanics*, **65**, 727–736.

Lichtenberg, A.J. and Lieberman, M.A., 1992, *Regular and Chaotic Dynamics*, 2nd Edition, Springer, New York.

Luo, A.C.J., 1995, Analytical modeling of bifurcations, chaos and multifractals in nonlinear dynamics, *Ph.D. Dissertation*, University of Manitoba, Winnipeg, Manitoba, Canada.

Luo, A.C.J., 2008, *Global Transversality, Resonance and Chaotic Dynamics*, World Scientific, Singapore.

Luo, A.C.J., Gu, K. and Han, R.P.S., 1999, Resonant-separatrix webs in stochastic layers of the twin-well Duffing oscillator, *Nonlinear Dynamics*, **19**, 37–48.

Luo, A.C.J. and Han, R.P.S., 2001, The resonance theory for stochastic layers in nonlinear dynamical systems, *Chaos, Solitons and Fractals*, **12**, 2493–2508.

Robert, I. M. and Pau, A., 1992, The accuracy of symplectic integrators, *Nonlinearity*, **5**, 541–565.

Zaslavskii and Filonenko, N.N., 1968, Stochastic instability of trapped particles and conditions of application of the quasi-linear approximation, *Soviet Physics JETP*, **27**, 851–857.

Chapter 3
Parametric Chaos in Pendulum

Parametric chaos is chaotic motions in the stochastic and resonant layers of a parametrically excited dynamical system. In this chapter, parametric chaos in pendulum will discuss the Hamiltonian chaos in the stochastic and resonant layers in a parametrically excited pendulum. The $(2M : 1)$ -librational and $(M : 1)$-rotational resonances are discovered in the stochastic layer for a parametrically excited pendulum. The analytical conditions for the onset of a resonance in the stochastic layer are derived. Numerical predictions of the appearance of resonance in the stochastic layer are also completed. Illustrations of the stochastic layer in the parametrically excited pendulums are given through the Poincaré mapping sections.

3.1 Resonance and energy increment

In this section, we will consider a parametrically excited pendulum for a further understanding the mechanism of the resonant separatrix in 2-dimesnional, periodically forced, nonlinear Hamiltonian systems. The parametric pendulum is

$$\ddot{x} + (\alpha + Q_0 \cos \Omega t) \sin x = 0, \tag{3.1}$$

where Q_0 and Ω are excitation strength and frequency, respectively. The Hamiltonian of Eq.(3.1) is

$$H(x, y, t) = \frac{1}{2}y^2 - \alpha \cos x - Q_0 \cos \Omega t \cos x$$
$$= H_0(x, y) + H_1(x, y, t) \tag{3.2}$$

where

$$H_0 = \frac{1}{2}y^2 - \alpha \cos x \text{ and}$$

$$H_1 = -Q_0 \cos \Omega t \cos x = -\frac{1}{2}Q_0[\cos(x + \Omega t) + \cos(x - \Omega t)]; \tag{3.3}$$

with $y = \dot{x}$. In the conservative system of Eq.(3.1), elliptic points $(\pm 2j\pi, 0)$ and hyperbolic points $(\pm(2j+1)\pi, 0)$ for $(j = 0, 1, \cdots)$ exist. Two homoclinic orbits connecting all the hyperbolic points separate the phase space in the conservative system of Eq.(3.1) into the librational and rotational motions. The resonant layers located in regions associated with the librational and rotational motion regions are termed *the librational and rotational resonant layers* in the parametrically excited pendulum.

3.1.1 Libration

For the conservative energy $H_0 = E_l$ $(-\alpha \leqslant E_l < \alpha)$, the time-dependent Hamiltonian H_1 in Eq.(3.3) is approximated by

$$H_1^l \approx -Q_0 \cos(\Omega t) \cos x_l^0, \tag{3.4}$$

where the subscript (or superscript) l denotes libration. As in Luo and Han (2000), x_l^0 is the librational solution of the unperturbed pendulum,

$$x_l^0 = \pm 2j\pi \pm 2\arcsin\left[k_l \mathrm{sn}\left(\frac{2K(k_l)\theta_l}{\pi}, k_l\right)\right], \tag{3.5}$$

where sn is the Jacobi elliptic functions, and $k_l = \sqrt{1 + E_l/\alpha}/\sqrt{2}$ is the elliptic modulus. $K(k_l)$ is the elliptic integral of the first kind. Note that in traditional whisker map, $K(k) \approx \log(4/\sqrt{1-k^2})$ is used (e.g., in Chirikov (1979), Zaslavskii and Filonenko (1968)). The phase angle θ_l and natural frequency ω_l (or period T_l) are

$$\varphi_l = \omega_l t + \varphi_{l0} \quad \text{and} \quad \omega_l = \frac{2\pi}{T_l} = \frac{\pi\sqrt{\alpha}}{2K(k_l)}, \tag{3.6}$$

where φ_{l0} is the initial phase angle.

For the libration in a potential well, substitution of the Fourier expansion of x_l^0 into Eq.(3.4) leads to

$$H_1^l \approx -Q_0 \cos \Omega t$$
$$+ \frac{2\pi^2 Q_0}{K_l^2} \sum_{M=1}^{\infty} \sum_{m=1}^{\infty} \sum_{n=1}^{\infty} \left(p_{2m-1}^l p_{2n-1}^l \delta_{m+n-1}^M - p_{2m-1}^l p_{2n-1}^l \delta_{|m-n|}^M\right)$$
$$\times [\sin(2M\omega_l + \Omega)t + \sin(2M\omega_l - \Omega)t], \tag{3.7}$$

where the Delta function $\delta_i^j = 1 (i = j)$ or $0 (i \neq j)$, $| \cdot |$ denotes the absolute value, and

$$p_{2i-1}^l = \operatorname{csch} \left[\frac{(2i-1)\pi K_{2i-1}^{\prime l}}{2K_{2i-1}^l} \right] \quad \text{for } i \in \{m, n\}, \tag{3.8}$$

where $K_{2i-1}^{\prime l} = K(k_l')$ and $k_l' = \sqrt{1 - k_l^2}$. As in Chirikov (1979) (or Zaslavskii and Filonenko (1968)), all the other time-dependent terms in H_1^l will average to zero over one period $2\pi/\Omega$ except for the $(2M\omega_l - \Omega)$ terms. Hence, from Eq.(3.7), the librational resonant condition and the corresponding modulus for $H_0 < \alpha$ are

$$2M\omega_l = \Omega \quad \text{and} \quad k_l^{(2M:1)} = \frac{\sqrt{\alpha + E_l^{(2M:1)}}}{\sqrt{2\alpha}}, \tag{3.9}$$

where $E_l^{(2M:1)}$ denotes the conservative energy related to the $(2M : 1)$-librational resonance. Note that in Luo and Han (2001), a resonance of $m\omega = n\Omega$ is termed the $(m : n)$-resonance.

Using Eq.(3.5), the approximate energy increment based on the unperturbed orbit of $(2M : 1)$-librational resonance is computed as in Melnikov (1963) and Chirikov (1979), i.e.,

$$\Delta H_0^l = \int_{t_0}^{T_l + t_0} [H_0, H_1] \, dt = \int_{t_0}^{T_l + t_0} (f_1 g_2 - f_2 g_1) \, dt \approx Q_0 Q_l^{(2M:1)} \sin \Omega t_0 \tag{3.10}$$

and

$$Q_l^{(2M:1)} \approx \frac{4\pi^2}{\sqrt{\alpha} K_l \Omega} \sum_{m=1}^{\infty} \sum_{n=1}^{\infty} (p_{2m-1}^l p_{2n-1}^l \delta_{m+n-1}^M - p_{2m-1}^l p_{2n-1}^l \delta_{|m-n|}^M), \tag{3.11}$$

where $[\cdot, \cdot]$ represents the Poisson bracket and

$$\left. \begin{array}{l} f_1 = y \approx y_l^0, \\ g_1 = 0; \\ f_2 = \alpha \sin x \approx \alpha \sin x_l^0, \\ g_2 = Q_0 \cos \Omega t \sin x \approx Q_0 \cos \Omega t \sin x_l^0; \end{array} \right\} \tag{3.12}$$

where $y = \dot{x}$.

3.1.2 Rotation

In a like fashion, for $H_0 = E_r > \alpha$, the time-dependent Hamiltonian H_1 in Eq.(3.3) for the rotation is approximated by

$$H_1^r \approx -Q_0 \cos \Omega t \cos x_r^0, \tag{3.13}$$

where the subscript (or superscript) r denotes the rotation. x_r^0 is the rotational solution of the unperturbed pendulum:

$$x_r^0 = \pm 2\mathrm{am}\left(\frac{K(k_r)\varphi_r}{\pi}, k_r\right); \tag{3.14}$$

where am is the amplitude of Jacobi elliptic function. The modulus $k_r = \sqrt{2\alpha}/\sqrt{E_r + \alpha}$ and phase $\varphi_r = \omega_r t + \varphi_{r0}$. The natural frequency ω_r (or period T_r) is

$$\omega_r = \frac{2\pi}{T_r} = \frac{\pi\sqrt{\alpha}}{k_r K(k_r)}. \tag{3.15}$$

Substitution of Eq.(3.14) into H_1 in Eq.(3.13) and the Fourier expansion of the elliptic function (am) gives

$$H_1^r \approx -Q_0 \cos \Omega t \tag{3.16}$$

$$+\frac{\pi^2 Q_0}{k_r^2 K_r^2} \sum_{M=1}^{\infty} \sum_{m=1}^{\infty} \sum_{n=1}^{\infty} (p_{2m-1}^r p_{2n-1}^r \delta_{m+n-1}^M - p_{2m-1}^r p_{2n-1}^r \delta_{|m-n|}^M)$$

$$\times [\sin(M\omega_r + \Omega)t + \sin(M\omega_r - \Omega)t].$$

Therefore, in the vicinity of the $(M:1)$ primary resonance, all other terms in H_1^r in Eq.(3.16) will average to zero over a period $2\pi/\Omega$ except the $(M\omega_r - \Omega)$ term. Thus, the rotational resonant condition and the corresponding modulus for $H_0 > \alpha$ are

$$M\omega_r = \Omega \quad \text{and} \quad k_r^{(M:1)} = \frac{\sqrt{2\alpha}}{\sqrt{\alpha + E_r^{(M:1)}}}, \tag{3.17}$$

where $E_r^{(M:1)}$ denotes the conservative energy associated with the $(M:1)$-rotational resonance.

The energy increment based on two unperturbed orbit of $(M:1)$-rotational resonance is approximately computed through Eq.(3.14), i.e.,

$$\Delta H_0^r = 2 \int_{t_0}^{T_r+t_0} [H_0, H_1]\, dt = 2 \int_{t_0}^{T_r+t_0} (f_1 g_2 - f_2 g_1)\, dt \approx 2 Q_0 Q_r^{(M:1)} \sin(\Omega t_0), \tag{3.18}$$

where

$$Q_r^{(M:1)} \approx \frac{2\pi^2}{\sqrt{\alpha} k_r K_r \Omega} \sum_{m=0}^{\infty} \sum_{n=1}^{\infty} (p_{2m-1}^r p_{2n-1}^r \delta_{m+n-1}^M - p_{2m-1}^r p_{2n-1}^r \delta_{|m-n|}^M). \tag{3.19}$$

3.2 Parametric stochastic layers

As in Luo (2001b), in this section, the analytical and numerical predictions of the onset of the $(2M : 1)$-librational and $(M : 1)$-rotational resonance in the stochastic layer are presented for the parametrically excited pendulum. A relationship between excitation frequency and amplitude will be developed for the onset of resonance.

3.2.1 Analytic predictions

As in Chirikov (1979), from the energy increments (ΔH_0^l and ΔH_0^r) and phase changes, the whisker maps are:

$$E_{i+1} \approx E_i + Q_0 Q_l^{(2M:1)} \sin \varphi_i \quad \text{and} \quad \varphi_{i+1} = \varphi_i + \frac{4\Omega K(k_l)}{\sqrt{\alpha}}, \qquad (3.20)$$

for the presence of the $(2M : 1)$-librational resonance in the stochastic layer, and

$$E_{i+1} \approx E_i + 2Q_0 Q_r^{(M:1)} \sin \varphi_i \quad \text{and} \quad \varphi_{i+1} = \varphi_i + \frac{2\Omega k_r K(k_r)}{\sqrt{\alpha}}, \qquad (3.21)$$

for the presence of the $(M : 1)$-rotational resonance in the stochastic layer, where $\varphi_i = \Omega t_i$, E_i is the conservative energy and $\Delta H_0 = E_{i+1} - E_i$, $\Delta \varphi_i = \varphi_{i+1} - \varphi_i$.

When the perturbation of the perturbed nonlinear Hamiltonian systems becomes stronger, the perturbed orbit will go through more and more unperturbed, periodic orbits. Therefore, the resonant interaction between the Hamiltonian system and its perturbation increases with perturbation strength. The resonant interaction leads to a new resonant overlap, and such a resonant overlap generates a new stochastic layer different from the old stochastic layer because a new, specific resonant-separatrix web is formed in the stochastic layer. Thus, in the second equation of Eq.(3.20), $\varphi_{i+1} - \varphi_i = 2(2M)\pi$ gives the resonant energy $E_l^{(2M:1)}$. Let $E_{i+1} = E_l^{(2M:1)}$ and $E_i = \alpha$, the excitation strength Q_0 for the onset of the $(2M : 1)$-order librational resonance in the stochastic layer, is estimated by

$$Q_0 = \frac{|E_l^{(2M:1)} - \alpha|}{Q_l^{(2M:1)}} \quad \text{and} \quad \Omega = \frac{2M\pi\sqrt{\alpha}}{K(k_l^{(2M:1)})} \qquad (3.22)$$

where

$$k_l^{(2M:1)} = \sqrt{\alpha + E_l^{(2M:1)}}/\sqrt{2\alpha}. \qquad (3.23)$$

From the standard mapping, the strength of the stochasticity parameter is $K = K^* \approx 0.9716 \cdots$ in Greene (1968, 1979) for the transition to global stochasticity. Therefore, for the special case, the excitation strength of the parametrically forced pendulum can be estimated. From the accurate standard mapping approach (e.g., Luo, 2008; Luo, 2001), an approximate critical condition for the onset of the librational resonance in the stochastic layer is

$$Q_0 \approx \frac{0.9716354}{Q_l^{(2M:1)}|G_l^{(2M:1)}|}, \tag{3.24}$$

where

$$G_l^{(2M:1)} = \frac{\Omega}{(\sqrt{\alpha})^3 (k_l^{(2M:1)})^2} \left[\frac{E(k_l^{(2M:1)})}{1 - (k_l^{(2M:1)})^2} - K(k_l^{(2M:1)}) \right]. \tag{3.25}$$

Similarly, when $E_{i+1} = E_r^{(M:1)}, E_i = \alpha$ and the second equation of Eq.(3.21) satisfies the resonant condition in Eq.(3.17), the excitation strength Q_0 for the onset of the $(M:1)$-rotational resonance in the stochastic layer is

$$Q_0 = \frac{|E_r^{(M:1)} - \alpha|}{2Q_r^{(M:1)}} \quad \text{and} \quad \Omega = \frac{M\pi\sqrt{\alpha}}{k_r^{(M:1)} K(k_r^{(M:1)})}, \tag{3.26}$$

where

$$k_r^{(M:1)} = \sqrt{2\alpha}/\sqrt{\alpha + E_r^{(M:1)}}. \tag{3.27}$$

Based on the accurate standard mapping approach, we have

$$Q_0 \approx \frac{0.9716354}{2Q_r^{(M:1)}|G_r^{(M:1)}|}, \tag{3.28}$$

where

$$G_r^{(M:1)} = -\frac{\Omega(k_r^{(M:1)})^3}{2(\sqrt{\alpha})^3[1 - (k_r^{(M:1)})^2]} E(k_r^{(M:1)}). \tag{3.29}$$

3.2.2 Numerical predictions

An energy spectrum approach developed in Luo et al. (1999) will be used for numerical prediction of the onset of resonance in the stochastic layer. In the energy spectrum, the maximum and minimum conservative energies are computed through the Poincaré mapping section. As in Luo and Han (2000), the Poincaré mapping section for the parametrically excited pendulum is

$$\Sigma = \{(\text{mod}(x(t_N), 2\pi), \dot{x}(t_N))|t_N = NT + t_0, T = 2\pi/\Omega, N = 0, 1, \cdots\}, \tag{3.30}$$

where $x(t_N) = x_N, \dot{x}(t_N) = \dot{x}_N$ and $x(t_0) = x_0, \dot{x}(t_0) = \dot{x}_0$ at $t = t_0$ are the initial conditions. The Poincaré map is: $\Sigma \to \Sigma$. The conservative energy for each Poincaré mapping point of the parametrically excited pendulum is

$$H_0^{(N)} = \frac{1}{2}\dot{x}_N^2 - \alpha \cos x_N \qquad (3.31)$$

and the corresponding minimum and maximum energies are defined by

$$E_{\max} = \max_{N \to \infty}\{H_0^{(N)}\} \quad \text{and} \quad E_{\min} = \min_{N \to \infty}\{H_0^{(N)}\}. \qquad (3.32)$$

For $Q_0 = 0.05$ and $\alpha = 1.0$ in Eq.(3.1), the maximum and minimum energy spectra are computed, as shown in Fig.3.1. The maximum and minimum energies are computed from 10000 iterations of Poincaré map for specific excitation strength. The critical excitation frequency $\Omega_l^{(2M:1)}$ (or $\Omega_r^{(M:1)}$) is located at energy jumps in the spectrum when appearance of the $(2M : 1)$-librational (or $(M : 1)$-rotational) resonance occurs in the stochastic layer. For a clear view of energy jump, four specific areas in the spectra are shown in detailed. The maximum energy jumps occur at $\Omega_r^{(1:1)} \approx 1.55$, $\Omega_r^{(2:1)} \approx 2.68$, $\Omega_r^{(3:1)} \approx 3.41$ and $\Omega_r^{(4:1)} \approx 4.15$ for the rotational stochastic layer, and the minimum energy jumps at $\Omega_l^{(2:1)} \approx 1.4$, $\Omega_l^{(4:1)} \approx 2.545$, $\Omega_l^{(6:1)} \approx 3.35$ and $\Omega_l^{(8:1)} \approx 4.05$ for the librational stochastic layer. If $\Omega > \Omega_r^{(2:1)} \approx 2.68$ for $Q_0 = 0.05$, then $\Omega > \Omega_l^{(2:1)}$ and $\Omega_l^{(4:1)}$. This implies that the $(1 : 1)$-rotational resonance and the $(2 : 1)$- and $(4 : 1)$-librational resonance will not appear in the stochastic layer.

The layer width is the minimum distance between inner and outer boundaries of the stochastic layer. As in Luo et al. (1999), the width is computed from the minimum and maximum energy spectra, i.e.,

$$w \equiv \min_{t \in [0,\infty)} ||\mathbf{x}(E^{\max}, t) - \mathbf{x}(E^{\min}, t)|| \equiv ||\mathbf{x}^{\max} - \mathbf{x}^{\min}|| \qquad (3.33)$$

where $|| \cdot ||$ is a norm. Two points $\mathbf{x}^{\max}$ and $\mathbf{x}^{\min}$ on the normal line of $f^{\pm}(\mathbf{x}_0) = (-f_2(\mathbf{x}_0), f_1(\mathbf{x}_0))^T$ of the tangential vector of separatrix at point $\mathbf{x}_0$ are the closest between the maximum and minimum energy orbits $\mathbf{x}(E_{\max}, t)$ and $\mathbf{x}(E_{\min}, t)$, which can be obtained by Eq. (3.3) with $E_{\max}$ and $E_{\min}$. The stochastic layer of the $(4 : 1)$-librational and $(2 : 1)$-rotational resonance is used for a better understanding of the above definition, as illustrated in Fig.3.2. The homoclinic orbit and the librational and rotational unperturbed orbits relative to the minimum $(E_{\min})$ and maximum $(E_{\max})$ energies are depicted in Fig.3.2, and the layer width (w) is shown as well. This is located at $x = \pm 2m\pi (m = 0, 1, 2, \cdots)$. The minimum and maximum energies in Fig.3.1 are used for computing the width of stochastic layer, as shown in Fig. 3.2. It is observed that the width of stochastic layer depends on the order of resonance. The layer width decreases with increasing resonance order for

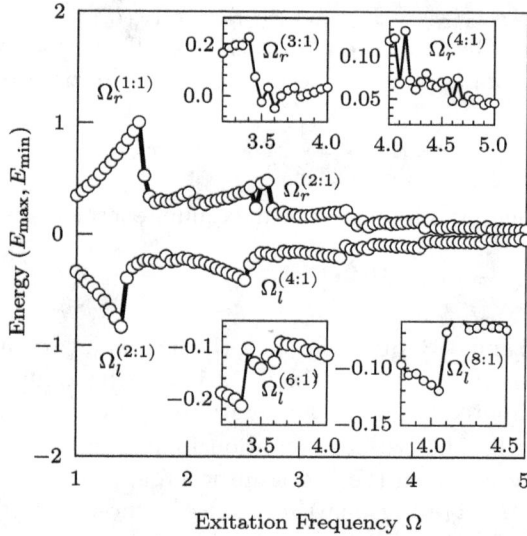

Fig. 3.1 Maximum and minimum energy spectra for stochastic layers in a parametrically forced pendulum with parameters ($\alpha = 1.0, Q_0 = 0.05$).

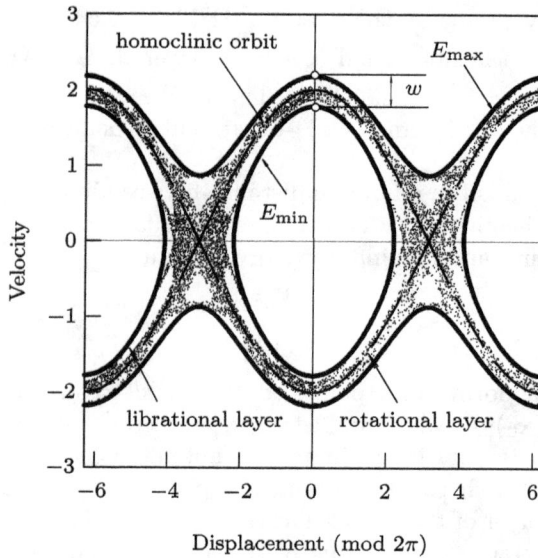

Fig. 3.2 Stochastic layer width (w) defined for a parametrically excited pendulum with parameters ($\alpha = 1.0$, $Q_0 = 0.05$, $\Omega = 2.5$).

a specific, excitation strength. Also, the magnitude of the stochastic layer width is not the same order of the magnitude of the excitation strength. Such a conclusion cannot be achieved from the traditional perturbation analysis.

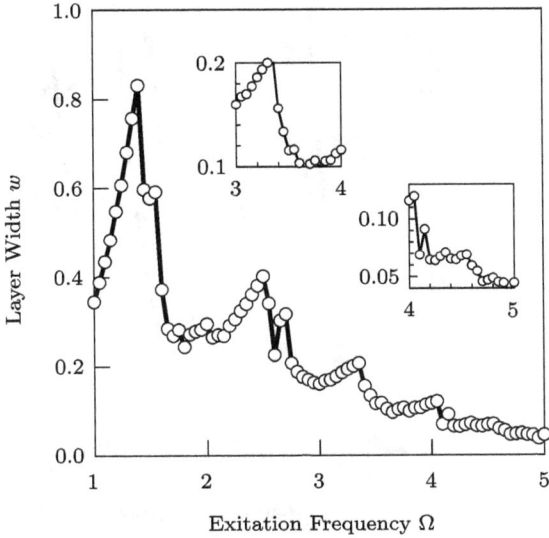

Fig. 3.3 The stochastic layer width of a parametrically excited pendulum with parameters ($\alpha = 1.0$, $Q_0 = 0.05$) at $x = \pm 2m\pi$ ($m = 0, 1, 2, \cdots$).

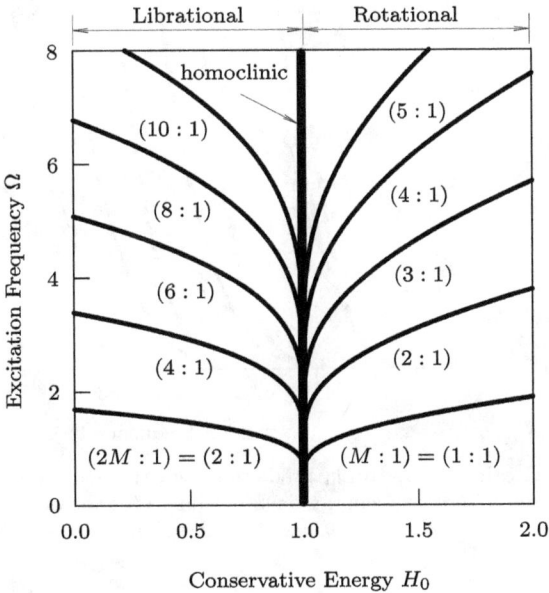

Fig. 3.4 Resonant conditions in the vicinity of homoclinic orbit for the parametrically excited pendulum at $\alpha = 1$.

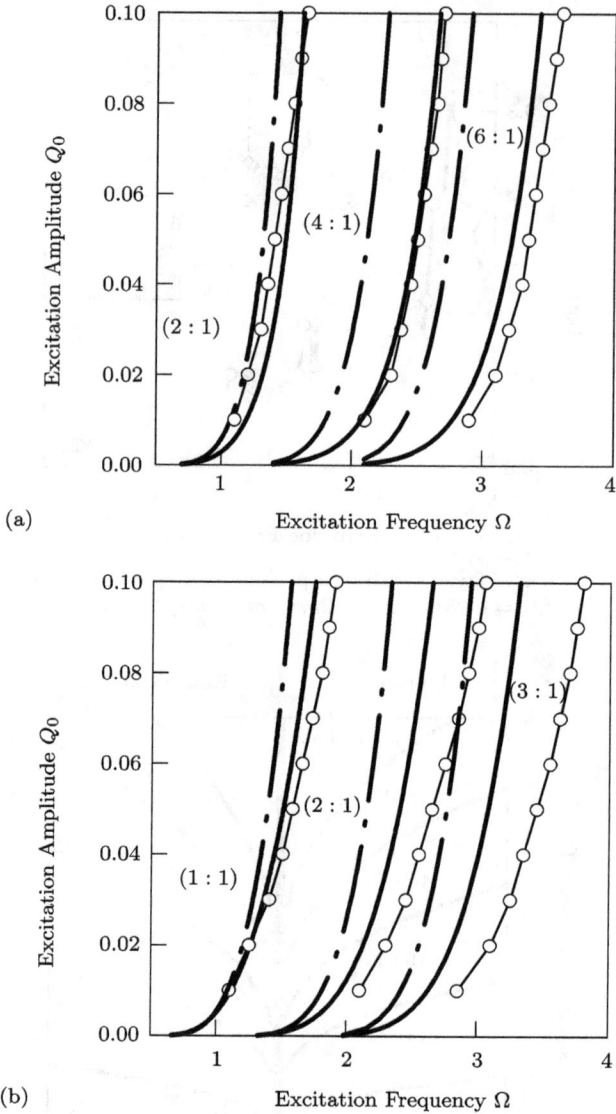

Fig. 3.5 The excitation strength Q_0 predicted from the accurate standard map by solid curves and the incremental energy dash-dot curve for the onset of (a) librational and (b) rotational resonance $\alpha = 1$. The circular symbol curves give the numerical prediction.

3.2.3 Illustrations

The $(2M : 1)$-librational and $(M : 1)$-rotational resonant conditions in the vicinity of homoclinic orbit are presented through the excitation frequency Ω versus the conservative energy H_0 in Fig.3.4. Accordingly, the conditions for the onset of a resonance in the stochastic layer are presented in Fig.3.5 through the excitation frequency and strength. The solid curves give the analytical predictions of excitation strength from Eqs.(3.22) and (3.26), and the dashed curves represent the analytical conditions from Eqs.(3.24) and (3.28). The upper and lower plots are for the librational and rotational stochastic layers, respectively. The numerical results for the onset of the resonance in stochastic layer are generated by the energy spectrum approach. Such results are also plotted in Fig.3.5 through the circular-symbol-curves for comparison of analytical predictions. In Fig.3.5, it is observed that the accurate standard map approach gives a better prediction than the incremental energy approach for the parametrically excited pendulum. Two analytical predictions are different from the numerical ones because the energy increments are computed approximately through the unperturbed orbits instead of the perturbed ones. The accuracy of energy increment approximation also decreases with increasing the resonance order, and effects of the sub-resonance in the above analysis of stochastic layers are not considered yet. Fig.3.4 gives a relationship between the conservative energy and excitation frequency when the specified resonance occurs. This relationship helps us determine the location of resonance. For the given excitation frequency, the conservative energies are determined for all the possible resonance in librational and rotational motions. Fig.3.5 shows a relationship between the excitation strength and frequency for the onset of a specified resonant separatrix in the stochastic layer. For a specified excitation frequency, the lower order resonant separatrices will appear with increasing excitation strength. For a given excitation strength, the lower order resonant separatrices will disappear with increasing excitation frequency.

3.2.4 Numerical simulations

Since the parametrically excited pendulum has infinite potential wells, its motions may move from one potential well into another through the rotation (see Eqs.(3.5) and (3.14)) and thus, the Poincaré mapping section in Eq.(3.30) is used for numerical simulations of the stochastic layer. All the input parameters must be appropriately selected from the above analysis. The resonant characteristics of the stochastic layer are simulated using the symplectic integration scheme. Therefore, consider the initial condition at $(\pi, 0)$ and excitation frequency and strength close to those critical values.

Consider a specific excitation frequency $\Omega = 1.65$. From the numerical prediction, the critical excitation strengths for the onset of the (1 : 1)-rotational and (2 : 1)-librational resonance in the stochastic layer are $Q_{0r}^{(1:1)} \approx 0.05$ and $Q_{0l}^{(2:1)} \approx 0.095$, respectively. For clear observation of such resonant characteristics in the stochastic layer, excitation strengths selected for numerical simulations are close to those critical values (i.e., $Q_0 = 0.06 > Q_{0r}^{(1:1)}$ and $Q_0 = 0.1 > Q_{0l}^{(2:1)}$). In the upper plot of Fig.3.6, the stochastic layer is characterized by the (1 : 1)-rotational resonance because of the excitation frequency greater than the critical value ($Q_0 = 0.06 > Q_{0r}^{(1:1)}$). Note that the specified resonant separatrix in the stochastic layer appears through one of the clockwise and counter clockwise motions. When excitation strength increases over some specific value, the resonant separatrix will appear through both of clockwise and counter clockwise motions. When excitation strength increases over the critical value ($Q_{0l}^{(2:1)} \approx 0.095$) for the (2:1)-librational resonance, this resonance will be involved in the stochastic layer. For instance, in the lower plot of Fig.3.6, $Q_0 = 0.1 > Q_{0l}^{(2:1)}$ is selected for the numerical simulation of the stochastic layer. Therefore, the stochastic layer characterized by the (2 : 1)-librational resonance is observed. It also is clearly observed that the layer in the vicinity of the (1 : 1)-rotational resonant separatrix becomes thicker due to the excitation strength greater than the critical values for the onset of (1 : 1)-rotational resonance ($Q_0 = 0.1 \gg Q_{0r}^{(1:1)} \approx 0.05$).

For a specific strength of excitation, the resonant characteristics of stochastic layers varying with excitation frequency are also simulated herein. At $Q = 0.1$, the numerical prediction gives the critical excitation frequency $\Omega_r^{(2:1)} \approx 3.05$ and $\Omega_l^{(4:1)} \approx 2.75$ for the onset of the (2 : 1)-rotational and (4 : 1)-librational resonance in the stochastic layer, respectively. As discussed in the previous section, excitation frequency is greater than critical values in which the corresponding resonance disappears in the stochastic layer. Therefore, consider the excitation frequency less than the aforementioned critical excitation frequencies ($\Omega = 2.7 < \Omega_l^{(4:1)}$ and $\Omega = 3.0 < \Omega_r^{(2:1)}$) for illustration of the stochastic layers. Due to the excitation frequency less than the critical values for the disappearance of the (4 : 1)-librational and (2 : 1)-rotational resonant separatrices ($\Omega = 2.7 < \Omega_l^{(4:1)}$ and $\Omega_r^{(2:1)}$), it indicates that the two resonant separatrices appear in the stochastic layers, and such a phenomenon are clearly observed in the upper plot of Fig.3.7. For $\Omega = 3.0$, the excitation frequency is located between the critical values ($\Omega_l^{(4:1)} < \Omega < \Omega_r^{(2:1)}$). From the analytical numerical predictions, only the (2 : 1)-rotational resonant separatrix can be observed. Such a phenomenon is simulated numerically and shown clearly in the lower plot of Fig.3.7. Besides, the other higher than (2 : 1)-librational and (4 : 1)-rotational resonant separatrices are involved in the stochastic layer and have been destroyed by the sub-resonance, which cannot clearly observed.

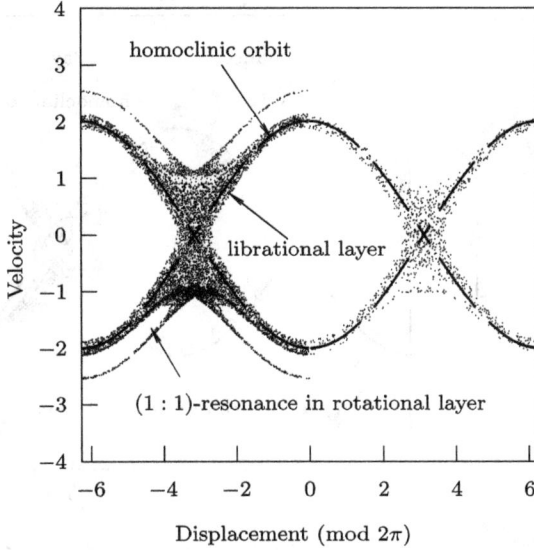

(a)

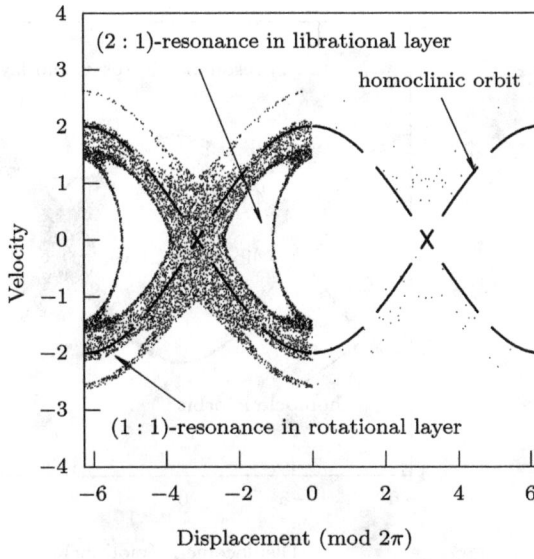

(b)

Fig. 3.6 (a) The $(1:1)$ rotational resonance (upper) for $\Omega = 1.65$ and $Q_0 = 0.06 > Q_{0r}^{(1:1)} \approx 0.05$ and (b) the $(2:1)$-librational and $(1:1)$-rotational resonance (lower) for $Q_0 = 0.1 > Q_{0l}^{(2:1)} \approx 0.95$ and $\Omega = 1.65$ in the stochastic layers.

(a)

(b)

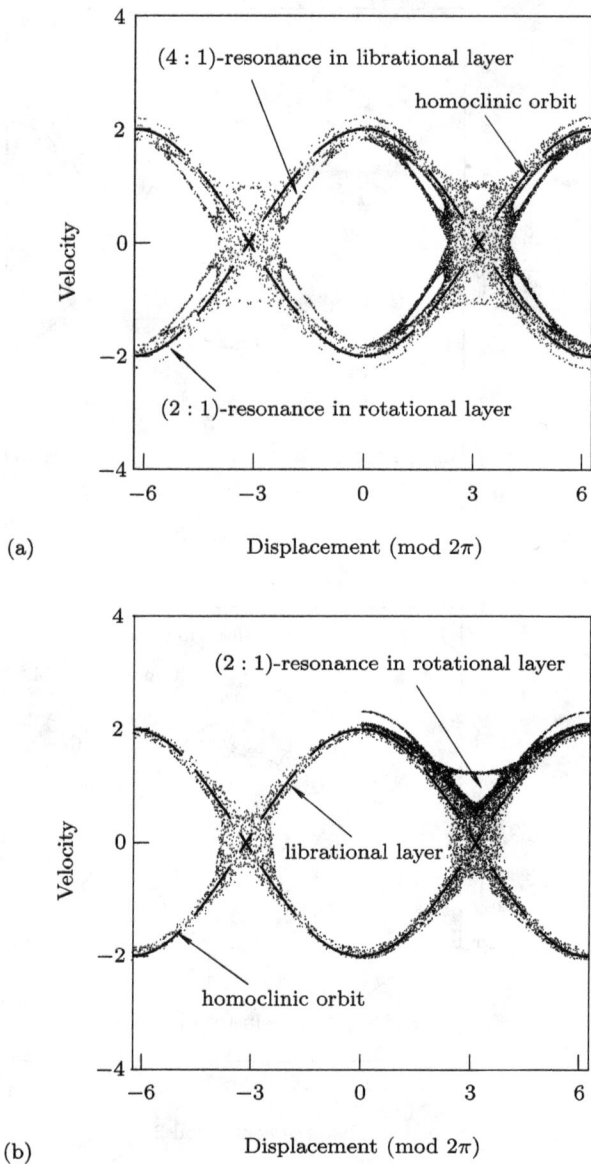

Fig. 3.7 The $(4 : 1)$-librational and $(2:1)$-rotational resonance (upper) for $\Omega = 2.7 < \Omega_l^{(4:1)} \approx 2.75$ and $Q_0 = 0.1$ and the $(2 : 1)$ rotational resonance (lower) for $\Omega = 3.0 < \Omega_r^{(2:1)} \approx 3.05$ and $Q_0 = 0.1$ in the stochastic layer.

3.3 Parametric resonant layers

As in Luo (2002), the stochastic and resonant layers in a parametric pendulum will be presented herein first since the parametric pendulum possesses qualitative, dynamic characteristics in other nonlinear Hamiltonian systems. For instance, the resonant layer in the periodically forced pendulum in Luo and Han (2000) possesses the similar characteristics of the stochastic layer in the parametric pendulum.

3.3.1 Approximate predictions

(A) *Librational Resonant Layer.* From a whisker map in Eq.(3.20) for the $(2M : 1)$-librational resonant layer, the accurate standard mapping approach (e.g., Luo, 2008; Luo, 2001a) gives an approximate, critical condition for onset of the librational resonant layer as

$$Q_0 \approx \frac{0.9716354}{2Q_l^{(2M:1)}|G_l^{(2M:1)}|},\tag{3.34}$$

where

$$G_l^{(2M:1)} = \frac{\Omega}{(\sqrt{\alpha})^3(k_l^{(2M:1)})^2}\left[\frac{K(k_l'^{(2M:1)})}{1-(k_l^{(2M:1)})^2}-K(k_l^{(2M:1)})\right],$$

$$k_l^{(2M:1)} = \sqrt{\alpha+E_l^{(2M:1)}}/\sqrt{2\alpha}\tag{3.35}$$

Let $E_i^l = E_l^{(2M:1)}$ and $E_{i+1}^l = E_l^{(2M+2:1)}$ and the excitation strength Q_0 for the $(2M : 1)$-librational resonant layer destroyed by the $(2M + 2 : 1)$ resonance is approximately predicted from Luo (2008) (also see, Luo, 1995; Han and Luo, 1998) by

$$Q_0 = \frac{|E_l^{(2M+2:1)} - E_l^{(2M:1)}|}{2Q_l^{(2M:1)}} \quad \text{and} \quad \Omega = \frac{2M\pi\sqrt{\alpha}}{K(k_l^{(2M:1)})}.\tag{3.36}$$

Setting $E_i^l = E_l^{(2M:1)}$ and $E_{i+1}^l = \alpha$, the excitation strength Q_0 for the $(2M : 1)$-order librational resonant layer involved in the stochastic layer is roughly estimated as in Eq.(3.22).

(B) *Rotational Resonant Layer.* Similarly, the energy increment (ΔH_0^r) and phase change $(\Delta\varphi_i^r = \Omega T_r)$ generate a whisker map in Eq.(3.21) for the $(M : 1)$-rotational resonant layer. From the accurate standard mapping approach in Luo (2001a), the condition for the appearance of the $(M : 1)$-rotational resonant layer is from Luo (2008), i.e.,

$$Q_0 \approx \frac{0.9716354}{2Q_r^{(M:1)}|G_r^{(M:1)}|},\tag{3.37}$$

where

$$G_r^{(M:1)} = -\frac{\Omega(k_r^{(M:1)})^3}{2(\sqrt{\alpha})^3[1-(k_r^{(M:1)})^2]}E(k_r^{(M:1)}),\tag{3.38}$$

$$k_r^{(M:1)} = \sqrt{2\alpha}/\sqrt{\alpha + E_r^{(M:1)}}.$$

When $E_i^r = E_l^{(M:1)}$ and $E_{i+1}^r = E_l^{(M+1:1)}$ are used, the excitation strength Q_0 for the $(M:1)$-rotational resonant layer destroyed by the $(M+1:1)$-resonance is approximated from Luo (2008) (also see Luo, 1995; Han and Luo,1998), i.e.,

$$Q_0 = \frac{|E_r^{(M+1:1)} - E_r^{(M:1)}|}{2Q_r^{(M:1)}} \quad \text{and} \quad \Omega = \frac{M\pi\sqrt{\alpha}}{k_r^{(M:1)}K(k_r^{(M:1)})}.\tag{3.39}$$

Setting $E_i^r = E_r^{(M:1)}$ and $E_{i+1}^r = \alpha$, the excitation strength Q_0 for the $(M:1)$-rotational resonant layer involved in the stochastic layer is estimated as in Eq.(3.26).

3.3.2 Numerical illustrations

(A) *Energy Increment Spectrum*: In the previous analytical predictions, the energy increment is approximated. As stated before, the exact energy increment can be numerically computed. Thus, Luo et al. (1999) developed the energy spectrum method, which is used to determine the onset of resonance in the stochastic layer. The stochastic and resonant layers for the parametrically excited pendulum are presented through the Poincaré mapping section. To detect the energy changes due to the sub-resonance in the specified resonant layer, the minimum and maximum energy increments with respect to the unperturbed resonant orbit are computed by

$$\Delta E_{\max} = \max_{N \to \infty}\{H_0^{(N)} - E_l^{(2M:1)}\} \quad \text{and} \quad \Delta E_{\min} = \min_{N \to \infty}\{H_0^{(N)} - E_l^{(2M:1)}\}\tag{3.40}$$

for the $(2M:1)$-librational resonant layer, and

$$\Delta E_{\max} = \max_{N \to \infty}\{H_0^{(N)} - E_r^{(M:1)}\} \quad \text{and} \quad \Delta E_{\min} = \min_{N \to \infty}\{H_0^{(N)} - E_r^{(M:1)}\}\tag{3.41}$$

for the $(M:1)$-rotational resonant layer. Such an energy increment versus excitation frequency is called the energy increment spectrum in Luo (2008). For $Q_0 = 0.05$ and $\alpha = 1.0$ in Eq.(3.1), the maximum and minimum energy

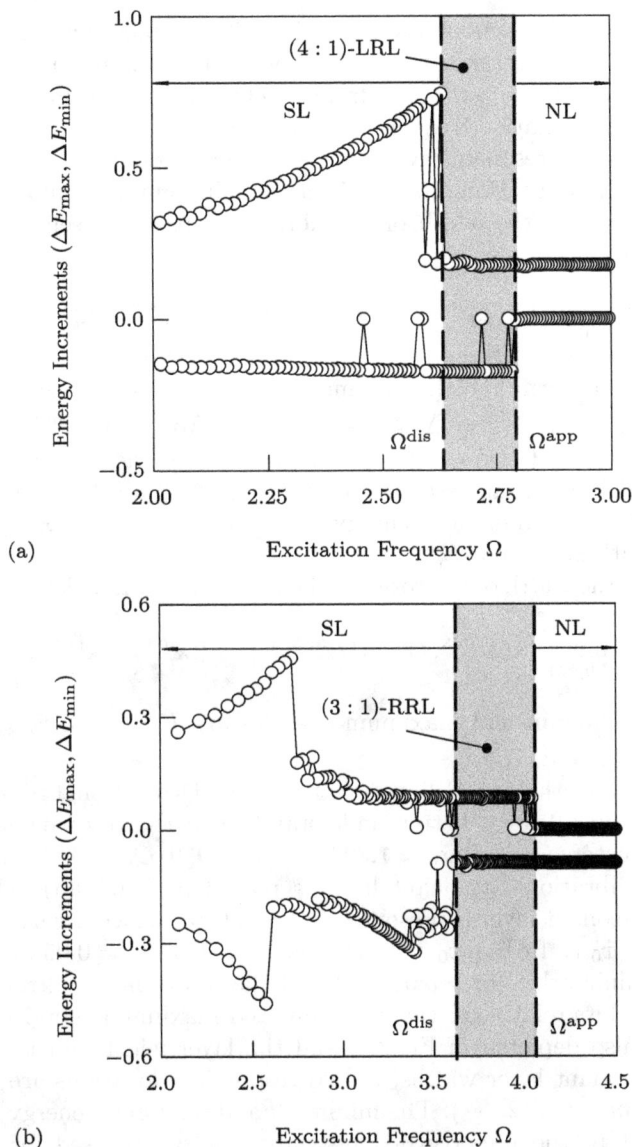

Fig. 3.8 The energy increment spectrum for (a) the (4 : 1)-librational and (b) (3 : 1)-rotational resonant layers at $Q_0 = 0.05$ and $\alpha = 1.0$. The acronyms "NL", "SL","LRL" and "RRL" represent "No Layer", "Stochastic Layer", "Librational Resonant Layer" and "Rotational Resonant Layer", respectively.

increment spectra are computed. The two plots of Fig.3.8 give the energy increment spectrum for the (4 : 1)-librational and (3 : 1)-rotational resonant layers. The critical values are $\Omega^{\text{app}} \approx 2.7695$ and $\Omega^{\text{dis}} \approx 2.6632$ for the

(4 : 1)-librational resonant layers, and $\Omega^{\text{app}} \approx 4.0381$ and $\Omega^{\text{dis}} \approx 3.6230$ are for the (3 : 1)-rotational resonant layers. The maximum and minimum energy increments are computed from 10000 iterations of Poincaré map for each specified excitation frequency and strength.

The width of a resonant layer can be computed as in references (e.g., Luo et al., 1999; Luo and Han, 2001). From the minimum and maximum energy increment spectra, the width of the librational resonant separatrix layer is computed by

$$w \equiv \min_{t \in [0,\infty)} ||\mathbf{x}(E_l^{\text{max}}, t) - \mathbf{x}(E_l^{\text{min}}, t)|| \equiv ||\mathbf{x}_l^{\text{max}} - \mathbf{x}_l^{\text{min}}|| \qquad (3.42)$$

where $|| \cdot ||$ is a norm and the minimum and maximum energies are $E_l^{\text{max}} = \Delta E_{\text{max}} + E_l^{(2M:1)}$, $E_l^{\text{min}} = \Delta E_{\text{min}} + E_l^{(2M:1)}$. Two points $\mathbf{x}_l^{\text{min}}$ and $\mathbf{x}_l^{\text{min}}$ on the normal vector $\mathbf{f}^{\perp}(\mathbf{x}_l^0) = (-f_2(\mathbf{x}_l^0), f_1(\mathbf{x}_l^0))^{\mathrm{T}}$ of the tangential vector of unperturbed librational, resonant orbit at point $\mathbf{x}_l^0$ are the closest between the maximum and minimum energy orbits $\mathbf{x}(E_l^{\text{max}}, t)$ and $\mathbf{x}(E_l^{\text{min}}, t)$ from Eq.(3.31) with E_l^{max} and E_l^{min}.

Similarly, the width of the rotational resonant separatrix layer is computed by

$$w \equiv \min_{t \in [0,\infty)} ||\mathbf{x}(E_r^{\text{max}}, t) - \mathbf{x}(E_r^{\text{min}}, t)|| \equiv ||\mathbf{x}_r^{\text{max}} - \mathbf{x}_r^{\text{min}}||, \qquad (3.43)$$

where the minimum and maximum energies are $E_r^{\text{max}} = \Delta E_{\text{max}} + E_r^{(M:1)}$, $E_r^{\text{min}} = \Delta E_{\text{min}} + E_r^{(M:1)}$.

Illustration of the (4 : 1)-librational and (3 : 1)-rotational resonant layers is given in Fig.3.9 for a better understanding of the resonant layer width. Using parameters ($x_0 = 0, \dot{x}_0 \approx 1.8221, \Omega \approx 2.6950, Q_0 = 0.05$ and $\alpha = 1.0$), the (4 : 1)-librational resonant layer is placed in Fig.3.9(a). The (3 : 1)-rotational resonant layer is presented in Fig.3.9(b), which are simulated with parameters ($x_0 \approx 1.6153, \dot{x}_0 \approx 1.4970, \Omega \approx 3.6230, Q_0 = 0.05$ and $\alpha = 1.0$). The homoclinic orbit (or separatrix) and the librational and rotational unperturbed orbits relative to the minimum and maximum energies (E^{min} and E^{max}) are also depicted in Fig.3.8, and the layer width (w) is sketched as well. For resonant layer widths, the corresponding locations are selected at $x = \pm 2m\pi (m = 0, 1, 2, \cdots)$. The minimum and maximum energy increments with respect to the unperturbed resonant energy are used for computing the width of stochastic layer. The width of resonant layers is illustrated in Fig.3.9. The ranges for non-layer and the stochastic layers are marked as well. The width of the resonant layer is almost constant instead of the exponential decaying with excitation frequency. Such a result is different from the asymptotic analysis (e.g., Melnikov, 1963).

(B) *Comparison with Analytical Prediction.* The analytical conditions for the appearance and destruction of the resonant layer have been presented and the energy spectrum approach has been given for the numerical prediction of the resonant layer. To make a comparison of two predictions, the excitation

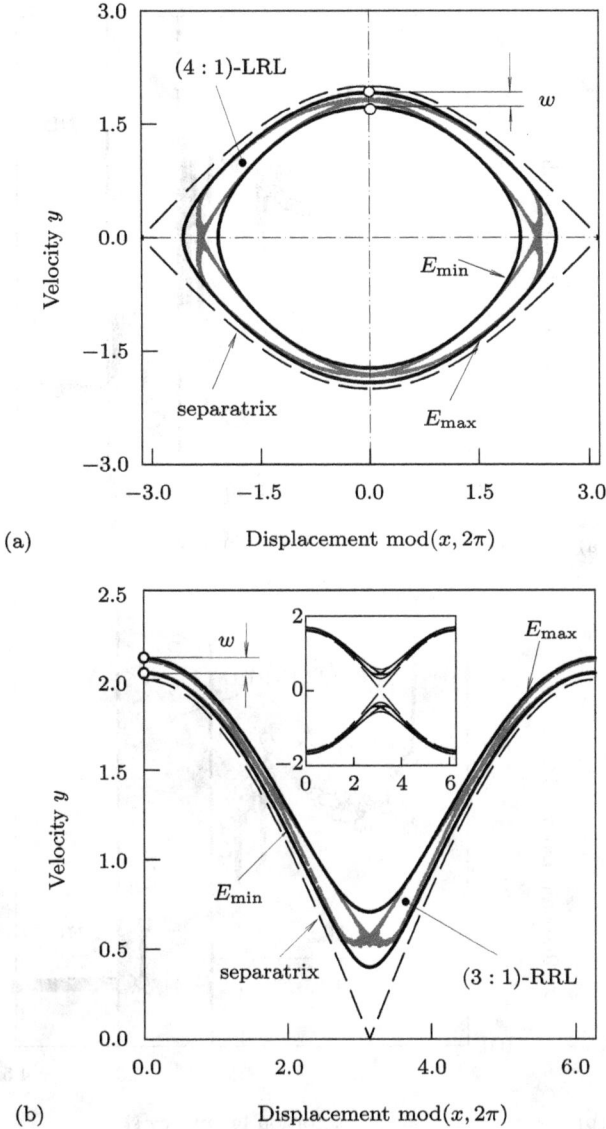

Fig. 3.9 The resonant layer width ($Q_0 = 0.05$ and $\alpha = 1.0$): (a) the $(4:1)$-librational resonant separatrix layer ($\Omega \approx 2.6950$, $x_0 = 0, \dot{x}_0 \approx 1.8221$) and (b) the $(3:1)$-rotational resonant separatrix layer ($\Omega \approx 3.6230$, $x_0 \approx 1.6153, \dot{x}_0 \approx 1.4970$). The acronyms "LRL" and "RRL" represent "Librational Resonant Layer" and "Rotational Resonant Layer", respectively.

frequency versus the excitation strength for a specific resonant layer will be presented in this section. To determine a specific resonant layer, the resonant condition is very important. Thus, for the parametrically excited pendulum,

(a)

(b)

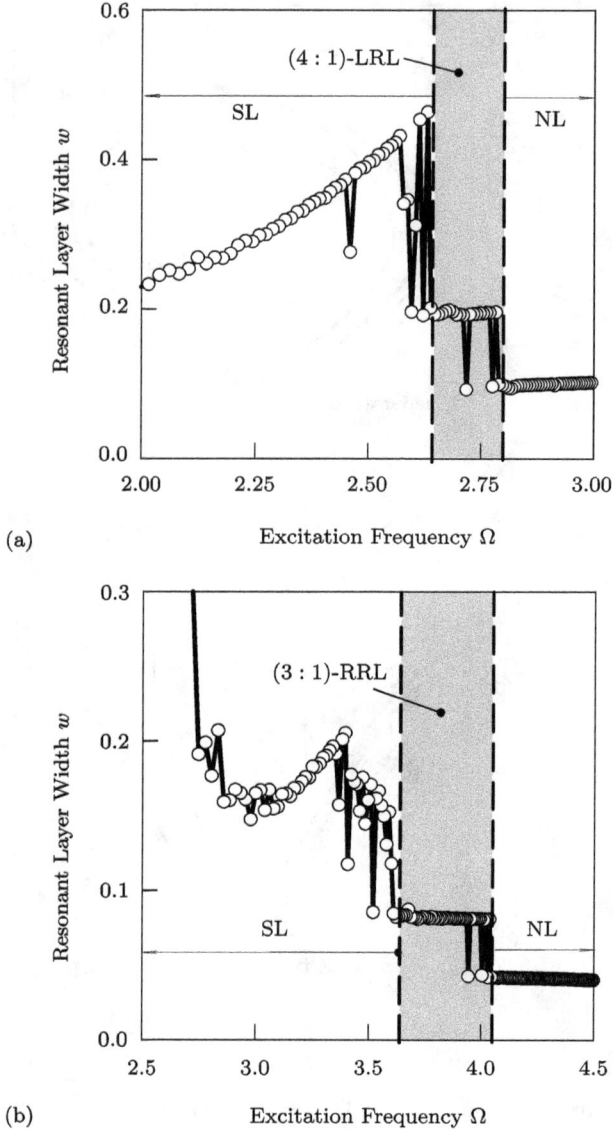

Fig. 3.10 The resonant layer widths for (a) the $(4:1)$-librational and (b) $(3:1)$-rotational (lower) resonant layers at $Q_0 = 0.05$ and $\alpha = 1.0$. The acronyms "NL", "SL","LRL" and "RRL" represent "No Layer", "Stochastic Layer", "Librational Resonant Layer" and "Rotational Resonant Layer", respectively.

the $(2M:1)$-librational and $(M:1)$-rotational resonant conditions are presented through the excitation frequency Ω versus the conservative energy H_0, as illustrated Figs.3.10(a) and 3.11(a). Based on such resonant conditions, the conditions for the appearance and destruction of the resonant layers are

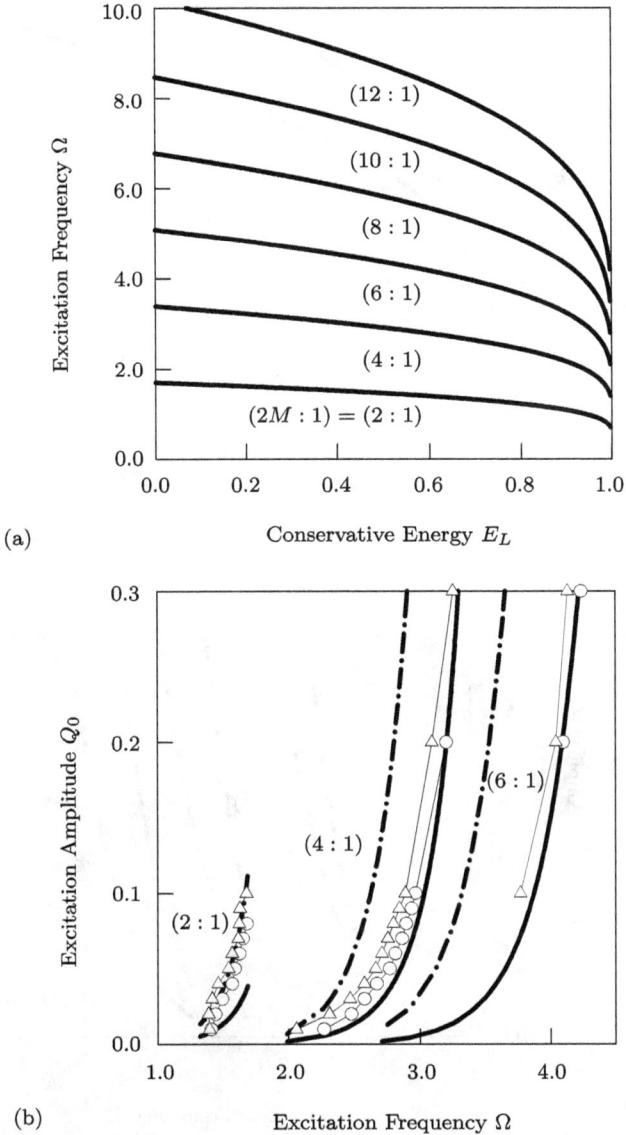

Fig. 3.11 (a) The $(2M : 1)$-librational resonant conditions and (b) the excitation strength conditions for the appearance (solid line) and disappearance (dash-dot) of the librational resonant layers at $\alpha = 1$. The circular and triangle symbol curves give the numerical predictions of the appearance and disappearance of the resonant layer.

computed and illustrated through the excitation frequency and strength in Figs.3.10(b) and 3.11(b). The solid curves give the analytical predictions of excitation strength for the appearance of the resonant layer from Eqs.(3.15) and (3.27), and the dashed curves represent the conditions for the destruc-

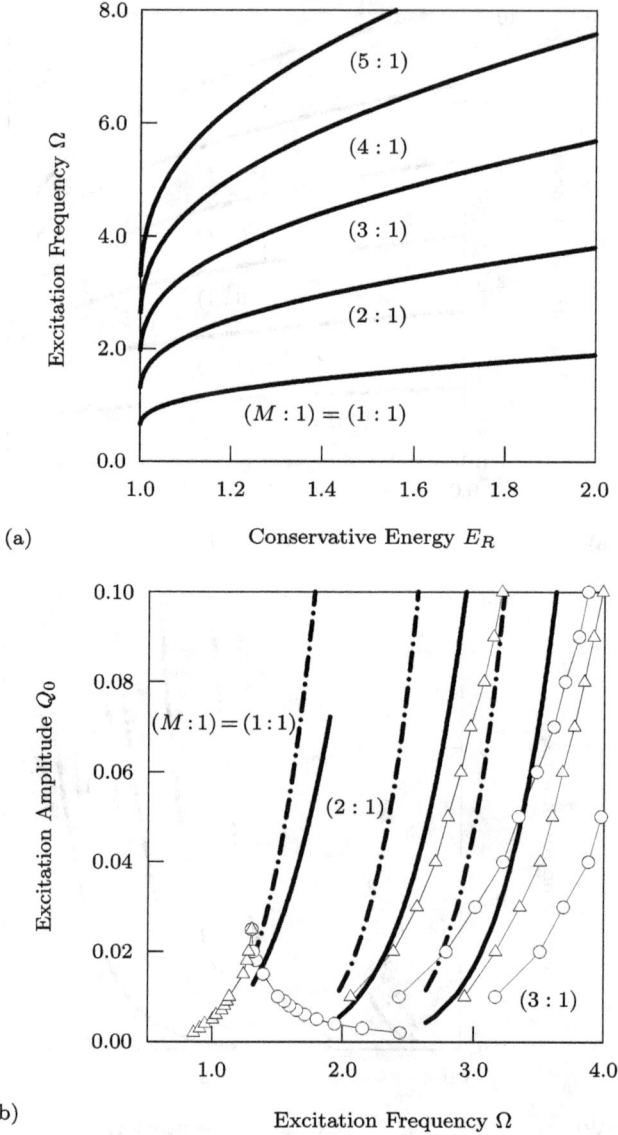

(a)

(b)

Fig. 3.12 (a) The $(M : 1)$-rotational resonant conditions, and (b) the excitation strength conditions for the appearance (solid) and disappearance (dash-dot) of the rotational resonant layers at $\alpha = 1$. The circular and triangle symbol curves give the numerical predictions of the appearance and disappearance of the resonant layers.

tion of the resonant layers, computed by Eqs.(3.17) and (3.29). The numerical predictions of the appearance and destruction of the resonant layers are given by the circular and triangle symbol-curves, which are generated from the energy increment spectrum technique. For the librational resonant layers, the

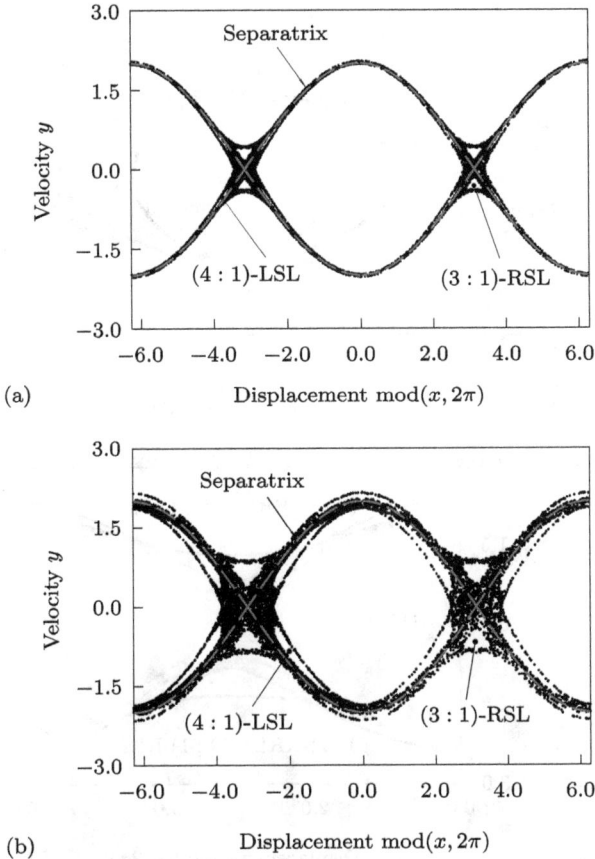

Fig. 3.13 Stochastic layers with resonance near separatrix ($\alpha = 1.0$): (a) thin layer ($Q_0 = 0.01, \Omega = 1.9$) and (b) thick layer ($Q_0 = 0.05, \Omega = 2.5$). Acronyms "LSL" and "RSL" represents librational and rotational stochastic layers, respectively.

numerical and analytical results are in a good agreement. But, the analytical results of the rotational resonant layer are different from the numerical predictions. Such a phenomenon may be caused by the sub-resonance. For the higher order resonant layer, the analytical prediction becomes poorer and poorer because the energy increments are computed by the unperturbed resonant orbit, instead of the perturbed resonant separatrix which can be obtained by renormalization.

(C) *Numerical Simulations.* To understand the mechanism of the resonant separatrix layer of nonlinear Hamiltonian systems, it is necessary to show the dynamic characteristics of the stochastic layer in Eq.(3.1). With the parameter ($\alpha = 1.0$), the stochastic layer of the parametric pendulum is shown in Fig.3.13. The thin stochastic layer with ($Q_0 = 0.01, \Omega = 1.9$) is presented in Fig.3.13(a). The initial condition for the stochastic layer is chosen

(a)

(b)

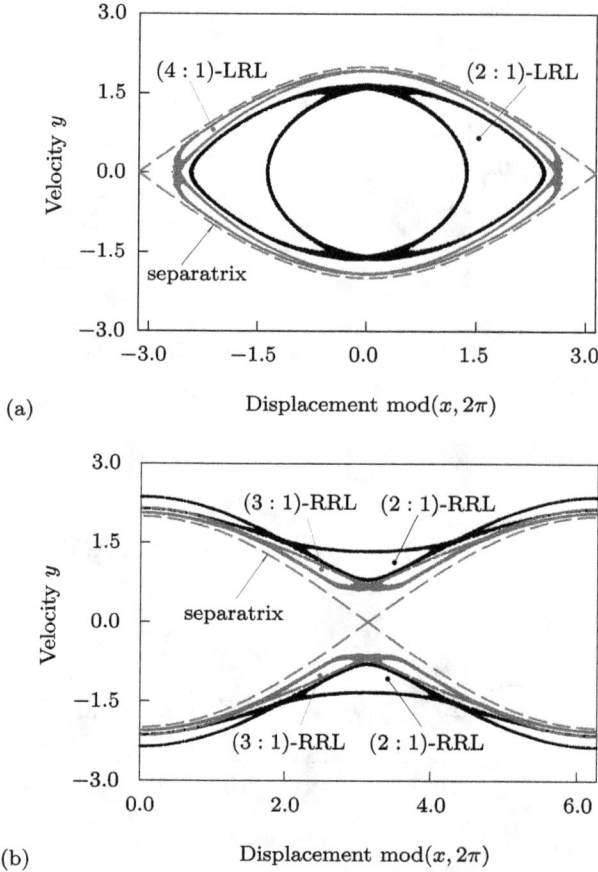

Fig. 3.14 (a) Two *librational* resonant layers (LRL) ((2:1)-layer: $\Omega \approx 1.572100$, $Q_0 = 0.06$, $x_0 = 0$, $y_0 \approx 1.603122$, and (4 : 1)-layer: $\Omega \approx 2.328332$, $Q_0 = 0.02$, $x_0 = 0$, $y_0 \approx 1.920937$); (b) the *rotational* resonant layers (RRL) ((2 : 1)-layer: $\Omega \approx 3.222313$, $Q_0 = 0.1$, $x_0 \approx 1.93871$, $y_0 \approx 1.555847$, and (3 : 1)-layer: $\Omega \approx 3.987793$, $Q_0 = 0.1$, $x_0 \approx 1.526633$, $y_0 \approx 1.615023$).

at hyperbolic points $(\pm(2j + 1)\pi, 0)$. The stochastic layer includes the (4 : 1) and higher order resonant separatrices in the librational stochastic layer and the (3:1) and higher order resonant separatrices in the rotational stochastic layer. With a similar resonance structure, the thick (3:1)-*rotational* stochastic layer is illustrated in Fig.3.13(b) with ($Q_0 = 0.05, \Omega = 2.5$). The stochastic layer with resonant separatrices becomes much thicker than in Fig.3.6(a). The investigation on dynamics in the stochastic layer of the parametric pendulum can be found in Luo (2001b). In Fig.3.14(a), the (2:1) and (4:1)-*librational* resonant layers inside the separatrix (or heteroclinic orbit) are presented with parameters ($\Omega \approx 1.5721, Q_0 = 0.06, x_0 = 0, y_0 \approx 1.603122$) and ($\Omega \approx 2.328332, Q_0 = 0.02, x_0 = 0, y_0 \approx 1.920937$), respectively. Fig.3.14(b) gives the (2:1)-primary resonant layers outside the separatrix with excitation strength ($Q_0 = 0.1$) and resonant excitation frequency and initial conditions ($\Omega \approx 3.222314, x_0 \approx 1.938710$ and $y_0 \approx 1.555847$) and ($\Omega \approx 3.987793$, $x_0 \approx 1.526633$ and $y_0 \approx 1.615023$), respectively.

References

Chirikov, B.V., 1979, A universal instability of many-dimensional oscillator systems, *Physics Report*, **52**, 263–379.

Greene, J.M., 1968, Two-dimensional measure-preserving mappings, *Journal of Mathematical Physics*, **9**, 760–768.

Greene, J.M., 1979, A method for computing the stochastic transition, *Journal of Mathematical Physics*, **20**, 1183–1201.

Han, R.P.S. and Luo, A.C.J., 1998, Resonant layers in nonlinear dynamics, *ASME Journal of Applied Mechanics*, **65**, 727–736.

Luo, A.C.J.,1995, *Analytical Modeling of Bifurcations, Chaos and Fractals in Nonlinear Dynamics*, Ph.D. Dissertation, University of Manitoba, Winnipeg, Canada.

Luo, A.C.J., 2001a, Resonant-overlap phenomena in stochastic layers of nonlinear Hamiltonian systems with periodical excitations, *Journal of Sound and Vibration*, **240**(5), 821–836.

Luo, A.C.J., 2001b, Resonance and stochastic layer in a parametrically excited pendulum, *Nonlinear Dynamics*, **26**(5), 355–367.

Luo, A.C.J., 2002, Resonant layers in a parametrically excited pendulum, *International Journal of Bifurcation and Chaos*, **12**, 409–419.

Luo, A.C.J., 2008, *Global Transversality, Resonance and Chaotic Dynamics*, World Scientific, Singapore.

Luo, A.C.J., Gu, K. and Han, R.P.S., 1999, Resonant-separatrix webs in stochastic layers of the twin-well Duffing oscillator, *Nonlinear Dynamics*, **19**, 37–48.

Luo, A.C.J. and Han, R.P.S., 2000, The dynamics of resonant and stochastic layers in a periodically-driven pendulum, *Chaos, Solitons and Fractals*, **11**, 2349–2359.

Luo, A.C.J. and Han, R.P.S., 2001, The resonance theory for stochastic layers in nonlinear dynamical systems, *Chaos, Solitons and Fractals*, **12**, 2493–2508.

Melnikov, V.K., 1963, On the stability of the center for time periodic perturbations, *Transactions Moscow Mathematics Society*, **12**, 1–57.

Zaslavskii and Filonenko, N.N., 1968, Stochastic instability of trapped particles and conditions of application of the quasi-linear approximation, *Soviet Physics JETP*, **27**, 851–857.

Chapter 4
Nonlinear Discrete Systems

In this chapter, a theory for nonlinear discrete systems will be presented. The local and global theory of stability and bifurcation for nonlinear discrete systems will be discussed. The stability switching and bifurcation on specific eigenvectors of the linearized system at fixed points under specific period will be presented. The higher order singularity and stability for nonlinear discrete systems on the specific eigenvectors will be presented.

4.1 Definitions

Definition 4.1 For $\Omega_\alpha \subseteq \mathscr{R}^n$ and $\Lambda \subseteq \mathscr{R}^m$ with $\alpha \in \mathbb{Z}$, consider a vector function $\mathbf{f}_\alpha : \Omega_\alpha \times \Lambda \to \Omega_\alpha$ which is $C^r (r \geqslant 1)$-continuous, and there is a discrete (or difference) equation in a form of

$$\mathbf{x}_{k+1} = \mathbf{f}_\alpha(\mathbf{x}_k, \mathbf{p}_\alpha) \text{ for } \mathbf{x}_k, \mathbf{x}_{k+1} \in \Omega_\alpha, \quad k \in \mathbb{Z} \quad \text{and} \quad \mathbf{p}_\alpha \in \Lambda. \tag{4.1}$$

With an initial condition of $\mathbf{x}_k = \mathbf{x}_0$, the solution of Eq.(4.1) is given by

$$\mathbf{x}_k = \underbrace{\mathbf{f}_\alpha(\mathbf{f}_\alpha(\cdots(\mathbf{f}_\alpha(\mathbf{x}_0, \mathbf{p}_\alpha))))}_{k}$$

$$\text{for } \mathbf{x}_k \in \Omega_\alpha, \quad k \in \mathbb{Z} \quad \text{and} \quad \mathbf{p} \in \Lambda. \tag{4.2}$$

(i) The difference equation with the initial condition is called a *discrete dynamical system*.
(ii) The vector function $\mathbf{f}_\alpha(\mathbf{x}_k, \mathbf{p}_\alpha)$ is called a *discrete vector field* on Ω_α.
(iii) The solution $\mathbf{x}_k$ for each $k \in \mathbb{Z}$ is called a *flow* of discrete system.
(iv) The solution $\mathbf{x}_k$ for all $k \in \mathbb{Z}$ on domain Ω_α is called the trajectory, phase curve or orbit of the discrete dynamical system, which is defined as

$$\Gamma = \{\mathbf{x}_k | \mathbf{x}_{k+1} = \mathbf{f}_\alpha(\mathbf{x}_k, \mathbf{p}_\alpha) \quad \text{for} \quad k \in \mathbb{Z} \text{ and } \mathbf{p}_\alpha \in \Lambda\} \subseteq \cup_\alpha \Omega_\alpha. \quad (4.3)$$

(v) The discrete dynamical system is called *a uniform discrete system* if

$$\mathbf{x}_{k+1} = \mathbf{f}_\alpha(\mathbf{x}_k, \mathbf{p}_\alpha) = \mathbf{f}(\mathbf{x}_k, \mathbf{p}) \quad \text{for} \quad k \in \mathbb{Z} \text{ and } \mathbf{x}_k \in \Omega_\alpha. \quad (4.4)$$

Otherwise, this discrete dynamical system is called a *non-uniform discrete system*.

Definition 4.2 For the discrete dynamical system in Eq.(4.1), the relation between state $\mathbf{x}_k$ and state $\mathbf{x}_{k+1}$ ($k \in \mathbb{Z}$) is called a discrete map if

$$P_\alpha : \mathbf{x}_k \xrightarrow{\mathbf{f}_\alpha} \mathbf{x}_{k+1} \quad \text{and} \quad \mathbf{x}_{k+1} = P_\alpha \mathbf{x}_k \quad (4.5)$$

with the following properties:

$$P_{(k,l)} : \mathbf{x}_k \xrightarrow{\mathbf{f}_{\alpha_1}, \mathbf{f}_{\alpha_2}, \cdots, \mathbf{f}_{\alpha_l}} \mathbf{x}_{k+l} \quad \text{and} \quad \mathbf{x}_{k+l} = P_{\alpha_l} \circ P_{\alpha_{l-1}} \circ \cdots \circ P_{\alpha_1} \mathbf{x}_k, \quad (4.6)$$

where

$$P_{(k;l)} = P_{\alpha_l} \circ P_{\alpha_{l-1}} \circ \cdots \circ P_{\alpha_1}. \quad (4.7)$$

If $P_{\alpha_l} = P_{\alpha_{l-1}} = \cdots = P_{\alpha_1} = P_\alpha$, then

$$P_{(\alpha;l)} \equiv P_\alpha^{(l)} = P_\alpha \circ P_\alpha \circ \cdots \circ P_\alpha \quad (4.8)$$

with

$$P_\alpha^{(n)} = P_\alpha \circ P_\alpha^{(n-1)} \quad \text{and} \quad P_\alpha^{(0)} = \mathbf{I}. \quad (4.9)$$

The total map with l-different sub-maps is shown in Fig.4.1. The map P_{α_k} with the relation function $\mathbf{f}_{\alpha_k}(\alpha_k \in \mathbb{Z})$ is given by Eq.(4.5). The total map $P_{(k,l)}$ is given in Eq.(4.7). The domains $\Omega_{\alpha_k}(\alpha_k \in \mathbb{Z})$ can fully overlap each other or can be completely separated without any intersection.

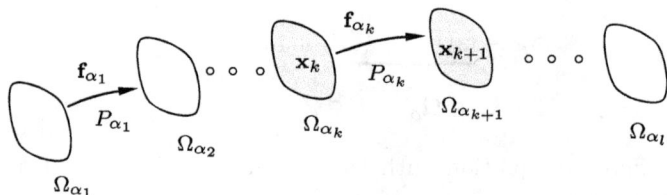

Fig. 4.1 Maps and vector functions on each sub-domain for discrete dynamical system.

Definition 4.3 For a vector function in $\mathbf{f}_\alpha \in \mathscr{R}^n, \mathbf{f}_\alpha : \mathscr{R}^n \to \mathscr{R}^n$. The operator norm of $\mathbf{f}_\alpha$ is defined by

$$\|\mathbf{f}_\alpha\| = \sum_{i=1}^{n} \max_{\|\mathbf{x}_k\| \leqslant 1, \mathbf{p}_\alpha} |f_{\alpha(i)}(\mathbf{x}_k, \mathbf{p}_\alpha)|. \tag{4.10}$$

For an $n \times n$ matrix $\mathbf{f}_\alpha(\mathbf{x}_k, \mathbf{p}_\alpha) = \mathbf{A}_\alpha \mathbf{x}_k$ and $\mathbf{A}_\alpha = (a_{ij})_{n \times n}$, the corresponding norm is defined by

$$\|\mathbf{A}_\alpha\| = \sum_{i,j=1}^{n} |a_{ij}|. \tag{4.11}$$

Definition 4.4 For $\Omega_\alpha \subseteq \mathscr{R}^n$ and $\Lambda \subseteq \mathscr{R}^m$ with $\alpha \in \mathbb{Z}$, the vector function $\mathbf{f}_\alpha(\mathbf{x}_k, \mathbf{p}_\alpha)$ with $\mathbf{f}_\alpha : \Omega_\alpha \times \Lambda \to \mathscr{R}^n$ is differentiable at $\mathbf{x}_k \in \Omega_\alpha$ if

$$\left. \frac{\partial \mathbf{f}_\alpha(\mathbf{x}_k, \mathbf{p}_\alpha)}{\partial \mathbf{x}_k} \right|_{(\mathbf{x}_k, \mathbf{p})} = \lim_{\Delta \mathbf{x}_k \to 0} \frac{\mathbf{f}_\alpha(\mathbf{x}_k + \Delta \mathbf{x}_k, \mathbf{p}_\alpha) - \mathbf{f}_\alpha(\mathbf{x}_k, \mathbf{p}_\alpha)}{\Delta \mathbf{x}_k}. \tag{4.12}$$

$\partial \mathbf{f}_\alpha / \partial \mathbf{x}_k$ is called the spatial derivative of $\mathbf{f}_\alpha(\mathbf{x}_k, \mathbf{p}_\alpha)$ at $\mathbf{x}_k$, and the derivative is given by the Jacobian matrix

$$\frac{\partial \mathbf{f}_\alpha(\mathbf{x}_k, \mathbf{p}_\alpha)}{\partial \mathbf{x}_k} = \left[\frac{\partial f_{\alpha i}}{\partial x_{kj}} \right]_{n \times n}. \tag{4.13}$$

Definition 4.5 For $\Omega_\alpha \subseteq \mathscr{R}^n$ and $\Lambda \subseteq \mathscr{R}^m$, consider a vector function $\mathbf{f}(\mathbf{x}_k, \mathbf{p})$ with $\mathbf{f} : \Omega_\alpha \times \Lambda \to \mathscr{R}^n$ where $\mathbf{x}_k \in \Omega_\alpha$ and $\mathbf{p} \in \Lambda$ with $k \in \mathbb{Z}$. The vector function $\mathbf{f}(\mathbf{x}_k, \mathbf{p})$ is said to satisfy the Lipschitz condition if

$$\|\mathbf{f}(\mathbf{y}_k, \mathbf{p}) - \mathbf{f}(\mathbf{x}_k, \mathbf{p})\| \leqslant L \|\mathbf{y}_k - \mathbf{x}_k\| \tag{4.14}$$

with $\mathbf{x}_k, \mathbf{y}_k \in \Omega_\alpha$ and L a constant. The constant L is called the Lipschitz constant.

4.2 Fixed points and stability

Definition 4.6 Consider a discrete, dynamical system $\mathbf{x}_{k+1} = \mathbf{f}_\alpha(\mathbf{x}_k, \mathbf{p}_\alpha)$ in Eq.(4.4).

(i) A point $\mathbf{x}_k^* \in \Omega_\alpha$ is called a fixed point or period-1 solution of a discrete nonlinear system $\mathbf{x}_{k+1} = \mathbf{f}_\alpha(\mathbf{x}_k, \mathbf{p}_\alpha)$ under a map P_α if for $\mathbf{x}_{k+1} = \mathbf{x}_k = \mathbf{x}_k^*$

$$\mathbf{x}_k^* = \mathbf{f}_\alpha(\mathbf{x}_k^*, \mathbf{p}_\alpha) \tag{4.15}$$

The linearized system of the nonlinear discrete system $\mathbf{x}_{k+1} = \mathbf{f}_\alpha(\mathbf{x}_k, \mathbf{p}_\alpha)$ in Eq.(4.4) at the fixed point $\mathbf{x}_k^*$ is given by

$$\mathbf{y}_{k+1} = DP_\alpha(\mathbf{x}_k^*, \mathbf{p}_\alpha)\mathbf{y}_k = D\mathbf{f}_\alpha(\mathbf{x}_k^*, \mathbf{p}_\alpha)\mathbf{y}_k \tag{4.16}$$

where

$$\mathbf{y}_k = \mathbf{x}_k - \mathbf{x}_k^* \quad \text{and} \quad \mathbf{y}_{k+1} = \mathbf{x}_{k+1} - \mathbf{x}_{k+1}^*. \tag{4.17}$$

(ii) A set of points $\mathbf{x}_j^* \in \Omega_{\alpha_j} (\alpha_j \in \mathbb{Z})$ is called the fixed point set or period-1 point set of the total map $P_{(k;l)}$ with l-different sub-maps in nonlinear discrete system of Eq.(4.2) if

$$
\begin{aligned}
\mathbf{x}_{k+j+1}^* &= \mathbf{f}_{\alpha_{j'}}(\mathbf{x}_{k+j}^*, \mathbf{p}_{\alpha_{j'}}) \quad \text{for} \quad j \in \mathbb{Z}_+ \text{ and } j' = \mathrm{mod}(j,l)+1; \\
\mathbf{x}_{k+\,\mathrm{mod}\,(j,l)}^* &= \mathbf{x}_k^*.
\end{aligned}
$$
$$\tag{4.18}$$

The linearized equation of the total map $P_{(k;l)}$ gives

$$
\begin{aligned}
\mathbf{y}_{k+j+1} &= DP_{\alpha_{j'}}(\mathbf{x}_{k+j}^*, \mathbf{p}_{\alpha_{j'}})\mathbf{y}_{k+j} = D\mathbf{f}_{\alpha_{j'}}(\mathbf{x}_{k+j}^*, \mathbf{p}_{\alpha_{j'}})\mathbf{y}_{k+j} \\
&\text{with } \mathbf{y}_{k+j+1} = \mathbf{x}_{k+j+1} - \mathbf{x}_{k+j+1}^* \quad \text{and} \quad \mathbf{y}_{k+j} = \mathbf{x}_{k+j} - \mathbf{x}_{k+j}^* \quad (4.19) \\
&\text{for } j \in \mathbb{Z}_+ \quad \text{and} \quad j' = \mathrm{mod}(j,l)+1.
\end{aligned}
$$

The resultant equation for each individual map is

$$\mathbf{y}_{k+j+1} = DP_{(k,l)}(\mathbf{x}_k^*, \mathbf{p})\mathbf{y}_{k+j} \quad \text{for} \quad j \in \mathbb{Z}_+ \tag{4.20}$$

where

$$
\begin{aligned}
DP_{(k,l)}(\mathbf{x}_k^*, \mathbf{p}) &= \prod_{j=l}^{1} DP_{\alpha_j}(\mathbf{x}_{k+j-1}^*, \mathbf{p}) \\
&= DP_{\alpha_l}(\mathbf{x}_{k+l-1}^*, \mathbf{p}_{\alpha_l}) \cdots \cdots DP_{\alpha_2}(\mathbf{x}_{k+1}^*, \mathbf{p}_{\alpha_2}) \cdot DP_{\alpha_1}(\mathbf{x}_k^*, \mathbf{p}_{\alpha_1}) \\
&= D\mathbf{f}_{\alpha_l}(\mathbf{x}_{k+l-1}^*, \mathbf{p}_{\alpha_l}) \cdots \cdots D\mathbf{f}_{\alpha_2}(\mathbf{x}_{k+1}^*, \mathbf{p}_{\alpha_2}) \cdot D\mathbf{f}_{\alpha_1}(\mathbf{x}_k^*, \mathbf{p}_{\alpha_1}).
\end{aligned}
$$
$$\tag{4.21}$$

The fixed point $\mathbf{x}_k^*$ lies in the intersected set of two domains Ω_k and Ω_{k+1}, as shown in Fig.4.2. In the vicinity of the fixed point $\mathbf{x}_k^*$, the incremental relations in the two domains Ω_k and Ω_{k+1} are different. In other words, setting $\mathbf{y}_k = \mathbf{x}_k - \mathbf{x}_k^*$ and $\mathbf{y}_{k+1} = \mathbf{x}_{k+1} - \mathbf{x}_{k+1}^*$, the corresponding linearization is generated as in Eq.(4.16). Similarly, The fixed point of the total map with n-different sub-maps requires the intersection set of two domains Ω_k and Ω_{k+n}, there are a set of equations to obtain the fixed points from Eq.(4.18). The other values of fixed points lie in different domains, i.e., $\mathbf{x}_j^* \in \Omega_j$ ($j = k+1, k+2, \cdots, k+n-1$), as shown in Fig.4.3.

The corresponding linearized equations are given in Eq.(4.19). From Eq.(4.20), the local characteristics of the total map can be discussed as a single map. Thus, the dynamical characteristics for the fixed point of the single map will be discussed comprehensively, and the fixed points for resultant map are applicable. The results can be extended to any period-m flows with $P^{(m)}$.

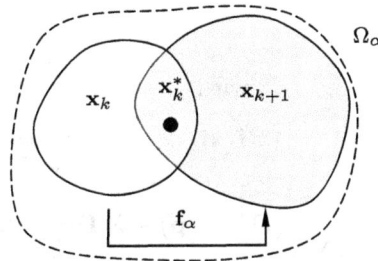

Fig. 4.2 A fixed point between domains Ω_k and Ω_{k+1} for a discrete dynamical system.

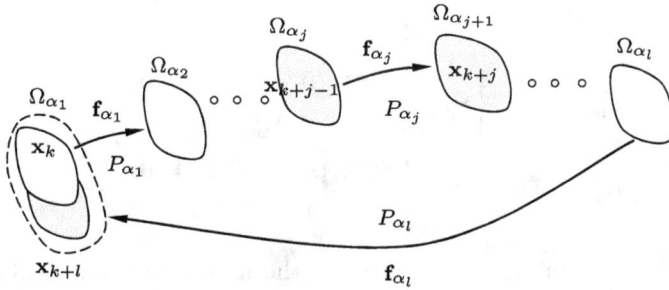

Fig. 4.3 Fixed points with l-maps for discrete dynamical system.

Definition 4.7 Consider a discrete, nonlinear dynamical system $\mathbf{x}_{k+1} = \mathbf{f}(\mathbf{x}_k, \mathbf{p})$ in Eq.(4.4) with a fixed point $\mathbf{x}_k^*$. The linearized system of the discrete nonlinear system in the neighborhood of $\mathbf{x}_k^*$ is $\mathbf{y}_{k+1} = D\mathbf{f}(\mathbf{x}_k^*, \mathbf{p})\mathbf{y}_k$ ($\mathbf{y}_l = \mathbf{x}_l - \mathbf{x}_k^*$ and $l = k, k+1$) in Eq.(4.16). The matrix $D\mathbf{f}(\mathbf{x}_k^*, \mathbf{p})$ possesses n_1 real eigenvalues $|\lambda_j| < 1$ ($j \in N_1$), n_2 real eigenvalues $|\lambda_j| > 1$ ($j \in N_2$), n_3 real eigenvalues $\lambda_j = 1$ ($j \in N_3$), and n_4 real eigenvalues $\lambda_j = -1$ ($j \in N_4$). $N = \{1, 2, \cdots, n\}$ and $N_i = \{i_1, i_2, \cdots, i_{n_i}\} \cup \varnothing$ ($i = 1, 2, 3, 4$) with $i_m \in N$ ($m = 1, 2, \cdots, n_i$) and $\sum_{i=1}^{4} n_i = n$. $N_i \subseteq N \cup \varnothing$, $\cup_{i=1}^{4} N_i = N$, $N_i \cap N_p = \varnothing$ ($p \neq i$). $N_i = \varnothing$ if $n_i = 0$. The corresponding eigenvectors for contraction, expansion, invariance and flip oscillation are $\{\mathbf{v}_j\}$ ($j \in N_i$) ($i = 1, 2, 3, 4$), respectively. The stable, unstable, invariant and flip subspaces of $\mathbf{y}_{k+1} = D\mathbf{f}(\mathbf{x}_k^*, \mathbf{p})\mathbf{y}_k$ in Eq.(4.16) are linear subspace spanned by $\{\mathbf{v}_j\}$ ($j \in N_i$) ($i = 1, 2, 3, 4$), respectively, i.e.,

$$
\begin{aligned}
\mathscr{E}^{\mathrm{s}} &= span \left\{ \mathbf{v}_j \left| \begin{array}{l} (D\mathbf{f}(\mathbf{x}_k^*, \mathbf{p}) - \lambda_j \mathbf{I})\mathbf{v}_j = \mathbf{0}, \\ |\lambda_j| < 1, j \in N_1 \subseteq N \cup \varnothing \end{array} \right. \right\}; \\
\mathscr{E}^{\mathrm{u}} &= span \left\{ \mathbf{v}_j \left| \begin{array}{l} (D\mathbf{f}(\mathbf{x}_k^*, \mathbf{p}) - \lambda_j \mathbf{I})\mathbf{v}_j = \mathbf{0}, \\ |\lambda_j| > 1, j \in N_2 \subseteq N \cup \varnothing \end{array} \right. \right\}; \\
\mathscr{E}^{\mathrm{i}} &= span \left\{ \mathbf{v}_j \left| \begin{array}{l} (D\mathbf{f}(\mathbf{x}_k^*, \mathbf{p}) - \lambda_j \mathbf{I})\mathbf{v}_j = \mathbf{0}, \\ \lambda_j = 1, j \in N_3 \subseteq N \cup \varnothing \end{array} \right. \right\}; \\
\mathscr{E}^{\mathrm{f}} &= span \left\{ \mathbf{v}_j \left| \begin{array}{l} (D\mathbf{f}(\mathbf{x}_k^*, \mathbf{p}) - \lambda_j \mathbf{I})\mathbf{v}_j = \mathbf{0}, \\ \lambda_j = -1, j \in N_4 \subseteq N \cup \varnothing \end{array} \right. \right\}.
\end{aligned}
\tag{4.22}
$$

where

$$\mathscr{E}^{\mathrm{s}} = \mathscr{E}^{\mathrm{s}}_{\mathrm{m}} \cup \mathscr{E}^{\mathrm{s}}_{\mathrm{o}} \cup \mathscr{E}^{\mathrm{s}}_{\mathrm{z}} \text{ with}$$

$$\mathscr{E}^{\mathrm{s}}_{\mathrm{m}} = span \left\{ \mathbf{v}_j \left| \begin{array}{l} (D\mathbf{f}(\mathbf{x}^*_k, \mathbf{p}) - \lambda_j \mathbf{I})\mathbf{v}_j = \mathbf{0}, \\ 0 < \lambda_j < 1, j \in N^{\mathrm{m}}_1 \subseteq N \cup \varnothing \end{array} \right. \right\};$$

$$\mathscr{E}^{\mathrm{s}}_{\mathrm{o}} = span \left\{ \mathbf{v}_j \left| \begin{array}{l} (D\mathbf{f}(\mathbf{x}^*_k, \mathbf{p}) - \lambda_j \mathbf{I})\mathbf{v}_j = \mathbf{0}, \\ -1 < \lambda_j < 0, j \in N^{\mathrm{o}}_1 \subseteq N \cup \varnothing \end{array} \right. \right\}; \qquad (4.23)$$

$$\mathscr{E}^{\mathrm{s}}_{\mathrm{z}} = span \left\{ \mathbf{v}_j \left| \begin{array}{l} (D\mathbf{f}(\mathbf{x}^*_k, \mathbf{p}) - \lambda_j \mathbf{I})\mathbf{v}_j = \mathbf{0}, \\ \lambda_j = 0, j \in N^{\mathrm{z}}_1 \subseteq N \cup \varnothing \end{array} \right. \right\}$$

$$\mathscr{E}^{\mathrm{u}} = \mathscr{E}^{\mathrm{u}}_{\mathrm{m}} \cup \mathscr{E}^{\mathrm{u}}_{\mathrm{o}} \text{ with}$$

$$\mathscr{E}^{\mathrm{u}}_{\mathrm{m}} = span \left\{ \mathbf{v}_j \left| \begin{array}{l} (D\mathbf{f}(\mathbf{x}^*_k, \mathbf{p}) - \lambda_j \mathbf{I})\mathbf{v}_j = \mathbf{0}, \\ \lambda_j > 1, j \in N^{\mathrm{m}}_2 \subseteq N \cup \varnothing \end{array} \right. \right\};$$

$$\mathscr{E}^{\mathrm{u}}_{\mathrm{o}} = span \left\{ \mathbf{v}_j \left| \begin{array}{l} (D\mathbf{f}(\mathbf{x}^*_k, \mathbf{p}) - \lambda_j \mathbf{I})\mathbf{v}_j = \mathbf{0}, \\ \lambda_j < -1, j \in N^{\mathrm{o}}_2 \subseteq N \cup \varnothing \end{array} \right. \right\}; \qquad (4.24)$$

where subscripts "m" and "o" represent the monotonic and oscillatory evolutions.

Definition 4.8 Consider a discrete, nonlinear dynamical system $\mathbf{x}_{k+1} = \mathbf{f}(\mathbf{x}_k, \mathbf{p})$ in Eq.(4.4) with a fixed point $\mathbf{x}^*_k$. The linearized system of the discrete nonlinear system in the neighborhood of $\mathbf{x}^*_k$ is $\mathbf{y}_{k+1} = D\mathbf{f}(\mathbf{x}^*_k, \mathbf{p})\mathbf{y}_k$ ($\mathbf{y}_l = \mathbf{x}_l - \mathbf{x}^*_k$ and $l = k, k+1$) in Eq.(4.16). The matrix $D\mathbf{f}(\mathbf{x}^*_k, \mathbf{p})$ has complex eigenvalues $\alpha_j \pm \mathbf{i}\beta_j$ with eigenvectors $\mathbf{u}_j \pm \mathbf{i}\mathbf{v}_j$ ($j \in \{1, 2, \cdots, n\}$) and the base of vector is

$$\mathbf{B} = \{\mathbf{u}_1, \mathbf{v}_1, \cdots, \mathbf{u}_j, \mathbf{v}_j, \cdots, \mathbf{u}_n, \mathbf{v}_n\}. \qquad (4.25)$$

The stable, unstable, center subspaces of $\mathbf{y}_{k+1} = D\mathbf{f}(\mathbf{x}^*_k, \mathbf{p})\mathbf{y}_k$ in Eq.(4.16) are linear subspaces spanned by $\{\mathbf{u}_j, \mathbf{v}_j\}$ ($j \in N_i, i = 1, 2, 3$), respectively. Set $N = \{1, 2, \cdots, n\}$ plus $N_i = \{i_1, i_2, \cdots, i_{n_i}\} \cup \varnothing \subseteq N \cup \varnothing$ with $i_m \in N$ ($m = 1, 2, \cdots, n_i$) and $\sum^3_{i=1} n_i = n$. $\cup^3_{i=1} N_i = N$ with $N_i \cap N_p = \varnothing (p \neq i)$. $N_i = \varnothing$ if $n_i = 0$. The stable, unstable, center subspaces of $\mathbf{y}_{k+1} = D\mathbf{f}(\mathbf{x}^*_k, \mathbf{p})\mathbf{y}_k$ in Eq.(4.16) are defined by

$$\mathscr{E}^{\mathrm{s}} = span \left\{ (\mathbf{u}_j, \mathbf{v}_j) \left| \begin{array}{l} r_j = \sqrt{\alpha^2_j + \beta^2_j} < 1, \\ (D\mathbf{f}(\mathbf{x}^*_k, \mathbf{p}) - (\alpha_j \pm \mathbf{i}\beta_j)\mathbf{I})(\mathbf{u}_j \pm \mathbf{i}\mathbf{v}_j) = \mathbf{0}, \\ j \in N_1 \subseteq \{1, 2, \cdots, n\} \cup \varnothing \end{array} \right. \right\};$$

$$\mathscr{E}^{\mathrm{u}} = span \left\{ (\mathbf{u}_j, \mathbf{v}_j) \left| \begin{array}{l} r_j = \sqrt{\alpha^2_j + \beta^2_j} > 1, \\ (D\mathbf{f}(\mathbf{x}^*_k, \mathbf{p}) - (\alpha_j \pm \mathbf{i}\beta_j)\mathbf{I})(\mathbf{u}_j \pm \mathbf{i}\mathbf{v}_j) = \mathbf{0}, \\ j \in N_2 \subseteq \{1, 2, \cdots, n\} \cup \varnothing \end{array} \right. \right\}; \qquad (4.26)$$

$$\mathscr{E}^{\mathrm{c}} = span \left\{ (\mathbf{u}_j, \mathbf{v}_j) \left| \begin{array}{l} r_j = \sqrt{\alpha_j^2 + \beta_j^2} = 1, \\ (D\mathbf{f}(\mathbf{x}_k^*, \mathbf{p}) - (\alpha_j \pm \mathrm{i}\beta_j)\mathbf{I})(\mathbf{u}_j \pm \mathrm{i}\mathbf{v}_j) = 0, \\ j \in N_3 \subseteq \{1, 2, \cdots, n\} \cup \varnothing \end{array} \right. \right\}.$$

Definition 4.9 Consider a discrete, nonlinear dynamical system $\mathbf{x}_{k+1} = \mathbf{f}(\mathbf{x}_k, \mathbf{p})$ in Eq.(4.4) with a fixed point $\mathbf{x}_k^*$. The linearized system of the discrete nonlinear system in the neighborhood of $\mathbf{x}_k^*$ is $\mathbf{y}_{k+1} = D\mathbf{f}(\mathbf{x}_k^*, \mathbf{p})\mathbf{y}_k$ ($\mathbf{y}_l = \mathbf{x}_l - \mathbf{x}_k^*$ and $l = k, k+1$) in Eq.(4.16). The fixed point or period-1point is *hyperbolic* if no any eigenvalues of $D\mathbf{f}(\mathbf{x}_k^*, \mathbf{p})$ are on the unit circle (i.e., $|\lambda_i| \neq 1$ for $i = 1, 2, \cdots, n$).

Theorem 4.1 *Consider a discrete, nonlinear dynamical system $\mathbf{x}_{k+1} = \mathbf{f}(\mathbf{x}_k, \mathbf{p})$ in Eq.(4.4) with a fixed point $\mathbf{x}_k^*$. The linearized system of the discrete nonlinear system in the neighborhood of $\mathbf{x}_k^*$ is $\mathbf{y}_{k+1} = D\mathbf{f}(\mathbf{x}_k^*, \mathbf{p})\mathbf{y}_k$ ($\mathbf{y}_j = \mathbf{x}_j - \mathbf{x}_k^*$ and $j = k, k+1$) in Eq.(4.16). The eigenspace of $D\mathbf{f}(\mathbf{x}_k^*, \mathbf{p})$ (i.e., $\mathscr{E} \subseteq \mathscr{R}^n$) in the linearized dynamical system is expressed by direct sum of three subspaces*

$$\mathscr{E} = \mathscr{E}^{\mathrm{s}} \oplus \mathscr{E}^{\mathrm{u}} \oplus \mathscr{E}^{\mathrm{c}}. \tag{4.27}$$

where $\mathscr{E}^{\mathrm{s}}, \mathscr{E}^{\mathrm{u}}$ and $\mathscr{E}^{\mathrm{c}}$ are the stable, unstable and center subspaces, respectively.

Proof. The proof can be referred to Luo (2011). ∎

Definition 4.10 Consider a discrete, nonlinear dynamical system $\mathbf{x}_{k+1} = \mathbf{f}(\mathbf{x}_k, \mathbf{p})$ in Eq.(4.4) with a fixed point $\mathbf{x}_k^*$. Suppose there is a neighborhood of the fixed point $\mathbf{x}_k^*$ as $U_k(\mathbf{x}_k^*) \subset \Omega_k$, and in the neighborhood,

$$\lim_{\|\mathbf{y}_k\| \to 0} \frac{\|\mathbf{f}(\mathbf{x}_k^* + \mathbf{y}_k, \mathbf{p}) - D\mathbf{f}(\mathbf{x}_k^*, \mathbf{p})\mathbf{y}_k\|}{\|\mathbf{y}_k\|} = 0, \tag{4.28}$$

and

$$\mathbf{y}_{k+1} = D\mathbf{f}(\mathbf{x}_k^*, \mathbf{p})\mathbf{y}_k. \tag{4.29}$$

(i) A C^r invariant manifold

$$\mathscr{S}_{loc}(\mathbf{x}_k, \mathbf{x}_k^*) = \{\mathbf{x}_k \in U(\mathbf{x}_k^*) | \lim_{j \to +\infty} \mathbf{x}_{k+j} = \mathbf{x}_k^* \text{ and}$$

$$\mathbf{x}_{k+j} \in U(\mathbf{x}_k^*) \text{ with } j \in \mathbb{Z}_+\} \tag{4.30}$$

is called the local stable manifold of $\mathbf{x}_k^*$, and the corresponding global stable manifold is defined as

$$\mathscr{S}(\mathbf{x}_k, \mathbf{x}_k^*) = \cup_{j \in \mathbb{Z}_-} \mathbf{f}(\mathscr{S}_{loc}(\mathbf{x}_{k+j}, \mathbf{x}_{k+j}^*)) = \cup_{j \in \mathbb{Z}_-} \mathbf{f}^{(j)}(\mathscr{S}_{loc}(\mathbf{x}_k, \mathbf{x}_k^*)). \tag{4.31}$$

(ii) A C^r invariant manifold

$$\mathscr{U}_{loc}(\mathbf{x}_k, \mathbf{x}_k^*) = \{\mathbf{x}_k \in U(\mathbf{x}_k^*) | \lim_{j \to -\infty} \mathbf{x}_{k+j} = \mathbf{x}_k^* \text{ and}$$

$$\mathbf{x}_{k+j} \in U(\mathbf{x}_k^*) \text{ with } j \in \mathbb{Z}_-\} \qquad (4.32)$$

is called the local unstable manifold of $\mathbf{x}^*$, and the corresponding global unstable manifold is defined as

$$\mathscr{U}(\mathbf{x}_k, \mathbf{x}_k^*) = \cup_{j \in \mathbb{Z}_+} \mathbf{f}(\mathscr{U}_{loc}(\mathbf{x}_{k+j}, \mathbf{x}_{k+j}^*)) = \cup_{j \in \mathbb{Z}_+} \mathbf{f}^{(j)}(\mathscr{U}_{loc}(\mathbf{x}_k, \mathbf{x}_k^*)). \qquad (4.33)$$

(iii) A C^{r-1} invariant manifold $\mathscr{C}_{loc}(\mathbf{x}_k, \mathbf{x}_k^*)$ is called the center manifold of $\mathbf{x}^*$ if $\mathscr{C}_{loc}(\mathbf{x}_k, \mathbf{x}_k^*)$ possesses the same dimension of $\mathscr{E}^c$ for $\mathbf{x}_k^* \in \mathscr{S}(\mathbf{x}_k, \mathbf{x}_k^*)$, and the tangential space of $\mathscr{C}_{loc}(\mathbf{x}_k, \mathbf{x}_k^*)$ is identical to $\mathscr{E}^c$.

As in continuous dynamical systems, the stable and unstable manifolds are unique, but the center manifold is not unique. If the nonlinear vector field $\mathbf{f}$ is C^∞-continuous, then a C^r center manifold can be found for any $r < \infty$.

Theorem 4.2 *Consider a discrete, nonlinear dynamical system $\mathbf{x}_{k+1} = \mathbf{f}(\mathbf{x}_k, \mathbf{p})$ in Eq.(4.4) with a hyperbolic fixed point $\mathbf{x}_k^*$. The corresponding solution is $\mathbf{x}_{k+j} = \mathbf{f}(\mathbf{x}_{k+j-1}, \mathbf{p})$ with $j \in \mathbb{Z}$. Suppose there is a neighborhood of the hyperbolic fixed point $\mathbf{x}_k^*$(i.e., $U_k(\mathbf{x}_k^*) \subset \Omega_\alpha$), and $\mathbf{f}(\mathbf{x}_k, \mathbf{p})$ is $C^r(r \geqslant 1)$-continuous in $U_k(\mathbf{x}_k^*)$. The linearized system is $\mathbf{y}_{k+j+1} = D\mathbf{f}(\mathbf{x}_k^*, \mathbf{p})\mathbf{y}_{k+j}(\mathbf{y}_{k+j} = \mathbf{x}_{k+j} - \mathbf{x}_k^*)$ in $U_k(\mathbf{x}_k^*)$. If the homeomorphism between the local invariant subspace $E(\mathbf{x}_k, \mathbf{x}_k^*) \subset U(\mathbf{x}_k^*)$ and the eigenspace $\mathscr{E}$ of the linearized system exists with the condition in Eq.(4.28), the local invariant subspace is decomposed by*

$$E(\mathbf{x}_k, \mathbf{x}_k^*) = \mathscr{S}_{loc}(\mathbf{x}_k, \mathbf{x}_k^*) \oplus \mathscr{U}_{loc}(\mathbf{x}_k, \mathbf{x}_k^*). \qquad (4.34)$$

(a) *The local stable invariant manifold $\mathscr{S}_{loc}(\mathbf{x}_k, \mathbf{x}_k^*)$ possesses the following properties:*

(i) *for $\mathbf{x}_k^* \in \mathscr{S}_{loc}(\mathbf{x}_k, \mathbf{x}_k^*)$, $\mathscr{S}_{loc}(\mathbf{x}_k, \mathbf{x}_k^*)$ possesses the same dimension of $\mathscr{E}^s$ and the tangential space of $\mathscr{S}_{loc}(\mathbf{x}_k, \mathbf{x}_k^*)$ is identical to $\mathscr{E}^s$;*
(ii) *for $\mathbf{x}_k \in \mathscr{S}_{loc}(\mathbf{x}_k, \mathbf{x}_k^*)$, $\mathbf{x}_{k+j} \in \mathscr{S}_{loc}(\mathbf{x}_k, \mathbf{x}_k^*)$ and $\lim_{j \to \infty} \mathbf{x}_{k+j} = \mathbf{x}_k^*$ for all $j \in \mathbb{Z}_+$;*
(iii) *for $\mathbf{x}_k \notin \mathscr{S}_{loc}(\mathbf{x}_k, \mathbf{x}_k^*)$, $\|\mathbf{x}_{k+j} - \mathbf{x}_k^*\| \geqslant \delta$ for $\delta > 0$ with $j, j_1 \in \mathbb{Z}_+$ and $j \geqslant j_1 \geqslant 0$.*

(b) *The local unstable invariant manifold $\mathscr{U}_{loc}(\mathbf{x}_k, \mathbf{x}_k^*)$ possesses the following properties:*

(i) *for $\mathbf{x}_k^* \in \mathscr{U}_{loc}(\mathbf{x}_k, \mathbf{x}_k^*)$, $\mathscr{U}_{loc}(\mathbf{x}_k, \mathbf{x}_k^*)$ possesses the same dimension of $\mathscr{E}^u$ and the tangential space of $\mathscr{U}_{loc}(\mathbf{x}_k, \mathbf{x}_k^*)$ is identical to $\mathscr{E}^u$;*
(ii) *for $\mathbf{x}_k \in \mathscr{U}_{loc}(\mathbf{x}_k, \mathbf{x}_k^*)$, $\mathbf{x}_{k+j} \in \mathscr{U}_{loc}(\mathbf{x}_k, \mathbf{x}_k^*)$ and $\lim_{j \to -\infty} \mathbf{x}_{k+j} = \mathbf{x}_k^*$ for all $j \in \mathbb{Z}_-$*
(iii) *for $\mathbf{x}_k \notin \mathscr{U}_{loc}(\mathbf{x}_k, \mathbf{x}_k^*)$, $\|\mathbf{x}_{k+j} - \mathbf{x}_k^*\| \geqslant \delta$ for $\delta > 0$ with $j_1, j \in \mathbb{Z}_-$ and $j \leqslant j_1 \leqslant 0$.*

Proof. See Nitecki (1971). ■

Theorem 4.3 *Consider a discrete, nonlinear dynamical system* $\mathbf{x}_{k+1} = \mathbf{f}(\mathbf{x}_k, \mathbf{p})$ *in Eq.(4.4) with a fixed point* $\mathbf{x}_k^*$. *The corresponding solution is* $\mathbf{x}_{k+j} = \mathbf{f}(\mathbf{x}_{k+j-1}, \mathbf{p})$ *with* $j \in \mathbb{Z}$. *Suppose there is a neighborhood of the fixed point* $\mathbf{x}_k^*$ *(i.e.,* $U_k(\mathbf{x}_k^*) \subset \Omega_\alpha$), *and* $\mathbf{f}(\mathbf{x}_k, \mathbf{p})$ *is* $C^r (r \geqslant 1)$-*continuous in* $U_k(\mathbf{x}_k^*)$. *The linearized system is* $\mathbf{y}_{k+j+1} = D\mathbf{f}(\mathbf{x}_k^*, \mathbf{p})\mathbf{y}_{k+j} (\mathbf{y}_{k+j} = \mathbf{x}_{k+j} - \mathbf{x}_k^*)$ *in* $U_k(\mathbf{x}_k^*)$. *If the homeomorphism between the local invariant subspace* $E(\mathbf{x}_k, \mathbf{x}_k^*) \subset U(\mathbf{x}_k^*)$ *and the eigenspace* $\mathscr{E}$ *of the linearized system exists with the condition in Eq.(4.28), in addition to the local stable and unstable invariant manifolds, there is a* C^{r-1} *center manifold* $\mathscr{C}_{loc}(\mathbf{x}_k, \mathbf{x}_k^*)$. *The center manifold possesses the same dimension of* $\mathscr{E}^c$ *for* $\mathbf{x}^* \in \mathscr{C}_{loc}(\mathbf{x}_k, \mathbf{x}_k^*)$, *and the tangential space of* $\mathscr{C}_{loc}(\mathbf{x}_k, \mathbf{x}_k^*)$ *is identical to* $\mathscr{E}^c$. *Thus, the local invariant subspace is decomposed by*

$$E(\mathbf{x}_k, \mathbf{x}_k^*) = \mathscr{S}_{loc}(\mathbf{x}_k, \mathbf{x}_k^*) \oplus \mathscr{U}_{loc}(\mathbf{x}_k, \mathbf{x}_k^*) \oplus \mathscr{C}_{loc}(\mathbf{x}_k, \mathbf{x}_k^*). \qquad (4.35)$$

Proof. See Guckenhiemer and Holmes (1990). ■

Definition 4.11 Consider a discrete, nonlinear dynamical system $\mathbf{x}_{k+1} = \mathbf{f}(\mathbf{x}_k, \mathbf{p}_\alpha)$ in Eq.(4.4) on domain $\Omega_\alpha \in \mathscr{R}^n$. Suppose there is a metric space (Ω_α, ρ), then the map P under the vector function $\mathbf{f}(\mathbf{x}_k, \mathbf{p}_\alpha)$ is called a contraction map if

$$\rho(\mathbf{x}_{k+1}^{(1)}, \mathbf{x}_{k+1}^{(2)}) = \rho(\mathbf{f}(\mathbf{x}_k^{(1)}, \mathbf{p}), \quad \mathbf{f}(\mathbf{x}_k^{(2)}, \mathbf{p})) \leqslant \lambda \rho(\mathbf{x}_k^{(1)}, \mathbf{x}_k^{(2)}) \qquad (4.36)$$

for $\lambda \in (0, 1)$ and $\mathbf{x}_k^{(1)}, \mathbf{x}_k^{(2)} \in \Omega_\alpha$ with $\rho(\mathbf{x}_k^{(1)}, \mathbf{x}_k^{(2)}) = \|\mathbf{x}_k^{(1)} - \mathbf{x}_k^{(2)}\|$.

Theorem 4.4 *Consider a discrete, nonlinear dynamical system* $\mathbf{x}_{k+1} = \mathbf{f}(\mathbf{x}_k, \mathbf{p})$ *in Eq.(4.4) on domain* $\Omega_\alpha \in \mathscr{R}^n$. *Suppose there is a metric space* (Ω_α, ρ), *if the map* P *under the vector function* $\mathbf{f}(\mathbf{x}_k, \mathbf{p})$ *is a contraction map, then there is an unique fixed point* $\mathbf{x}_k^*$ *which is globally stable.*

Proof. The proof can be referred to Luo (2011). ■

Definition 4.12 Consider a discrete, nonlinear dynamical system $\mathbf{x}_{k+1} = \mathbf{f}(\mathbf{x}_k, \mathbf{p}_\alpha)$ in Eq.(4.4) with a fixed point $\mathbf{x}_k^*$. The corresponding solution is given by $\mathbf{x}_{k+j} = \mathbf{f}(\mathbf{x}_{k+j-1}, \mathbf{p})$ with $j \in \mathbb{Z}$. Suppose there is a neighborhood of the fixed point $\mathbf{x}_k^*$ (i.e., $U_k(\mathbf{x}_k^*) \subset \Omega_\alpha$), and $\mathbf{f}(\mathbf{x}_k, \mathbf{p})$ is C^r $(r \geqslant 1)$-continuous in $U_k(\mathbf{x}_k^*)$. The linearized system is $\mathbf{y}_{k+j+1} = D\mathbf{f}(\mathbf{x}_k^*, \mathbf{p})\mathbf{y}_{k+j}$ $(\mathbf{y}_{k+j} = \mathbf{x}_{k+j} - \mathbf{x}_k^*)$ in $U_k(\mathbf{x}_k^*)$. Consider a real eigenvalue λ_i of matrix $D\mathbf{f}(\mathbf{x}_k^*, \mathbf{p})$ $(i \in N = \{1, 2, \cdots, n\})$ and there is a corresponding eigenvector $\mathbf{v}_i$. On the invariant eigenvector $\mathbf{v}_k^{(i)} = \mathbf{v}_i$, consider $\mathbf{y}_k^{(i)} = c_k^{(i)}\mathbf{v}_i$ and $\mathbf{y}_{k+1}^{(i)} = c_{k+1}^{(i)}\mathbf{v}_i = \lambda_i c_k^{(i)}\mathbf{v}_i$, thus, $c_{k+1}^{(i)} = \lambda_i c_k^{(i)}$.

(i) $\mathbf{x}_k^{(i)}$ on the direction $\mathbf{v}_i$ is stable if

$$\lim_{k \to \infty} |c_k^{(i)}| = \lim_{k \to \infty} |(\lambda_i)^k| \times |c_0^{(i)}| = 0 \text{ for } |\lambda_i| < 1. \qquad (4.37)$$

(ii) $\mathbf{x}_k^{(i)}$ on the direction $\mathbf{v}_i$ is unstable if

$$\lim_{k\to\infty} |c_k^{(i)}| = \lim_{k\to\infty} |(\lambda_i)^k| \times |c_0^{(i)}| = \infty \text{ for } |\lambda_i| > 1. \tag{4.38}$$

(iii) $\mathbf{x}_k^{(i)}$ on the direction $\mathbf{v}_i$ is invariant if

$$\lim_{k\to\infty} c_k^{(i)} = \lim_{k\to\infty} (\lambda_i)^k c_0^{(i)} = c_0^{(i)} \text{ for } \lambda_i = 1. \tag{4.39}$$

(iv) $\mathbf{x}_k^{(i)}$ on the direction $\mathbf{v}_i$ is flipped if

$$\left.\begin{array}{l}\lim\limits_{2k\to\infty} c_{2k}^{(i)} = \lim\limits_{2k\to\infty} (\lambda_i)^{2k} \times c_0^{(i)} = c_0^{(i)} \\[4pt] \lim\limits_{2k+1\to\infty} c_{2k+1}^{(i)} = \lim\limits_{2k+1\to\infty} (\lambda_i)^{2k+1} \times c_0^{(i)} = -c_0^{(i)}\end{array}\right\} \text{ for } \lambda_i = -1. \tag{4.40}$$

(v) $\mathbf{x}_k^{(i)}$ on the direction $\mathbf{v}_i$ is degenerate if

$$c_k^{(i)} = (\lambda_i)^k c_0^{(i)} = 0 \text{ for } \lambda_i = 0. \tag{4.41}$$

Definition 4.13 Consider a discrete, nonlinear dynamical system $\mathbf{x}_{k+1} = \mathbf{f}(\mathbf{x}_k, \mathbf{p})$ in Eq.(4.4) with a fixed point $\mathbf{x}_k^*$. The corresponding solution is given by $\mathbf{x}_{k+j} = \mathbf{f}(\mathbf{x}_{k+j-1}, \mathbf{p})$ with $j \in \mathbb{Z}$. Suppose there is a neighborhood of the fixed point $\mathbf{x}_k^*$ (i.e., $U_k(\mathbf{x}_k^*) \subset \Omega_\alpha$), and $\mathbf{f}(\mathbf{x}_k, \mathbf{p})$ is $C^r (r \geqslant 1)$-continuous in $U_k(\mathbf{x}_k^*)$. Consider a pair of complex eigenvalue $\alpha_i \pm \mathbf{i}\beta_i$ of matrix $D\mathbf{f}(\mathbf{x}_k^*, \mathbf{p})$ ($i \in N = \{1, 2, \cdots, n\}, \mathbf{i} = \sqrt{-1}$) and there is a corresponding eigenvector $\mathbf{u}_i \pm \mathbf{i}\mathbf{v}_i$. On the invariant plane of $(\mathbf{u}_k^{(i)}, \mathbf{v}_k^{(i)}) = (\mathbf{u}_i, \mathbf{v}_i)$, consider $\mathbf{x}_k^{(i)} = \mathbf{x}_{k+}^{(i)} + \mathbf{x}_{k-}^{(i)}$ with

$$\mathbf{x}_k^{(i)} = c_k^{(i)} \mathbf{u}_i + d_k^{(i)} \mathbf{v}_i, \quad \mathbf{x}_{k+1}^{(i)} = c_{k+1}^{(i)} \mathbf{u}_i + d_{k+1}^{(i)} \mathbf{v}_i. \tag{4.42}$$

Thus, $\mathbf{c}_k^{(i)} = (c_k^{(i)}, d_k^{(i)})^{\mathrm{T}}$ with

$$\mathbf{c}_{k+1}^{(i)} = \mathbf{E}_i \mathbf{c}_k^{(i)} = r_i \mathbf{R}_i \mathbf{c}_k^{(i)} \tag{4.43}$$

where

$$\mathbf{E}_i = \begin{bmatrix} \alpha_i & \beta_i \\ -\beta_i & \alpha_i \end{bmatrix} \quad \text{and} \quad \mathbf{R}_i = \begin{bmatrix} \cos\theta_i & \sin\theta_i \\ -\sin\theta_i & \cos\theta_i \end{bmatrix},$$
$$r_i = \sqrt{\alpha_i^2 + \beta_i^2}, \quad \cos\theta_i = \alpha_i/r_i \quad \text{and} \quad \sin\theta_i = \beta_i/r_i; \tag{4.44}$$

and

$$\mathbf{E}_i^k = \begin{bmatrix} \alpha_i & \beta_i \\ -\beta_i & \alpha_i \end{bmatrix}^k \quad \text{and} \quad \mathbf{R}_i^k = \begin{bmatrix} \cos k\theta_i & \sin k\theta_i \\ -\sin k\theta_i & \cos k\theta_i \end{bmatrix}. \tag{4.45}$$

(i) $\mathbf{x}_k^{(i)}$ on the plane of $(\mathbf{u}_i, \mathbf{v}_i)$ is spirally stable if

$$\lim_{k\to\infty} \|\mathbf{c}_k^{(i)}\| = \lim_{k\to\infty} r_i^k \|\mathbf{R}_i^k\| \times \|\mathbf{c}_0^{(i)}\| = 0 \text{ for } r_i = |\lambda_i| < 1. \qquad (4.46)$$

(ii) $\mathbf{x}_k^{(i)}$ on the plane of $(\mathbf{u}_i, \mathbf{v}_i)$ is spirally unstable if

$$\lim_{k\to\infty} \|\mathbf{c}_k^{(i)}\| = \lim_{k\to\infty} r_i^k \|\mathbf{R}_i^k\| \times \|\mathbf{c}_0^{(i)}\| = \infty \text{ for } r_i = |\lambda_i| > 1. \qquad (4.47)$$

(iii) $\mathbf{x}_k^{(i)}$ on the plane of $(\mathbf{u}_i, \mathbf{v}_i)$ is on the invariant circles if

$$\|\mathbf{c}_k^{(i)}\| = r_i^k \|\mathbf{R}_i^k\| \times \|\mathbf{c}_0^{(i)}\| = \|\mathbf{c}_0^{(i)}\| \text{ for } r_i = |\lambda_i| = 1. \qquad (4.48)$$

(iv) $\mathbf{x}_k^{(i)}$ on the plane of $(\mathbf{u}_i, \mathbf{v}_i)$ is degenerate in the direction of $\mathbf{u}_i$ if $\beta_i = 0$.

Definition 4.14 Consider a discrete, nonlinear dynamical system $\mathbf{x}_{k+1} = \mathbf{f}(\mathbf{x}_k, \mathbf{p})$ in Eq.(4.4) with a fixed point $\mathbf{x}_k^*$. The corresponding solution is given by $\mathbf{x}_{k+j} = \mathbf{f}(\mathbf{x}_{k+j-1}, \mathbf{p})$ with $j \in \mathbb{Z}$. Suppose there is a neighborhood of the fixed point $\mathbf{x}_k^*$ (i.e., $U_k(\mathbf{x}_k^*) \subset \Omega_\alpha$), and $\mathbf{f}(\mathbf{x}_k, \mathbf{p})$ is $C^r (r \geqslant 1)$-continuous in $U_k(\mathbf{x}_k^*)$ with Eq.(4.28). The linearized system is $\mathbf{y}_{k+j+1} = D\mathbf{f}(\mathbf{x}_k^*, \mathbf{p})\mathbf{y}_{k+j} (\mathbf{y}_{k+j} = \mathbf{x}_{k+j} - \mathbf{x}_k^*)$ in $U_k(\mathbf{x}_k^*)$. The matrix $D\mathbf{f}(\mathbf{x}_k^*, \mathbf{p})$ possesses n eigenvalues $\lambda_i (i = 1, 2, \cdots, n)$.

(i) The fixed point $\mathbf{x}_k^*$ is called a hyperbolic point if $|\lambda_i| \neq 1$ $(i = 1, 2, \cdots, n)$.
(ii) The fixed point $\mathbf{x}_k^*$ is called a sink if $|\lambda_i| < 1$ $(i = 1, 2, \cdots, n)$.
(iii) The fixed point $\mathbf{x}_k^*$ is called a source if $|\lambda_i| > 1$ $(i = 1, 2, \cdots, n)$.
(iv) The fixed point $\mathbf{x}_k^*$ is called a center if $|\lambda_i| = 1$ $(i = 1, 2, \cdots, n)$ with distinct eigenvalues.

Definition 4.15 Consider a discrete, nonlinear dynamical system $\mathbf{x}_{k+1} = \mathbf{f}(\mathbf{x}_k, \mathbf{p})$ in Eq.(4.4) with a fixed point $\mathbf{x}_k^*$. The corresponding solution is given by $\mathbf{x}_{k+j} = \mathbf{f}(\mathbf{x}_{k+j-1}, \mathbf{p})$ with $j \in \mathbb{Z}$. Suppose there is a neighborhood of the fixed point $\mathbf{x}_k^*$ (i.e., $U_k(\mathbf{x}_k^*) \subset \Omega_\alpha$), and $\mathbf{f}(\mathbf{x}_k, \mathbf{p})$ is $C^r (r \geqslant 1)$-continuous in $U_k(\mathbf{x}_k^*)$ with Eq.(4.28). The linearized system is $\mathbf{y}_{k+j+1} = D\mathbf{f}(\mathbf{x}_k^*, \mathbf{p})\mathbf{y}_{k+j} (\mathbf{y}_{k+j} = \mathbf{x}_{k+j} - \mathbf{x}_k^*)$ in $U_k(\mathbf{x}_k^*)$. The matrix $D\mathbf{f}(\mathbf{x}_k^*, \mathbf{p})$ possesses n eigenvalues λ_i $(i = 1, 2, \cdots, n)$.

(i) The fixed point $\mathbf{x}_k^*$ is called a stable node if $|\lambda_i| < 1$ $(i = 1, 2, \cdots, n)$.
(ii) The fixed point $\mathbf{x}_k^*$ is called an unstable node if $|\lambda_i| > 1$ $(i = 1, 2, \cdots, n)$.
(iii) The fixed point $\mathbf{x}_k^*$ is called an $(l_1 : l_2)$-saddle if at least one $|\lambda_i| > 1$ $(i \in L_1 \subset \{1, 2, \cdots, n\})$ and the other $|\lambda_j| < 1$ $(j \in L_2 \subset \{1, 2, \cdots, n\})$ with $L_1 \cup L_2 = \{1, 2, \cdots, n\}$ and $L_1 \cap L_2 = \varnothing$.
(iv) The fixed point $\mathbf{x}_k^*$ is called an lth-order degenerate case if $\lambda_i = 0$ $(i \in L \subseteq \{1, 2, \cdots, n\})$.

Definition 4.16 Consider a discrete, nonlinear dynamical system $\mathbf{x}_{k+1} = \mathbf{f}(\mathbf{x}_k, \mathbf{p})$ in Eq.(4.4) with a fixed point $\mathbf{x}_k^*$. The corresponding solution is given by $\mathbf{x}_{k+j} = \mathbf{f}(\mathbf{x}_{k+j-1}, \mathbf{p})$ with $j \in \mathbb{Z}$. Suppose there is a neighborhood of the fixed point $\mathbf{x}_k^*$ (i.e., $U_k(\mathbf{x}_k^*) \subset \Omega_\alpha$), and $\mathbf{f}(\mathbf{x}_k, \mathbf{p})$ is $C^r (r \geqslant 1)$-continuous in $U_k(\mathbf{x}_k^*)$ with Eq.(4.28). The linearized system is $\mathbf{y}_{k+j+1} = D\mathbf{f}(\mathbf{x}_k^*, \mathbf{p})\mathbf{y}_{k+j}$ $(\mathbf{y}_{k+j} = \mathbf{x}_{k+j} - \mathbf{x}_k^*)$ in $U_k(\mathbf{x}_k^*)$. The matrix $D\mathbf{f}(\mathbf{x}_k^*, \mathbf{p})$ possesses n-pairs of complex eigenvalues λ_i $(i = 1, 2, \cdots, n)$.

(i) The fixed point $\mathbf{x}_k^*$ is called a spiral sink if $|\lambda_i| < 1$ $(i = 1, 2, \cdots, n)$ and $\mathrm{Im}\lambda_j \neq 0$ $(j \in \{1, 2, \cdots, n\})$.

(ii) fixed point $\mathbf{x}_k^*$ is called a spiral source if $|\lambda_i| > 1$ $(i = 1, 2, \cdots, n)$ with $\mathrm{Im}\lambda_j \neq 0$ $(j \in \{1, 2, \cdots, n\})$.

(iii) fixed point $\mathbf{x}_k^*$ is called a center if $|\lambda_i| = 1$ with distinct $\mathrm{Im}\lambda_i \neq 0$ $(i = 1, 2, \cdots, n)$.

The generalized stability and bifurcation of flows in linearized, nonlinear dynamical systems in Eq.(4.4) will be discussed as follows.

Definition 4.17 Consider a discrete, nonlinear dynamical system $\mathbf{x}_{k+1} = \mathbf{f}(\mathbf{x}_k, \mathbf{p})$ in Eq.(4.4) with a fixed point $\mathbf{x}_k^*$. The corresponding solution is given by $\mathbf{x}_{k+s} = \mathbf{f}(\mathbf{x}_{k+s-1}, \mathbf{p})$ with $s \in \mathbb{Z}$. Suppose there is a neighborhood of the fixed point $\mathbf{x}_k^*$ (i.e., $U_k(\mathbf{x}_k^*) \subset \Omega_\alpha$), and $\mathbf{f}(\mathbf{x}_k, \mathbf{p})$ is $C^r (r \geqslant 1)$-continuous in $U_k(\mathbf{x}_k^*)$ with Eq.(4.28). The linearized system is $\mathbf{y}_{k+s+1} = D\mathbf{f}(\mathbf{x}_k^*, \mathbf{p})\mathbf{y}_{k+s}$ $(\mathbf{y}_{k+s} = \mathbf{x}_{k+s} - \mathbf{x}_k^*)$ in $U_k(\mathbf{x}_k^*)$. The matrix $D\mathbf{f}(\mathbf{x}_k^*, \mathbf{p})$ possesses n eigenvalues λ_i $(i = 1, 2, \cdots, n)$. Set $N = \{1, 2, \cdots, m, m + 1, \cdots, (n + m)/2\}$, $N_j = \{j_1, j_2, \cdots, j_{n_j}\} \cup \varnothing$ with $j_p \in N$ $(p = 1, 2, \cdots, n_j; j = 1, 2, \cdots, 7)$, $\Sigma_{j=1}^4 n_j = m$ and $2\Sigma_{j=5}^7 n_j = n - m$. $\cup_{j=1}^7 N_j = N$ with $N_j \cap N_l = \varnothing(l \neq j)$. $N_j = \varnothing$ if $n_j = 0$. $N_\alpha = N_\alpha^{\mathrm{m}} \cup N_\alpha^{\mathrm{o}}$ $(\alpha = 1, 2)$ and $N_\alpha^{\mathrm{m}} \cap N_\alpha^{\mathrm{o}} = \varnothing$ with $n_\alpha^{\mathrm{m}} + n_\alpha^{\mathrm{o}} = n_\alpha$ where superscripts "m" and "o" represent monotonic and oscillatory evolutions. The matrix $D\mathbf{f}(\mathbf{x}_k^*, \mathbf{p})$ possesses n_1-stable, n_2-unstable, n_3-invariant, and n_4-flip real eigenvectors plus n_5-stable, n_6-unstable and n_7-center pairs of complex eigenvectors. Without repeated complex eigenvalues of $|\lambda_i| = 1 (i \in N_3 \cup N_4 \cup N_7)$, an iterative response of $\mathbf{x}_{k+1} = \mathbf{f}(\mathbf{x}_k, \mathbf{p})$ is an $([n_1^{\mathrm{m}}, n_1^{\mathrm{o}}] : [n_2^{\mathrm{m}}, n_2^{\mathrm{o}}] : [n_3; \kappa_3] : [n_4; \kappa_4] | n_5 : n_6 : n_7)$ flow in the neighborhood of the fixed point $\mathbf{x}_k^*$. With repeated complex eigenvalues of $|\lambda_i| = 1$ $(i \in N_3 \cup N_4 \cup N_7)$, an iterative response of $\mathbf{x}_{k+1} = \mathbf{f}(\mathbf{x}_k, \mathbf{p})$ is an $([n_1^{\mathrm{m}}, n_1^{\mathrm{o}}] : [n_2^{\mathrm{m}}, n_2^{\mathrm{o}}] : [n_3; \kappa_3] : [n_4; \kappa_4] | n_5 : n_6 : [n_7, l; \boldsymbol{\kappa}_7])$ flow in the neighborhood of the fixed point $\mathbf{x}_k^*$, where $\kappa_3 \in \{\varnothing, m_3\}$, $\boldsymbol{\kappa}_7 = (\kappa_{71}, \kappa_{72}, \cdots, \kappa_{7l})^{\mathrm{T}}$ with $\kappa_{7s} \in \{\varnothing, m_{7s}\}$ $(s = 1, 2, \cdots, l)$. The meanings of notations in the aforementioned structures are defined as follows:

(i) $[n_1^{\mathrm{m}}, n_1^{\mathrm{o}}]$ represents n_1-sinks with n_1^{m}-monotonic convergence and n_1^{o}-oscillatory convergence among n_1-directions of $\mathbf{v}_i$ $(i \in N_1)$ if $|\lambda_i| < 1$ $(i \in N_1$ and $1 \leqslant n_1 \leqslant m)$ with distinct or repeated eigenvalues.

(ii) $[n_2^{\mathrm{m}}, n_2^{\mathrm{o}}]$ represents n_2-sources with n_2^{m}-monotonic divergence and n_2^{o}-oscillatory divergence among n_2-directions of $\mathbf{v}_i$ $(i \in N_2)$ if $|\lambda_i| > 1$ $(i \in N_2$ and $1 \leqslant n_2 \leqslant m)$ with distinct or repeated eigenvalues.

(iii) $n_3 = 1$ represents an invariant center on 1-direction of $\mathbf{v}_i$ $(i \in N_3)$ if $\lambda_i = 1$ $(i \in N_3$ and $n_3 = 1)$.

(iv) $n_4 = 1$ represents an flip center on 1-direction of $\mathbf{v}_i$ $(i \in N_4)$ if $\lambda_i = -1$ $(i \in N_4$ and $n_4 = 1)$.

(v) n_5 represents n_5-spiral sinks on n_5-pairs of $(\mathbf{u}_i, \mathbf{v}_i)$ $(i \in N_5)$ if $|\lambda_i| < 1$ and $\mathrm{Im}\lambda_i \neq 0$ $(i \in N_5$ and $1 \leqslant n_5 \leqslant (n - m)/2)$ with distinct or repeated eigenvalues.

(vi) n_6 represents n_6-spiral sources on n_6-directions of $(\mathbf{u}_i, \mathbf{v}_i)$ $(i \in N_6)$ if $|\lambda_i| > 1$ and $\mathrm{Im}\lambda_i \neq 0$ $(i \in N_6$ and $1 \leqslant n_6 \leqslant (n - m)/2)$ with distinct or repeated eigenvalues.

(vii) n_7 represents n_7-invariant centers on n_7-pairs of $(\mathbf{u}_i, \mathbf{v}_i)$ $(i \in N_7)$ if $|\lambda_i| = 1$ and $\mathrm{Im}\lambda_i \neq 0$ $(i \in N_7$ and $1 \leqslant n_7 \leqslant (n - m)/2)$ with distinct eigenvalues.

(viii) $\varnothing$ represents none if $n_j = 0$ $(j \in \{1, 2, \cdots, 7\})$.

(ix) $[n_3; \kappa_3]$ represents $(n_3 - \kappa_3)$ invariant centers on $(n_3 - \kappa_3)$ directions of $\mathbf{v}_{i_3}$ $(i_3 \in N_3)$ and κ_3-sources in κ_3-directions of $\mathbf{v}_{j_3}$ $(j_3 \in N_3$ and $j_3 \neq i_3)$ if $\lambda_i = 1$ $(i \in N_3$ and $n_3 \leqslant m)$ with the $(\kappa_3 + 1)$th-order nilpotent matrix $\mathbf{N}_3^{\kappa_3+1} = \mathbf{0}$ $(0 < \kappa_3 \leqslant n_3 - 1)$.

(x) $[n_3; \varnothing]$ represents n_3 invariant centers on n_3-directions of $\mathbf{v}_i$ $(i \in N_3)$ if $\lambda_i = 1$ $(i \in N_3$ and $1 < n_3 \leqslant m)$ with a nilpotent matrix $\mathbf{N}_3 = \mathbf{0}$.

(xi) $[n_4; \kappa_4]$ represents $(n_4 - \kappa_4)$ flip oscillatory centers on $(n_4 - \kappa_4)$ directions of $\mathbf{v}_{i_4}$ $(i_4 \in N_4)$ and κ_4-sources in κ_4-directions of $\mathbf{v}_{j_4}$ $(j_4 \in N_4$ and $j_4 \neq i_4)$ if $\lambda_i = -1$ $(i \in N_4$ and $n_4 \leqslant m)$ with the $(\kappa_4 + 1)$th-order nilpotent matrix $\mathbf{N}_4^{\kappa_4+1} = \mathbf{0}$ $(0 < \kappa_4 \leqslant n_4 - 1)$.

(xii) $[n_4; \varnothing]$ represents n_4 flip oscillatory centers on n_4-directions of $\mathbf{v}_i$ $(i \in N_3)$ if $\lambda_i = -1$ $(i \in N_4$ and $1 < n_4 \leqslant m)$ with a nilpotent matrix $\mathbf{N}_4 = \mathbf{0}$.

(xiii) $[n_7, l; \kappa_7]$ represents $(n_7 - \Sigma_{s=1}^{l} \kappa_{7s})$ invariant centers on $(n_7 - \Sigma_{s=1}^{l} \kappa_{7s})$ pairs of $(\mathbf{u}_{i_7}, \mathbf{v}_{i_7})(i_7 \in N_7)$ and $\Sigma_{s=1}^{l} \kappa_{7s}$ sources on $\Sigma_{s=1}^{l} \kappa_{7s}$ pairs of $(\mathbf{u}_{j_7}, \mathbf{v}_{j_7})$ $(j_7 \in N_7$ and $j_7 \neq i_7)$ if $|\lambda_i| = 1$ and $\mathrm{Im}\lambda_i \neq 0$ $(i \in N_7$ and $n_7 \leqslant (n - m)/2)$ for $\Sigma_{s=1}^{l} \kappa_{7s}$-pairs of repeated eigenvalues with the $(\kappa_{7s} + 1)$th-order nilpotent matrix $\mathbf{N}_7^{\kappa_{7s}+1} = \mathbf{0}$ $(0 < \kappa_{7s} \leqslant l, s = 1, 2, \cdots, l)$.

(xiv) $[n_7, l; \varnothing]$ represents n_7-invariant centers on n_7-pairs of $(\mathbf{u}_i, \mathbf{v}_i)$ $(i \in N_6)$ if $|\lambda_i| = 1$ and $\mathrm{Im}\lambda_i \neq 0$ $(i \in N_7$ and $1 \leqslant n_7 \leqslant (n - m)/2)$ for $\Sigma_{s=1}^{l} \kappa_{7s}$-pairs of repeated eigenvalues with a nilpotent matrix $\mathbf{N}_7 = \mathbf{0}$.

4.3 Stability switching theory

To extend the idea of Definitions 4.11 and 4.12, a new function will be defined to determine the stability and the stability state switching.

Definition 4.18 Consider a discrete, nonlinear dynamical system $\mathbf{x}_{k+1} = \mathbf{f}(\mathbf{x}_k, \mathbf{p}) \in \mathscr{R}^n$ in Eq.(4.4) with a fixed point $\mathbf{x}_k^*$. The corresponding solution is given by $\mathbf{x}_{k+j} = \mathbf{f}(\mathbf{x}_{k+j-1}, \mathbf{p})$ with $j \in \mathbb{Z}$. Suppose there is a neighborhood of the fixed point $\mathbf{x}_k^*$ (i.e., $U_k(\mathbf{x}_k^*) \subset \Omega_\alpha$), and $\mathbf{f}(\mathbf{x}_k, \mathbf{p})$ is $C^r (r \geqslant 1)$-continuous in $U_k(\mathbf{x}_k^*)$ with Eq.(4.28). The linearized system is $\mathbf{y}_{k+j+1} = D\mathbf{f}(\mathbf{x}_k^*, \mathbf{p})\mathbf{y}_{k+j}$ $(\mathbf{y}_{k+j} = \mathbf{x}_{k+j} - \mathbf{x}_k^*)$ in $U_k(\mathbf{x}_k^*)$ and there are n linearly independent vectors

$\mathbf{v}_i$ $(i = 1, 2, \cdots, n)$. For a perturbation of fixed point $\mathbf{y}_k = \mathbf{x}_k - \mathbf{x}_k^*$, let $\mathbf{y}_k^{(i)} = c_k^{(i)}\mathbf{v}_i$ and $\mathbf{y}_{k+1}^{(i)} = c_{k+1}^{(i)}\mathbf{v}_i$,

$$s_k^{(i)} = \mathbf{v}_i^{\mathrm{T}} \cdot \mathbf{y}_k = \mathbf{v}_i^{\mathrm{T}} \cdot (\mathbf{x}_k - \mathbf{x}_k^*) \tag{4.49}$$

where $s_k^{(i)} = c_k^{(i)}\|\mathbf{v}_i\|^2$. Define the following functions

$$G_i(\mathbf{x}_k, \mathbf{p}) = \mathbf{v}_i^{\mathrm{T}} \cdot [\mathbf{f}(\mathbf{x}_k, \mathbf{p}) - \mathbf{x}_k^*] \tag{4.50}$$

and

$$G_{s_k^{(i)}}^{(1)}(\mathbf{x}, \mathbf{p}) = \mathbf{v}_i^{\mathrm{T}} \cdot D_{c_k^{(i)}}\mathbf{f}(\mathbf{x}_k(s_k^{(i)}), \mathbf{p}) = \mathbf{v}_i^{\mathrm{T}} \cdot D_{\mathbf{x}_k}\mathbf{f}(\mathbf{x}_k(s_k^{(i)}), \mathbf{p})\partial_{c_k^{(i)}}\mathbf{x}_k\partial_{s_k}c_k^{(i)}$$

$$= \mathbf{v}_i^{\mathrm{T}} \cdot D_{\mathbf{x}_k}\mathbf{f}(\mathbf{x}_k(s_k^{(i)}), \mathbf{p})\mathbf{v}_i\|\mathbf{v}_i\|^{-2} \tag{4.51}$$

$$G_{s_k^{(i)}}^{(m)}(\mathbf{x}, \mathbf{p}) = \mathbf{v}_i^{\mathrm{T}} \cdot D_{s_k^{(i)}}^{(m)}\mathbf{f}(\mathbf{x}_k(s_k^{(i)}), \mathbf{p}) = \mathbf{v}_i^{\mathrm{T}} \cdot D_{s_k^{(i)}}(D_{s_k^{(i)}}^{(m-1)}\mathbf{f}(\mathbf{x}_k(s_k^{(i)}), \mathbf{p})) \tag{4.52}$$

where $D_{s_k^{(i)}}(\cdot) = \partial(\cdot)/\partial s_k^{(i)}$ and $D_{s_k^{(i)}}^{(m)}(\cdot) = D_{s_k^{(i)}}(D_{s_k^{(i)}}^{(m-1)}(\cdot))$.

Definition 4.19 Consider a discrete, nonlinear dynamical system $\mathbf{x}_{k+1} = \mathbf{f}(\mathbf{x}_k, \mathbf{p}) \in \mathscr{R}^n$ in Eq.(4.4) with a fixed point $\mathbf{x}_k^*$. The corresponding solution is given by $\mathbf{x}_{k+j} = \mathbf{f}(\mathbf{x}_{k+j-1}, \mathbf{p})$ with $j \in \mathbb{Z}$. Suppose there is a neighborhood of the fixed point $\mathbf{x}_k^*$ (i.e., $U_k(\mathbf{x}_k^*) \subset \Omega_\alpha$), and $\mathbf{f}(\mathbf{x}_k, \mathbf{p})$ is $C^r (r \geqslant 1)$-continuous in $U_k(\mathbf{x}_k^*)$ with Eq.(4.28). The linearized system is $\mathbf{y}_{k+j+1} = D\mathbf{f}(\mathbf{x}_k^*, \mathbf{p})\mathbf{y}_{k+j}$ ($\mathbf{y}_{k+j} = \mathbf{x}_{k+j} - \mathbf{x}_k^*$) in $U_k(\mathbf{x}_k^*)$ and there are n linearly independent vectors $\mathbf{v}_i$ $(i = 1, 2, \cdots, n)$. For a perturbation of fixed point $\mathbf{y}_k = \mathbf{x}_k - \mathbf{x}_k^*$, let $\mathbf{y}_k^{(i)} = c_k^{(i)}\mathbf{v}_i$ and $\mathbf{y}_{k+1}^{(i)} = c_{k+1}^{(i)}\mathbf{v}_i$.

(i) $\mathbf{x}_{k+j}(j \in \mathbb{Z})$ at fixed point $\mathbf{x}_k^*$ on the direction $\mathbf{v}_i$ is stable if

$$|\mathbf{v}_i^{\mathrm{T}} \cdot (\mathbf{x}_{k+1} - \mathbf{x}_k^*)| < |\mathbf{v}_i^{\mathrm{T}} \cdot (\mathbf{x}_k - \mathbf{x}_k^*)| \tag{4.53}$$

for $\mathbf{x}_k \in U_k(\mathbf{x}_k^*) \subset \Omega_\alpha$. The fixed point $\mathbf{x}_k^*$ is called the sink (or stable node) on the direction $\mathbf{v}_i$.

(ii) $\mathbf{x}_{k+j}(j \in \mathbb{Z})$ at fixed point $\mathbf{x}_k^*$ on the direction $\mathbf{v}_i$ is unstable if

$$|\mathbf{v}_i^{\mathrm{T}} \cdot (\mathbf{x}_{k+1} - \mathbf{x}_k^*)| > |\mathbf{v}_i^{\mathrm{T}} \cdot (\mathbf{x}_k - \mathbf{x}_k^*)| \tag{4.54}$$

for $\mathbf{x}_k \in U(\mathbf{x}_k^*) \subset \Omega_\alpha$. The fixed point $\mathbf{x}_k^*$ is called the source (or unstable node) on the direction $\mathbf{v}_i$.

(iii) $\mathbf{x}_{k+j}(j \in \mathbb{Z})$ at fixed point $\mathbf{x}_k^*$ on the direction $\mathbf{v}_i$ is invariant if

$$\mathbf{v}_i^{\mathrm{T}} \cdot (\mathbf{x}_{k+1} - \mathbf{x}_k^*) = \mathbf{v}_i^{\mathrm{T}} \cdot (\mathbf{x}_k - \mathbf{x}_k^*) \tag{4.55}$$

for $\mathbf{x}_k \in U(\mathbf{x}_k^*) \subset \Omega_\alpha$. The fixed point $\mathbf{x}_k^*$ is called to be degenerate on the direction $\mathbf{v}_i$.

(iv) $\mathbf{x}_{k+j}^{(i)}(j \in \mathbb{Z})$ at fixed point $\mathbf{x}_k^*$ on the direction $\mathbf{v}_i$ is symmetrically flipped if

$$\mathbf{v}_i^{\mathrm{T}} \cdot (\mathbf{x}_{k+1} - \mathbf{x}_k^*) = -\mathbf{v}_i^{\mathrm{T}} \cdot (\mathbf{x}_k - \mathbf{x}_k^*) \tag{4.56}$$

for $\mathbf{x}_k \in U(\mathbf{x}_k^*) \subset \Omega_\alpha$. The equilibrium $\mathbf{x}^*$ is called to be degenerate on the direction $\mathbf{v}_i$.

The stability of fixed points for a specific eigenvector is presented in Fig.4.4. The solid curve is $\mathbf{v}_i^{\mathrm{T}} \cdot \mathbf{x}_{k+1} = \mathbf{v}_i^{\mathrm{T}} \cdot \mathbf{f}(\mathbf{x}_k, \mathbf{p})$. The circular symbol is a fixed point. The shaded regions are stable. The horizontal solid line is for a degenerate case. The vertical solid line is for a line with infinite slope. The monotonically stable node (sink) is presented in Fig.4.4(a). From the fixed point $\mathbf{x}_k^*$, let $\mathbf{y}_k = \mathbf{x}_k - \mathbf{x}_k^*$ and $\mathbf{y}_{k+1} = \mathbf{x}_{k+1} - \mathbf{x}_k^*$. $\mathbf{v}_i^{\mathrm{T}} \cdot \mathbf{x}_k = \mathbf{v}_i^{\mathrm{T}} \cdot \mathbf{x}_{k+1}$ and $\mathbf{v}_i^{\mathrm{T}} \cdot \mathbf{y}_{k+1} = -\mathbf{v}_i^{\mathrm{T}} \cdot \mathbf{y}_k$ are represented by dashed and dotted lines, respectively. The iterative responses approach the fixed point. However, the monotonically unstable (source) is presented in Fig.4.4(b). The iterative responses go away from the fixed point. Similarly, the oscillatory stable node (sink) after iteration with a flip $\mathbf{v}_i^{\mathrm{T}} \cdot \mathbf{y}_k = -\mathbf{v}_i^{\mathrm{T}} \cdot \mathbf{y}_{k+1}$ is presented in Fig.4.4(c). The dashed and dotted lines are used for two lines $\mathbf{v}_i^{\mathrm{T}} \cdot \mathbf{y}_{k+1} = -\mathbf{v}_i^{\mathrm{T}} \cdot \mathbf{y}_k$ and $\mathbf{v}_i^{\mathrm{T}} \cdot \mathbf{x}_k = \mathbf{v}_i^{\mathrm{T}} \cdot \mathbf{x}_{k+1}$, respectively. In a similar fashion, the oscillatory unstable node (source) is presented in Fig.4.4(d). This illustration can be easily observed the stability of fixed points. In Fig.4.4(e) and (f), the oscillatory stable and unstable nodes are presented as usual through the two time iterations.

Theorem 4.5 *Consider a discrete, nonlinear dynamical system $\mathbf{x}_{k+1} = \mathbf{f}(\mathbf{x}_k, \mathbf{p}) \in \mathscr{R}^n$ in Eq.(4.4) with a fixed point $\mathbf{x}_k^*$. The corresponding solution is given by $\mathbf{x}_{k+j} = \mathbf{f}(\mathbf{x}_{k+j-1}, \mathbf{p})$ with $j \in \mathbb{Z}$. Suppose there is a neighborhood of the fixed point $\mathbf{x}_k^*$ (i.e., $U_k(\mathbf{x}_k^*) \subset \Omega_\alpha$), and $\mathbf{f}(\mathbf{x}_k, \mathbf{p})$ is $C^r (r \geqslant 1)$-continuous in $U_k(\mathbf{x}_k^*)$ with Eq.(4.28). The linearized system is $\mathbf{y}_{k+j+1} = D\mathbf{f}(\mathbf{x}_k^*, \mathbf{p})\mathbf{y}_{k+j}$ $(\mathbf{y}_{k+j} = \mathbf{x}_{k+j} - \mathbf{x}_k^*)$ in $U_k(\mathbf{x}_k^*)$ and there are n linearly independent vectors $\mathbf{v}_i(i = 1, 2, \cdots, n)$. For a perturbation of fixed point $\mathbf{y}_k = \mathbf{x}_k - \mathbf{x}_k^*$, let $\mathbf{y}_k^{(i)} = c_k^{(i)}\mathbf{v}_i$ and $\mathbf{y}_{k+1}^{(i)} = c_{k+1}^{(i)}\mathbf{v}_i$.*

(i) *$\mathbf{x}_{k+j}(j \in \mathbb{Z})$ at fixed point $\mathbf{x}_k^*$ on the direction $\mathbf{v}_i$ is stable if and only if*

$$G_{s_k^{(i)}}^{(1)}(\mathbf{x}_k^*, \mathbf{p}) = \lambda_i \in (-1, 1) \tag{4.57}$$

for $\mathbf{x}_k \in U_k(\mathbf{x}_k^*) \subset \Omega_\alpha$.

(ii) *$\mathbf{x}_{k+j}(j \in \mathbb{Z})$ at fixed point $\mathbf{x}_k^*$ on the direction $\mathbf{v}_i$ is unstable if and only if*

$$G_{s_k^{(i)}}^{(1)}(\mathbf{x}_k^*, \mathbf{p}) = \lambda_i \in (1, \infty) \quad and \quad (-\infty, -1) \tag{4.58}$$

for $\mathbf{x}_k \in U_k(\mathbf{x}_k^*) \subset \Omega_\alpha$.

(iii) *$\mathbf{x}_{k+j}(j \in \mathbb{Z})$ at fixed point $\mathbf{x}_k^*$ on the direction $\mathbf{v}_i$ is invariant if and only if*

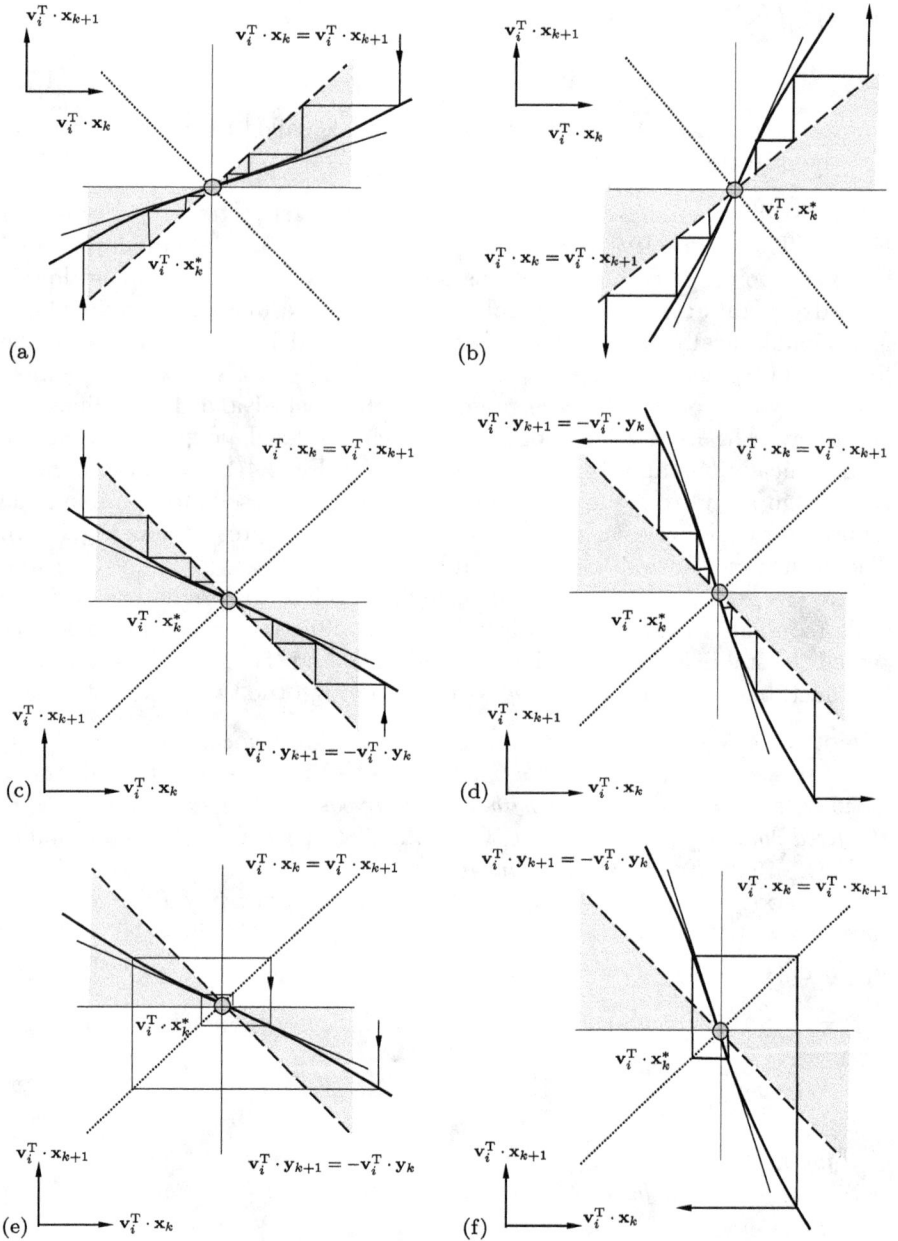

Fig. 4.4 Stability of fixed points: (a) monotonically stable node (sink), (b) monotonically unstable node (source); (c) oscillatory stable node (sink) and (d) oscillatory unstable node (sink); (e) oscillatory stable node (sink) and (f) oscillatory unstable node (sink). Shaded areas are stable zones. ($\mathbf{y}_k = \mathbf{x}_k - \mathbf{x}_k^*$ and $\mathbf{y}_{k+1} = \mathbf{x}_{k+1} - \mathbf{x}_k^*$).

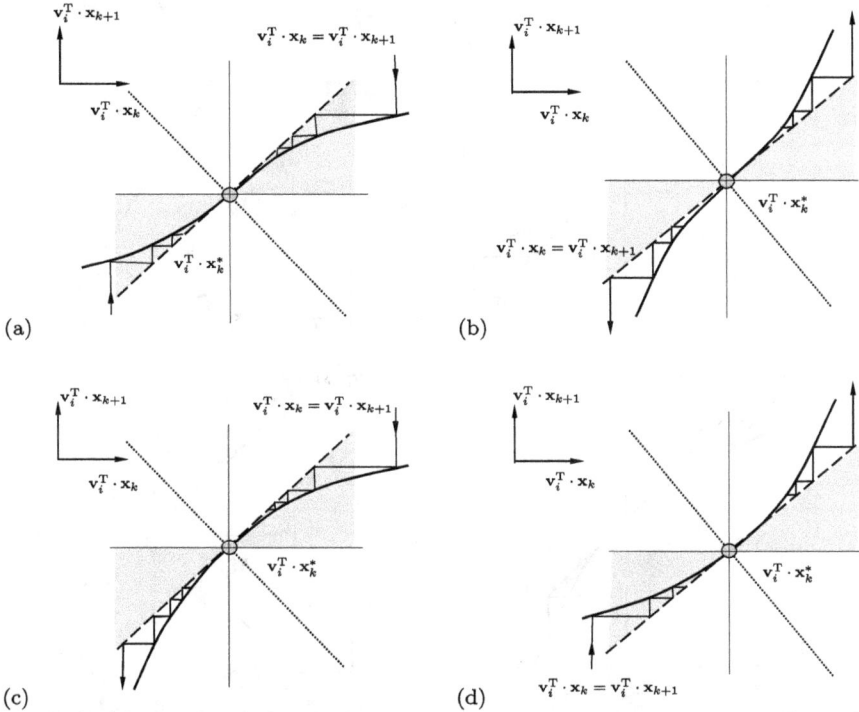

Fig. 4.5 Monotonic stability of fixed points with higher-order singularity: (a) monotonically stable node (sink) of $(2m_i + 1)$th-order, (b) monotonically unstable node (source) of $(2m_i + 1)$th-order, (c) monotonically lower saddle of $(2m_i)$th-order and (d) monotonically upper saddle of $(2m_i)$th-order. Shaded areas are stable zones. ($\mathbf{y}_k = \mathbf{x}_k - \mathbf{x}_k^*$ and $\mathbf{y}_{k+1} = \mathbf{x}_{k+1} - \mathbf{x}_k^*$).

$$G_{s_k^{(i)}}^{(1)}(\mathbf{x}_k^*, \mathbf{p}) = \lambda_i = 1 \quad and \quad G_{s_k^{(i)}}^{(m_i)}(\mathbf{x}_k^*, \mathbf{p}) = 0 \quad m_i = 2, 3, \cdots \quad (4.59)$$

for $\mathbf{x}_k \in U_k(\mathbf{x}_k^*) \subset \Omega_\alpha$.

(iv) $\mathbf{x}_{k+j}^{(i)} (j \in \mathbb{Z})$ *at fixed point* $\mathbf{x}_k^*$ *on the direction* $\mathbf{v}_k$ *is symmetrically flipped if and only if*

$$G_{s_k^{(i)}}^{(1)}(\mathbf{x}_k^*, \mathbf{p}) = \lambda_i = -1 \quad and \quad G_{s_k^{(i)}}^{(m_i)}(\mathbf{x}_k^*, \mathbf{p}) = 0 \quad m_i = 2, 3, \cdots \quad (4.60)$$

for $\mathbf{x}_k \in U_k(\mathbf{x}_k^*) \subset \Omega_\alpha$.

Proof. The proof can be referred to Luo (2012).   ∎

The monotonic stability of fixed points with higher order singularity for a specific eigenvector is presented in Fig.4.5. The solid curve is $\mathbf{v}_i^{\mathrm{T}} \cdot \mathbf{x}_{k+1} = \mathbf{v}_i^{\mathrm{T}} \cdot \mathbf{f}(\mathbf{x}_k, \mathbf{p})$. The circular symbol is a fixed point. The shaded regions are stable. The horizontal solid line is also for the degenerate case. The vertical solid line is for a line with infinite slope. The monotonically stable node (sink) of the $(2m_i+1)$th order is sketched in Fig.4.5(a). The dashed and dotted lines

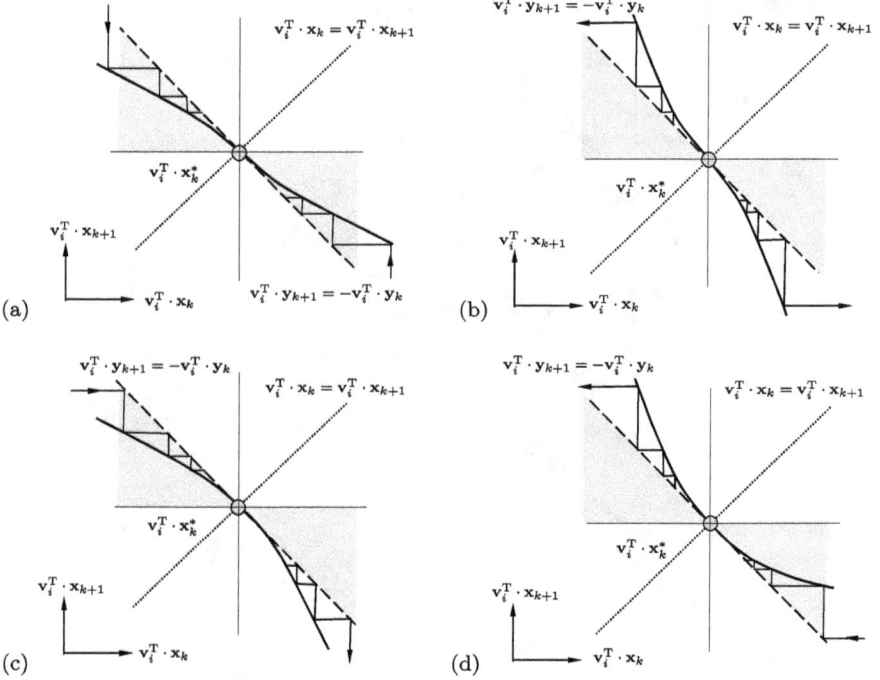

Fig. 4.6 Oscillatory stability of fixed points with higher-order singularity after iteration with a flip $\mathbf{v}_i^T \cdot \mathbf{y}_k = -\mathbf{v}_i^T \cdot \mathbf{y}_{k+1}$: (a) oscillatory stable node (sink) of $(2m_i + 1)$th-order, (b) oscillatory unstable node (source) of $(2m_i + 1)$th-order, (c) oscillatory lower saddle of $(2m_i)$th-order and (d) oscillatory upper saddle of $(2m_i)$th-order. Shaded areas are stable zones. ($\mathbf{y}_k = \mathbf{x}_k - \mathbf{x}_k^*$ and $\mathbf{y}_{k+1} = \mathbf{x}_{k+1} - \mathbf{x}_k^*$).

are for $\mathbf{v}_i^T \cdot \mathbf{x}_k = \mathbf{v}_i^T \cdot \mathbf{x}_{k+1}$ and $\mathbf{v}_i^T \cdot \mathbf{y}_{k+1} = -\mathbf{v}_i^T \cdot \mathbf{y}_k$, respectively. The nonlinear curve lies in the stable zone, and the iterative responses approach the fixed point. However, the monotonically unstable (source) of the $(2m_i + 1)$th order is presented in Fig.4.5(b). The nonlinear curve lies in the unstable zone, and the iterative responses go away from the fixed point. The monotonically lower saddle of the $(2m_i)$th order is presented in Fig.4.5(c). The nonlinear curve is tangential to the line of $\mathbf{v}_i^T \cdot \mathbf{x}_k = \mathbf{v}_i^T \cdot \mathbf{x}_{k+1}$ with the $(2m_i)$th order, and the upper one branch is in the stable zone and the lower branch is in the unstable zone. Similarly, the monotonically upper saddle of the $(2m_i)$th order is presented in Fig.4.5(d). The oscillatory stability of fixed points with higher order singularity for a specific eigenvector after iteration with a flip $\mathbf{v}_i^T \cdot \mathbf{y}_k = -\mathbf{v}_i^T \cdot \mathbf{y}_{k+1}$ is presented in Fig.4.6. The oscillatory stable node (sink) of the $(2m_i + 1)$th order is sketched in Fig.4.6(a). The dashed and dotted lines are for $\mathbf{v}_i^T \cdot \mathbf{y}_{k+1} = -\mathbf{v}_i^T \cdot \mathbf{y}_k$ and $\mathbf{v}_i^T \cdot \mathbf{x}_k = \mathbf{v}_i^T \cdot \mathbf{x}_{k+1}$, respectively. The nonlinear curve lies in the stable zone, and the iterative responses approach the fixed point. However, the oscillatory unstable (source) of the $(2m_i + 1)$th order is presented in Fig.4.6(b). The nonlinear curve lies in the unstable zone, and the iterative responses go away from the fixed point. The oscillatory lower

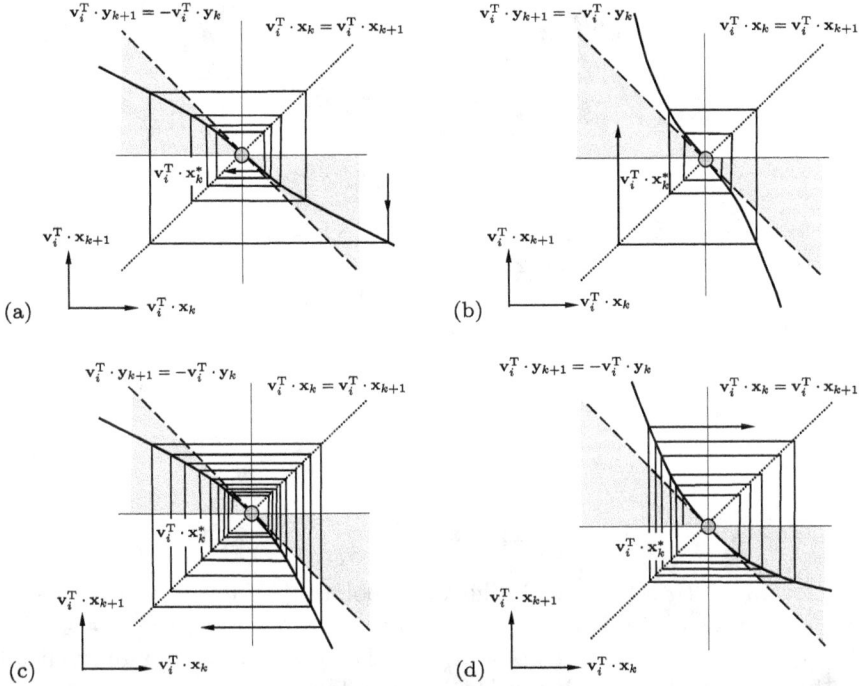

Fig. 4.7 Oscillatory stability of fixed points with higher-order singularity for the two-time iterations: (a) oscillatory stable node (sink) of $(2m_i + 1)$th-order, (b) oscillatory unstable node (source) of $(2m_i + 1)$th-order, (c) oscillatory lower saddle of $(2m_i)$th-order and (d) oscillatory upper saddle of $(2m_i)$th-order. Shaded areas are stable zones. ($\mathbf{y}_k = \mathbf{x}_k - \mathbf{x}_k^*$ and $\mathbf{y}_{k+1} = \mathbf{x}_{k+1} - \mathbf{x}_k^*$).

saddle of the $(2m_i)$th order is presented in Fig.4.6(c). The nonlinear curve is tangential to and below the line of $\mathbf{v}_i^{\mathrm{T}} \cdot \mathbf{y}_{k+1} = -\mathbf{v}_i^{\mathrm{T}} \cdot \mathbf{y}_k$ with the $(2m_i)$th order, and the upper branch is in the stable zone and the lower branch is in the unstable zone. Finally, the oscillatory upper saddle of the $(2m_i)$th order is presented in Fig.4.6(d). For clear illustrations, oscillatory stability of fixed points with higher-order singularity for the two-time iterations are presented in Fig.4.7.

Definition 4.20 Consider a discrete, nonlinear dynamical system $\mathbf{x}_{k+1} = \mathbf{f}(\mathbf{x}_k, \mathbf{p}) \in \mathscr{R}^n$ in Eq.(4.4) with a fixed point $\mathbf{x}_k^*$. The corresponding solution is given by $\mathbf{x}_{k+j} = \mathbf{f}(\mathbf{x}_{k+j-1}, \mathbf{p})$ with $j \in \mathbb{Z}$. Suppose there is a neighborhood of the fixed point $\mathbf{x}_k^*$ (i.e., $U_k(\mathbf{x}_k^*) \subset \Omega_\alpha$), and $\mathbf{f}(\mathbf{x}_k, \mathbf{p})$ is $C^r (r \geqslant 1)$-continuous in $U_k(\mathbf{x}_k^*)$ with Eq.(4.28). The linearized system is $\mathbf{y}_{k+j+1} = D\mathbf{f}(\mathbf{x}_k^*, \mathbf{p})\mathbf{y}_{k+j}$ ($\mathbf{y}_{k+j} = \mathbf{x}_{k+j} - \mathbf{x}_k^*$) in $U_k(\mathbf{x}_k^*)$ and there are n linearly independent vectors $\mathbf{v}_i (i = 1, 2, \cdots, n)$. For a perturbation of fixed point $\mathbf{y}_k = \mathbf{x}_k - \mathbf{x}_k^*$, let $\mathbf{y}_k^{(i)} = c_k^{(i)} \mathbf{v}_i$ and $\mathbf{y}_{k+1}^{(i)} = c_{k+1}^{(i)} \mathbf{v}_i$.

(i) $\mathbf{x}_{k+j}(j \in \mathbb{Z})$ at fixed point $\mathbf{x}_k^*$ on the direction $\mathbf{v}_i$ is monotonically stable of the $(2m_i + 1)$th-order if

$$G^{(1)}_{s^{(i)}_k}(\mathbf{x}^*_k, \mathbf{p}) = \lambda_i = 1,$$
$$G^{(r_i)}_{s^{(i)}_k}(\mathbf{x}^*_k, \mathbf{p}) = 0 \text{ for } r_i = 2, 3, \cdots, 2m_i,$$
$$G^{(2m_i+1)}_{s^{(i)}_k}(\mathbf{x}^*_k, \mathbf{p}) \neq 0, \tag{4.61}$$
$$|\mathbf{v}^{\mathrm{T}}_i \cdot (\mathbf{x}_{k+1} - \mathbf{x}^*_k)| < |\mathbf{v}^{\mathrm{T}}_i \cdot (\mathbf{x}_k - \mathbf{x}^*_k)|$$

for $\mathbf{x}_k \in U_k(\mathbf{x}^*_k) \subset \Omega_\alpha$. The fixed point $\mathbf{x}^*_k$ is called the monotonic sink (or stable node) of the $(2m_i+1)$th-order on the direction $\mathbf{v}_i$.

(ii) $\mathbf{x}_{k+j}(j \in \mathbb{Z})$ at fixed point $\mathbf{x}^*_k$ on the direction $\mathbf{v}_i$ is monotonically unstable of the $(2m_i+1)$th-order if

$$G^{(1)}_{s^{(i)}_k}(\mathbf{x}^*_k, \mathbf{p}) = \lambda_i = 1,$$
$$G^{(r_i)}_{s^{(i)}_k}(\mathbf{x}^*_k, \mathbf{p}) = 0 \text{ for } r_i = 2, 3, \cdots, 2m_i;$$
$$G^{(2m_i+1)}_{s^{(i)}_k}(\mathbf{x}^*_k, \mathbf{p}) \neq 0; \tag{4.62}$$
$$|\mathbf{v}^{\mathrm{T}}_i \cdot (\mathbf{x}_{k+1} - \mathbf{x}^*_k)| > |\mathbf{v}^{\mathrm{T}}_i \cdot (\mathbf{x}_k - \mathbf{x}^*_k)|$$

for $\mathbf{x}_k \in U_k(\mathbf{x}^*_k) \subset \Omega_\alpha$. The fixed point $\mathbf{x}^*_k$ is called the monotonic source (or unstable node) of the $(2m_i+1)$th-order on the direction $\mathbf{v}_i$.

(iii) $\mathbf{x}_{k+j}(j \in \mathbb{Z})$ at fixed point $\mathbf{x}^*_k$ on the direction $\mathbf{v}_i$ is monotonically unstable of the $(2m_i)$th-order, lower saddle if

$$G^{(1)}_{s^{(i)}_k}(\mathbf{x}^*_k, \mathbf{p}) = \lambda_i = 1,$$
$$G^{(r_i)}_{s^{(i)}_k}(\mathbf{x}^*_k, \mathbf{p}) = 0 \text{ for } r_i = 2, 3, \cdots, 2m_i - 1;$$
$$G^{(2m_i)}_{s^{(i)}_k}(\mathbf{x}^*_k, \mathbf{p}) \neq 0, \tag{4.63}$$
$$|\mathbf{v}^{\mathrm{T}}_i \cdot (\mathbf{x}_{k+1} - \mathbf{x}^*_k)| < |\mathbf{v}^{\mathrm{T}}_i \cdot (\mathbf{x}_k - \mathbf{x}^*_k)| \text{ for } s^{(i)}_k > 0,$$
$$|\mathbf{v}^{\mathrm{T}}_i \cdot (\mathbf{x}_{k+1} - \mathbf{x}^*_k)| > |\mathbf{v}^{\mathrm{T}}_i \cdot (\mathbf{x}_k - \mathbf{x}^*_k)| \text{ for } s^{(i)}_k < 0$$

for $\mathbf{x}_k \in U_k(\mathbf{x}^*_k) \subset \Omega_\alpha$. The fixed point $\mathbf{x}^*_k$ is called the monotonic, lower saddle of the $(2m_i)$th-order on the direction $\mathbf{v}_i$.

(iv) $\mathbf{x}_{k+j}(j \in \mathbb{Z})$ at fixed point $\mathbf{x}^*_k$ on the direction $\mathbf{v}_i$ is monotonically unstable of the $(2m_i)$th-order, upper saddle if

$$G^{(1)}_{s^{(i)}_k}(\mathbf{x}^*_k, \mathbf{p}) = \lambda_i = 1,$$
$$G^{(r_i)}_{s^{(i)}_k}(\mathbf{x}^*_k, \mathbf{p}) = 0 \text{ for } r_i = 2, 3, \cdots, 2m_i - 1;$$
$$G^{(2m_i)}_{s^{(i)}_k}(\mathbf{x}^*_k, \mathbf{p}) \neq 0, \tag{4.64}$$
$$|\mathbf{v}^{\mathrm{T}}_i \cdot (\mathbf{x}_{k+1} - \mathbf{x}^*_k)| > |\mathbf{v}^{\mathrm{T}}_i \cdot (\mathbf{x}_k - \mathbf{x}^*_k)| \text{ for } s^{(i)}_k > 0,$$
$$|\mathbf{v}^{\mathrm{T}}_i \cdot (\mathbf{x}_{k+1} - \mathbf{x}^*_k)| < |\mathbf{v}^{\mathrm{T}}_i \cdot (\mathbf{x}_k - \mathbf{x}^*_k)| \text{ for } s^{(i)}_k < 0$$

for $\mathbf{x}_k \in U_k(\mathbf{x}^*_k) \subset \Omega_\alpha$. The fixed point $\mathbf{x}^*_k$ is called the monotonic, upper saddle of the $(2m_i)$th-order on the direction $\mathbf{v}_i$.

(v) $\mathbf{x}_{k+j}(j \in \mathbb{Z})$ at fixed point $\mathbf{x}_k^*$ on the direction $\mathbf{v}_i$ is oscillatory stable of the $(2m_i + 1)$th-order if

$$
\begin{aligned}
&G^{(1)}_{s_k^{(i)}}(\mathbf{x}_k^*, \mathbf{p}) = \lambda_i = -1, \\
&G^{(r_i)}_{s_k^{(i)}}(\mathbf{x}_k^*, \mathbf{p}) = 0 \text{ for } r_i = 2, 3, \cdots, 2m_i; \\
&G^{(2m_i+1)}_{s_k^{(i)}}(\mathbf{x}_k^*, \mathbf{p}) \neq 0; \\
&|\mathbf{v}_i^{\mathrm{T}} \cdot (\mathbf{x}_{k+1} - \mathbf{x}_k^*)| < |\mathbf{v}_i^{\mathrm{T}} \cdot (\mathbf{x}_k - \mathbf{x}_k^*)|
\end{aligned}
\tag{4.65}
$$

for $\mathbf{x}_k \in U_k(\mathbf{x}_k^*) \subset \Omega_\alpha$. The fixed point $\mathbf{x}_k^*$ is called the oscillatory sink (or stable node) of the $(2m_i + 1)$th-order on the direction $\mathbf{v}_i$.

(vi) $\mathbf{x}_{k+j}(j \in \mathbb{Z})$ at fixed point $\mathbf{x}_k^*$ on the direction $\mathbf{v}_i$ is oscillatory unstable of the $(2m_i + 1)$th-order if

$$
\begin{aligned}
&G^{(1)}_{s_k^{(i)}}(\mathbf{x}_k^*, \mathbf{p}) = \lambda_i = -1; \\
&G^{(r_i)}_{s_k^{(i)}}(\mathbf{x}_k^*, \mathbf{p}) = 0 \text{ for } r_i = 2, 3, \cdots, 2m_i; \\
&G^{(2m_i+1)}_{s_k^{(i)}}(\mathbf{x}_k^*, \mathbf{p}) \neq 0, \\
&|\mathbf{v}_i^{\mathrm{T}} \cdot (\mathbf{x}_{k+1} - \mathbf{x}_k^*)| > |\mathbf{v}_i^{\mathrm{T}} \cdot (\mathbf{x}_k - \mathbf{x}_k^*)|
\end{aligned}
\tag{4.66}
$$

for $\mathbf{x}_k \in U_k(\mathbf{x}_k^*) \subset \Omega_\alpha$. The fixed point $\mathbf{x}_k^*$ is called the oscillatory source (or unstable node) of the $(2m_i + 1)$th-order on the direction $\mathbf{v}_i$.

(vii) $\mathbf{x}_{k+j}(j \in \mathbb{Z})$ at fixed point $\mathbf{x}_k^*$ on the direction $\mathbf{v}_i$ is oscillatory unstable of the $(2m_i)$th-order, lower saddle if

$$
\begin{aligned}
&G^{(1)}_{s_k^{(i)}}(\mathbf{x}_k^*, \mathbf{p}) = \lambda_i = -1, \\
&G^{(r_i)}_{s_k^{(i)}}(\mathbf{x}_k^*, \mathbf{p}) = 0 \text{ for } r_i = 2, 3, \cdots, 2m_i - 1; \\
&G^{(2m_i)}_{s_k^{(i)}}(\mathbf{x}_k^*, \mathbf{p}) \neq 0, \\
&|\mathbf{v}_i^{\mathrm{T}} \cdot (\mathbf{x}_{k+1} - \mathbf{x}_k^*)| > |\mathbf{v}_i^{\mathrm{T}} \cdot (\mathbf{x}_k - \mathbf{x}_k^*)| \text{ for } s_k^{(i)} > 0, \\
&|\mathbf{v}_i^{\mathrm{T}} \cdot (\mathbf{x}_{k+1} - \mathbf{x}_k^*)| < |\mathbf{v}_i^{\mathrm{T}} \cdot (\mathbf{x}_k - \mathbf{x}_k^*)| \text{ for } s_k^{(i)} < 0
\end{aligned}
\tag{4.67}
$$

for $\mathbf{x}_k \in U_k(\mathbf{x}_k^*) \subset \Omega_\alpha$. The fixed point $\mathbf{x}_k^*$ is called the oscillatory, lower saddle of the $(2m_i)$th-order on the direction $\mathbf{v}_i$.

(viii) $\mathbf{x}_{k+j}(j \in \mathbb{Z})$ at fixed point $\mathbf{x}_k^*$ on the direction $\mathbf{v}_i$ is oscillatory unstable of the $(2m_i)$th-order, upper saddle if

$$
\begin{aligned}
&G^{(1)}_{s_k^{(i)}}(\mathbf{x}_k^*, \mathbf{p}) = \lambda_i = -1, \\
&G^{(r_i)}_{s_k^{(i)}}(\mathbf{x}_k^*, \mathbf{p}) = 0 \text{ for } r_i = 2, 3, \cdots, 2m_i - 1; \\
&G^{(2m_i)}_{s_k^{(i)}}(\mathbf{x}_k^*, \mathbf{p}) \neq 0, \\
&|\mathbf{v}_i^{\mathrm{T}} \cdot (\mathbf{x}_{k+1} - \mathbf{x}_k^*)| < |\mathbf{v}_i^{\mathrm{T}} \cdot (\mathbf{x}_k - \mathbf{x}_k^*)| \text{ for } s_k^{(i)} > 0, \\
&|\mathbf{v}_i^{\mathrm{T}} \cdot (\mathbf{x}_{k+1} - \mathbf{x}_k^*)| > |\mathbf{v}_i^{\mathrm{T}} \cdot (\mathbf{x}_k - \mathbf{x}_k^*)| \text{ for } s_k^{(i)} < 0
\end{aligned}
\tag{4.68}
$$

for $\mathbf{x}_k \in U_k(\mathbf{x}_k^*) \subset \Omega_\alpha$. The fixed point $\mathbf{x}_k^*$ is called the oscillatory, upper saddle of the $(2m_i)$th-order on the direction $\mathbf{v}_i$.

Theorem 4.6 *Consider a discrete, nonlinear dynamical system $\mathbf{x}_{k+1} = \mathbf{f}(\mathbf{x}_k, \mathbf{p}) \in \mathscr{R}^n$ in Eq.(4.4) with a fixed point $\mathbf{x}_k^*$. The corresponding solution is given by $\mathbf{x}_{k+j} = \mathbf{f}(\mathbf{x}_{k+j-1}, \mathbf{p})$ with $j \in \mathbb{Z}$. Suppose there is a neighborhood of the fixed point $\mathbf{x}_k^*$ (i.e., $U_k(\mathbf{x}_k^*) \subset \Omega_\alpha$), and $\mathbf{f}(\mathbf{x}_k, \mathbf{p})$ is C^r ($r \geqslant 1$)-continuous in $U_k(\mathbf{x}_k^*)$ with Eq.(4.28). The linearized system is $\mathbf{y}_{k+j+1} = D\mathbf{f}(\mathbf{x}_k^*, \mathbf{p})\mathbf{y}_{k+j}$ ($\mathbf{y}_{k+j} = \mathbf{x}_{k+j} - \mathbf{x}_k^*$) in $U_k(\mathbf{x}_k^*)$ and there are n linearly independent vectors $\mathbf{v}_i$ ($i = 1, 2, \cdots, n$). For a perturbation of fixed point $\mathbf{y}_k = \mathbf{x}_k - \mathbf{x}_k^*$, let $\mathbf{y}_k^{(i)} = c_k^{(i)} \mathbf{v}_i$ and $\mathbf{y}_{k+1}^{(i)} = c_{k+1}^{(i)} \mathbf{v}_i$.*

(i) *$\mathbf{x}_{k+j}$ ($j \in \mathbb{Z}$) at fixed point $\mathbf{x}_k^*$ on the direction $\mathbf{v}_i$ is monotonically stable of the $(2m_i + 1)$th-order if and only if*

$$
\begin{aligned}
&G_{s_k^{(i)}}^{(1)}(\mathbf{x}_k^*, \mathbf{p}) = \lambda_i = 1, \\
&G_{s_k^{(i)}}^{(r_i)}(\mathbf{x}_k^*, \mathbf{p}) = 0 \text{ for } r_i = 2, 3, \cdots, 2m_i, \\
&G_{s_k^{(i)}}^{(2m_i+1)}(\mathbf{x}_k^*, \mathbf{p}) < 0
\end{aligned}
\tag{4.69}
$$

for $\mathbf{x}_k \in U_k(\mathbf{x}_k^*) \subset \Omega_\alpha$.

(ii) *$\mathbf{x}_{k+j}$ ($j \in \mathbb{Z}$) at fixed point $\mathbf{x}_k^*$ on the direction $\mathbf{v}_i$ is monotonically unstable of the $(2m_i + 1)$th-order if and only if*

$$
\begin{aligned}
&G_{s_k^{(i)}}^{(1)}(\mathbf{x}_k^*, \mathbf{p}) = \lambda_i = 1, \\
&G_{s_k^{(i)}}^{(r_i)}(\mathbf{x}_k^*, \mathbf{p}) = 0 \text{ for } r_i = 2, 3, \cdots, 2m_i, \\
&G_{s_k^{(i)}}^{(2m_i+1)}(\mathbf{x}_k^*, \mathbf{p}) > 0
\end{aligned}
\tag{4.70}
$$

for $\mathbf{x}_k \in U_k(\mathbf{x}_k^*) \subset \Omega_\alpha$.

(iii) *$\mathbf{x}_{k+j}$ ($j \in \mathbb{Z}$) at fixed point $\mathbf{x}_k^*$ on the direction $\mathbf{v}_i$ is monotonically unstable of the $(2m_i)$th-order, lower saddle if and only if*

$$
\begin{aligned}
&G_{s_k^{(i)}}^{(1)}(\mathbf{x}_k^*, \mathbf{p}) = \lambda_i = 1, \\
&G_{s_k^{(i)}}^{(r_i)}(\mathbf{x}_k^*, \mathbf{p}) = 0 \text{ for } r_i = 2, 3, \cdots, 2m_i - 1, \\
&G_{s_k^{(i)}}^{(2m_i)}(\mathbf{x}_k^*, \mathbf{p}) < 0 \text{ stable for } s_k^{(i)} > 0; \\
&G_{s_k^{(i)}}^{(2m_i)}(\mathbf{x}_k^*, \mathbf{p}) < 0 \text{ unstable for } s_k^{(i)} < 0
\end{aligned}
\tag{4.71}
$$

for $\mathbf{x}_k \in U_k(\mathbf{x}_k^*) \subset \Omega_\alpha$.

(iv) *$\mathbf{x}_{k+j}$ ($j \in \mathbb{Z}$) at fixed point $\mathbf{x}_k^*$ on the direction $\mathbf{v}_i$ is monotonically unstable of the $(2m_i)$th-order, upper saddle if and only if*

$$G^{(1)}_{s^{(i)}_k}(\mathbf{x}^*_k, \mathbf{p}) = \lambda_i = 1,$$
$$G^{(r_i)}_{s^{(i)}_k}(\mathbf{x}^*_k, \mathbf{p}) = 0 \text{ for } r_i = 2, 3, \cdots, 2m_i - 1,$$
$$G^{(2m_i)}_{s^{(i)}_k}(\mathbf{x}^*_k, \mathbf{p}) > 0 \text{ unstable for } s^{(i)}_k > 0;$$
$$G^{(2m_i)}_{s^{(i)}_k}(\mathbf{x}^*_k, \mathbf{p}) > 0 \text{ stable for } s^{(i)}_k < 0$$

(4.72)

for $\mathbf{x}_k \in U_k(\mathbf{x}^*_k) \subset \Omega_\alpha$.

(v) $\mathbf{x}_{k+j}(j \in \mathbb{Z})$ *at fixed point* $\mathbf{x}^*_k$ *on the direction* $\mathbf{v}_i$ *is oscillatory stable of the* $(2m_i + 1)$*th-order if and only if*

$$G^{(1)}_{s^{(i)}_k}(\mathbf{x}^*_k, \mathbf{p}) = \lambda_i = -1,$$
$$G^{(r_i)}_{s^{(i)}_k}(\mathbf{x}^*_k, \mathbf{p}) = 0 \text{ for } r_i = 2, 3, \cdots, 2m_i,$$
$$G^{(2m_i+1)}_{s^{(i)}_k}(\mathbf{x}^*_k, \mathbf{p}) > 0$$

(4.73)

for $\mathbf{x}_k \in U_k(\mathbf{x}^*_k) \subset \Omega_\alpha$.

(vi) $\mathbf{x}_{k+j}(j \in \mathbb{Z})$ *at fixed point* $\mathbf{x}^*_k$ *on the direction* $\mathbf{v}_i$ *is oscillatory unstable of the* $(2m_i + 1)$*th-order if and only if*

$$G^{(1)}_{s^{(i)}_k}(\mathbf{x}^*_k, \mathbf{p}) = \lambda_i = -1,$$
$$G^{(r_i)}_{s^{(i)}_k}(\mathbf{x}^*_k, \mathbf{p}) = 0 \text{ for } r_i = 2, 3, \cdots, 2m_i,$$
$$G^{(2m_i+1)}_{s^{(i)}_k}(\mathbf{x}^*_k, \mathbf{p}) < 0$$

(4.74)

for $\mathbf{x}_k \in U_k(\mathbf{x}^*_k) \subset \Omega_\alpha$.

(vii) $\mathbf{x}_{k+j}(j \in \mathbb{Z})$ *at fixed point* $\mathbf{x}^*_k$ *on the direction* $\mathbf{v}_i$ *is oscillatory unstable of the* $(2m_i)$*th-order, lower saddle if and only if*

$$G^{(1)}_{s^{(i)}_k}(\mathbf{x}^*_k, \mathbf{p}) = \lambda_i = -1,$$
$$G^{(r_i)}_{s^{(i)}_k}(\mathbf{x}^*_k, \mathbf{p}) = 0 \text{ for } r_i = 2, 3, \cdots, 2m_i - 1,$$
$$G^{(2m_i)}_{s^{(i)}_k}(\mathbf{x}^*_k, \mathbf{p}) > 0 \text{ stable for } s^{(i)}_k > 0;$$
$$G^{(2m_i)}_{s^{(i)}_k}(\mathbf{x}^*_k, \mathbf{p}) > 0 \text{ unstable for } s^{(i)}_k < 0$$

(4.75)

for $\mathbf{x}_k \in U_k(\mathbf{x}^*_k) \subset \Omega_\alpha$.

(viii) $\mathbf{x}_{k+j}(j \in \mathbb{Z})$ *at fixed point* $\mathbf{x}^*_k$ *on the direction* $\mathbf{v}_i$ *is oscillatory unstable of the* $(2m_i)$*th-order, upper saddle if and only if*

$$G^{(1)}_{s^{(i)}_k}(\mathbf{x}^*_k, \mathbf{p}) = \lambda_i = -1,$$
$$G^{(r_i)}_{s^{(i)}_k}(\mathbf{x}^*_k, \mathbf{p}) = 0 \text{ for } r_i = 2, 3, \cdots, 2m_i - 1,$$
$$G^{(2m_i)}_{s^{(i)}_k}(\mathbf{x}^*_k, \mathbf{p}) < 0 \text{ stable for } s^{(i)}_k < 0;$$
$$G^{(2m_i)}_{s^{(i)}_k}(\mathbf{x}^*_k, \mathbf{p}) < 0 \text{ unstable for } s^{(i)}_k > 0$$

(4.76)

for $\mathbf{x}_k \in U_k(\mathbf{x}_k^*) \subset \Omega_\alpha$.

Proof. The proof can be referred to Luo (2012). ∎

Definition 4.21 Consider a discrete, nonlinear dynamical system $\mathbf{x}_{k+1} = \mathbf{f}(\mathbf{x}_k, \mathbf{p}) \in \mathscr{R}^n$ in Eq.(4.4) with a fixed point $\mathbf{x}_k^*$. The corresponding solution is given by $\mathbf{x}_{k+j} = \mathbf{f}(\mathbf{x}_{k+j-1}, \mathbf{p})$ with $j \in \mathbb{Z}$. Suppose there is a neighborhood of the fixed point $\mathbf{x}_k^*$ (i.e., $U_k(\mathbf{x}_k^*) \subset \Omega_\alpha$), and $\mathbf{f}(\mathbf{x}_k, \mathbf{p})$ is $C^r(r \geqslant 1)$-continuous in $U_k(\mathbf{x}_k^*)$ with Eq.(4.28). The linearized system is $\mathbf{y}_{k+j+1} = D\mathbf{f}(\mathbf{x}_k^*, \mathbf{p})\mathbf{y}_{k+j}$ $(\mathbf{y}_{k+j} = \mathbf{x}_{k+j} - \mathbf{x}_k^*)$ in $U_k(\mathbf{x}_k^*)$. Consider a pair of complex eigenvalues $\alpha_i \pm \mathbf{i}\beta_i$ $(i \in N = \{1, 2, \cdots, n\}, \mathbf{i} = \sqrt{-1})$ of matrix $D\mathbf{f}(\mathbf{x}^*, \mathbf{p})$ with a pair of eigenvectors $\mathbf{u}_i \pm \mathbf{i}\mathbf{v}_i$. On the invariant plane of $(\mathbf{u}_i, \mathbf{v}_i)$, consider $\mathbf{r}_k^{(i)} = \mathbf{y}_k^{(i)} = \mathbf{y}_{k+}^{(i)} + \mathbf{y}_{k-}^{(i)}$ with

$$\mathbf{r}_k^{(i)} = c_k^{(i)}\mathbf{u}_i + d_k^{(i)}\mathbf{v}_i,$$
$$\mathbf{r}_{k+1}^{(i)} = c_{k+1}^{(i)}\mathbf{u}_i + d_{k+1}^{(i)}\mathbf{v}_i \tag{4.77}$$

and

$$c_k^{(i)} = \frac{1}{\Delta}[\Delta_2(\mathbf{u}_i^{\mathrm{T}} \cdot \mathbf{y}_k) - \Delta_{12}(\mathbf{v}_i^{\mathrm{T}} \cdot \mathbf{y}_k)],$$

$$d_k^{(i)} = \frac{1}{\Delta}[\Delta_1(\mathbf{v}_i^{\mathrm{T}} \cdot \mathbf{y}_k) - \Delta_{12}(\mathbf{u}_i^{\mathrm{T}} \cdot \mathbf{y}_k)]; \tag{4.78}$$

$$\Delta_1 = \|\mathbf{u}_i\|^2, \quad \Delta_2 = \|\mathbf{v}_i\|^2, \quad \Delta_{12} = \mathbf{u}_i^{\mathrm{T}} \cdot \mathbf{v}_i;$$

$$\Delta = \Delta_1\Delta_2 - \Delta_{12}^2.$$

Consider a polar coordinate of (r_k, θ_k) defined by

$$c_k^{(i)} = r_k^{(i)} \cos\theta_k^{(i)}, \text{ and } d_k^{(i)} = r_k^{(i)} \sin\theta_k^{(i)};$$

$$r_k^{(i)} = \sqrt{(c_k^{(i)})^2 + (d_k^{(i)})^2}, \text{ and } \theta_k^{(i)} = \arctan\left(d_k^{(i)}/c_k^{(i)}\right). \tag{4.79}$$

Thus

$$c_{k+1}^{(i)} = \frac{1}{\Delta}[\Delta_2 G_{c_k^{(i)}}(\mathbf{x}_k, \mathbf{p}) - \Delta_{12}G_{d_k^{(i)}}(\mathbf{x}_k, \mathbf{p})],$$

$$d_{k+1}^{(i)} = \frac{1}{\Delta}[\Delta_1 G_{d_k^{(i)}}(\mathbf{x}_k, \mathbf{p}) - \Delta_{12}G_{c_k^{(i)}}(\mathbf{x}_k, \mathbf{p})] \tag{4.80}$$

where

$$G_{c_k^{(i)}}(\mathbf{x}_k, \mathbf{p}) = \mathbf{u}_i^{\mathrm{T}} \cdot [\mathbf{f}(\mathbf{x}_k, \mathbf{p}) - \mathbf{x}_k^*] = \sum_{m_i=1}^{\infty} \frac{1}{m_i!}G_{c_k^{(i)}}^{(m_i)}(\theta_k^{(i)})(r_k^{(i)})^{m_i},$$

$$G_{d_k^{(i)}}(\mathbf{x}_k, \mathbf{p}) = \mathbf{v}_i^{\mathrm{T}} \cdot [\mathbf{f}(\mathbf{x}_k, \mathbf{p}) - \mathbf{x}_k^*] = \sum_{m_i=1}^{\infty} \frac{1}{m_i!}G_{d_k^{(i)}}^{(m_i)}(\theta_k^{(i)})(r_k^{(i)})^{m_i}; \tag{4.81}$$

$$G_{c_k^{(i)}}^{(m_i)}(\theta_k^{(i)}) = \mathbf{u}_i^{\mathrm{T}} \cdot \partial_{\mathbf{x}_k}^{(m_i)}\mathbf{f}(\mathbf{x}_k, \mathbf{p})[\mathbf{u}_i \cos\theta_k^{(i)} + \mathbf{v}_i \sin\theta_k^{(i)}]^{m_i}\Big|_{(\mathbf{x}_k^*, \mathbf{p})},$$

$$G_{d_k^{(i)}}^{(m_i)}(\theta_k^{(i)}) = \mathbf{v}_i^{\mathrm{T}} \cdot \partial_{\mathbf{x}_k}^{(m_i)}\mathbf{f}(\mathbf{x}_k, \mathbf{p})[\mathbf{u}_i \cos\theta_k^{(i)} + \mathbf{v}_i \sin\theta_k^{(i)}]^{m_i}\Big|_{(\mathbf{x}_k^*, \mathbf{p})}. \tag{4.82}$$

Thus

$$r_{k+1}^{(i)} = \sqrt{(c_{k+1}^{(i)})^2 + (d_{k+1}^{(i)})^2} = \sqrt{\sum_{m=2}^{\infty} (r_k^{(i)})^{m_i} G_{r_{k+1}^{(i)}}^{(m_i)}(\theta_k^{(i)})}$$

$$= \sqrt{G_{r_{k+1}^{(i)}}^{(2)} r_k^{(i)} \sqrt{1 + (G_{r_{k+1}^{(i)}}^{(2)})^{-1} \sum_{m=3}^{\infty} (r_k^{(i)})^{m_i-2} G_{r_{k+1}^{(i)}}^{(m_i)}(\theta_k^{(i)})}} \qquad (4.83)$$

$$\theta_{k+1}^{(i)} = \arctan(d_{k+1}^{(i)}/c_{k+1}^{(i)})$$

where

$$G_{r_{k+1}^{(i)}}^{(m_i)}(\theta_k^{(i)})$$

$$= \sum_{r_i=1}^{\infty} \sum_{s_i=1}^{\infty} \frac{1}{r_i!} \frac{1}{s_i!} [G_{c_{k+1}^{(i)}}^{(r_i)}(\theta_k^{(i)}) G_{c_{k+1}^{(i)}}^{(s_i)}(\theta_k^{(i)}) + G_{d_{k+1}^{(i)}}^{(r_i)}(\theta_k^{(i)}) G_{d_{k+1}^{(i)}}^{(s_i)}(\theta_k^{(i)})] \delta_{m_i}^{r_i+s_i}$$

$$= \frac{1}{m_i!} \sum_{r_i=1}^{m_i-1} [C_{m_i}^{r_i} G_{c_{k+1}^{(i)}}^{(r_i)}(\theta_k^{(i)}) G_{c_{k+1}^{(i)}}^{(m_i-r_i)}(\theta_k^{(i)}) + G_{d_{k+1}^{(i)}}^{(r_i)}(\theta_k^{(i)}) G_{d_{k+1}^{(i)}}^{(m_i-r_i)}(\theta_k^{(i)})]$$

$$(4.84)$$

and

$$G_{c_{k+1}^{(i)}}^{(m_i)}(\theta_k^{(i)}) = \frac{1}{\Delta}[\Delta_2 G_{c_k^{(i)}}^{(m_i)}(\theta_k^{(i)}) - \Delta_{12} G_{d_k^{(i)}}^{(m_i)}(\theta_k^{(i)})],$$

$$G_{d_{k+1}^{(i)}}^{(m_i)}(\theta_k^{(i)}) = \frac{1}{\Delta}[\Delta_1 G_{d_k^{(i)}}^{(m_i)}(\theta_k^{(i)}) - \Delta_{12} G_{c_k^{(i)}}^{(m_i)}(\theta_k^{(i)})]. \qquad (4.85)$$

From the foregoing definition, consider the first order terms of G-function

$$G_{c_k^{(i)}}^{(1)}(\mathbf{x}_k, \mathbf{p}) = G_{c_k^{(i)}1}^{(1)}(\mathbf{x}_k, \mathbf{p}) + G_{c_k^{(i)}2}^{(1)}(\mathbf{x}_k, \mathbf{p}),$$

$$G_{d_k^{(i)}}^{(1)}(\mathbf{x}_k, \mathbf{p}) = G_{d_k^{(i)}1}^{(1)}(\mathbf{x}_k, \mathbf{p}) + G_{d_k^{(i)}2}^{(1)}(\mathbf{x}_k, \mathbf{p}) \qquad (4.86)$$

where

$$G_{c_k^{(i)}1}^{(1)}(\mathbf{x}_k, \mathbf{p}) = \mathbf{u}_i^T \cdot D_{\mathbf{x}_k} \mathbf{f}(\mathbf{x}_k, \mathbf{p}) \partial_{c_k^{(i)}} \mathbf{x}_k = \mathbf{u}_i^T \cdot D_{\mathbf{x}_k} \mathbf{f}(\mathbf{x}_k, \mathbf{p}) \mathbf{u}_i$$

$$= \mathbf{u}_i^T \cdot (-\beta_i \mathbf{v}_i + \alpha_i \mathbf{u}_i) = \alpha_i \Delta_1 - \beta_i \Delta_{12},$$

$$G_{c_k^{(i)}2}^{(1)}(\mathbf{x}_k, \mathbf{p}) = \mathbf{u}_i^T \cdot D_{\mathbf{x}_k} \mathbf{f}(\mathbf{x}_k, \mathbf{p}) \partial_{d_k^{(i)}} \mathbf{x}_k = \mathbf{u}_i^T \cdot D_{\mathbf{x}_k} \mathbf{f}(\mathbf{x}_k, \mathbf{p}) \mathbf{v}_i \qquad (4.87)$$

$$= \mathbf{u}_i^T \cdot (\beta_i \mathbf{u}_i + \alpha_i \mathbf{v}_i) = \alpha_i \Delta_{12} + \beta_i \Delta_1;$$

and

$$G_{d_k^{(i)}1}^{(1)}(\mathbf{x}_k, \mathbf{p}) = \mathbf{v}_i^T \cdot D_{\mathbf{x}_k} \mathbf{f}(\mathbf{x}_k, \mathbf{p}) \partial_{c_k^{(i)}} \mathbf{x}_k = \mathbf{v}_i^T \cdot D_{\mathbf{x}_k} \mathbf{f}(\mathbf{x}_k, \mathbf{p}) \mathbf{u}_i$$

$$= \mathbf{v}_i^T \cdot (-\beta_i \mathbf{v}_i + \alpha_i \mathbf{u}_i) = -\beta_i \Delta_2 + \alpha_i \Delta_{12},$$

$$G_{d_k^{(i)}2}^{(1)}(\mathbf{x}, \mathbf{p}) = \mathbf{v}_i^T \cdot D_{\mathbf{x}_k} \mathbf{f}(\mathbf{x}_k, \mathbf{p}) \partial_{d_k^{(i)}} \mathbf{x}_k = \mathbf{v}_i^T \cdot D_{\mathbf{x}_k} \mathbf{f}(\mathbf{x}_k, \mathbf{p}) \mathbf{v}_i \qquad (4.88)$$

$$= \mathbf{v}_i^T \cdot (\beta_i \mathbf{u}_i + \alpha_i \mathbf{v}_i) = \alpha_i \Delta_2 + \beta_i \Delta_{12}.$$

Substitution of Eqs.(4.86)–(4.88) into Eq.(4.82) gives

$$
\begin{aligned}
G^{(1)}_{c^{(i)}_k}(\theta^{(i)}_k) &= G^{(1)}_{c^{(i)}_k 1}(\mathbf{x}_k, \mathbf{p}) \cos\theta^{(i)}_k + G^{(1)}_{c^{(i)}_k 2}(\mathbf{x}_k, \mathbf{p}) \sin\theta^{(i)}_k \\
&= (\alpha_i \Delta_1 - \beta_i \Delta_{12}) \cos\theta^{(i)}_k + (\alpha_i \Delta_{12} + \beta_i \Delta_1) \sin\theta^{(i)}_k, \\
G^{(1)}_{d^{(i)}_k}(\theta^{(i)}_k) &= G^{(1)}_{d^{(i)}_k 1}(\mathbf{x}_k, \mathbf{p}) \cos\theta^{(i)}_k + G^{(1)}_{d^{(i)}_k 2}(\mathbf{x}_k, \mathbf{p}) \sin\theta^{(i)}_k \\
&= (-\beta_i \Delta_2 + \alpha_i \Delta_{12}) \cos\theta^{(i)}_k + (\alpha_i \Delta_2 + \beta_i \Delta_{12}) \sin\theta^{(i)}_k.
\end{aligned}
\tag{4.89}
$$

From Eq.(4.85), we have

$$
\begin{aligned}
G^{(1)}_{c^{(i)}_{k+1}}(\theta^{(i)}_k) &= \frac{1}{\Delta}[\Delta_2 G^{(1)}_{c^{(i)}_k}(\theta^{(i)}_k) - \Delta_{12} G^{(1)}_{d^{(i)}_k}(\theta^{(i)}_k)] \\
&= \alpha_i \cos\theta^{(i)}_k + \beta_i \sin\theta^{(i)}_k, \\
G^{(1)}_{d^{(i)}_{k+1}}(\theta^{(i)}_k) &= \frac{1}{\Delta}[\Delta_1 G^{(1)}_{d^{(i)}_k}(\theta^{(i)}_k) - \Delta_{12} G^{(1)}_{c^{(i)}_k}(\theta^{(i)}_k)] \\
&= \alpha_i \sin\theta^{(i)}_k - \beta_i \cos\theta^{(i)}_k.
\end{aligned}
\tag{4.90}
$$

Thus

$$
\begin{aligned}
G^{(2)}_{r^{(i)}_{k+1}}(\theta^{(i)}_k) &= [G^{(1)}_{c^{(i)}_{k+1}}(\theta^{(i)}_k) G^{(1)}_{c^{(i)}_{k+1}}(\theta^{(i)}_k) + G^{(1)}_{d^{(i)}_{k+1}}(\theta^{(i)}_k) G^{(1)}_{d^{(i)}_{k+1}}(\theta^{(i)}_k)] \\
&= \alpha_i^2 + \beta_i^2.
\end{aligned}
\tag{4.91}
$$

Furthermore, equation (4.83) gives

$$
r^{(i)}_{k+1} = \rho_i r^{(i)}_k + o(r^{(i)}_k) \quad \text{and} \quad \theta^{(i)}_{k+1} = \theta^{(i)}_k - \vartheta_i + o(r^{(i)}_k).
\tag{4.92}
$$

where

$$
\vartheta_i = \arctan(\beta_i/\alpha_i) \quad \text{and} \quad \rho_i = \sqrt{\alpha_i^2 + \beta_i^2}.
\tag{4.93}
$$

As $r^{(i)}_k \ll 1$ and $r^{(i)}_k \to 0$, we have

$$
r^{(i)}_{k+1} = \rho_i r^{(i)}_k \quad \text{and} \quad \theta^{(i)}_{k+1} = \vartheta_i - \theta^{(i)}_k.
\tag{4.94}
$$

With an initial condition of $r^{(i)}_k = r^0_k$ and $\theta^{(i)}_k = \theta^{(i)}_k$, the corresponding solution of Eq.(4.94) is

$$
r^{(i)}_{k+j} = (\rho_i)^j r^0_k \quad \text{and} \quad \theta^{(i)}_{k+j} = j\vartheta_i - \theta^{(i)}_k.
\tag{4.95}
$$

From Eqs.(4.80), (4.81) and (4.90), we have

$$
\begin{aligned}
c^{(i)}_{k+1} &= \alpha_i r^{(i)}_k \cos\theta^{(i)}_k + \beta_i r^{(i)}_k \sin\theta^{(i)}_k = \alpha_i c^{(i)}_k + \beta_i d^{(i)}_k, \\
d^{(i)}_{k+1} &= \alpha_i r^{(i)}_k \sin\theta^{(i)}_k - \beta_i r^{(i)}_k \cos\theta^{(i)}_k = -\beta_i c^{(i)}_k + \alpha_i d^{(i)}_k.
\end{aligned}
\tag{4.96}
$$

That is,

$$\left\{ \begin{matrix} c_{k+1}^{(i)} \\ d_{k+1}^{(i)} \end{matrix} \right\} = \begin{bmatrix} \alpha_i & \beta_i \\ -\beta_i & \alpha_i \end{bmatrix} \left\{ \begin{matrix} c_k^{(i)} \\ d_k^{(i)} \end{matrix} \right\} = \rho_i \begin{bmatrix} \cos \vartheta_i & \sin \vartheta_i \\ -\sin \vartheta_i & \cos \vartheta_i \end{bmatrix} \left\{ \begin{matrix} c_k^{(i)} \\ d_k^{(i)} \end{matrix} \right\}. \quad (4.97)$$

From the foregoing equation, we have

$$\left\{ \begin{matrix} c_{k+j}^{(i)} \\ d_{k+j}^{(i)} \end{matrix} \right\} = \begin{bmatrix} \alpha_i & \beta_i \\ -\beta_i & \alpha_i \end{bmatrix}^j \left\{ \begin{matrix} c_k^{(i)} \\ d_k^{(i)} \end{matrix} \right\} = (\rho_i)^j \begin{bmatrix} \cos j\vartheta_i & \sin j\vartheta_i \\ -\sin j\vartheta_i & \cos j\vartheta_i \end{bmatrix} \left\{ \begin{matrix} c_k^{(i)} \\ d_k^{(i)} \end{matrix} \right\}.$$
$$(4.98)$$

Definition 4.22 Consider a discrete, nonlinear dynamical system $\mathbf{x}_{k+1} = \mathbf{f}(\mathbf{x}_k, \mathbf{p}) \in \mathscr{R}^n$ in Eq.(4.4) with a fixed point $\mathbf{x}_k^*$. The corresponding solution is given by $\mathbf{x}_{k+j} = \mathbf{f}(\mathbf{x}_{k+j-1}, \mathbf{p})$ with $j \in \mathbb{Z}$. Suppose there is a neighborhood of the fixed point $\mathbf{x}_k^*$ (i.e., $U_k(\mathbf{x}_k^*) \subset \Omega_\alpha$), and $\mathbf{f}(\mathbf{x}_k, \mathbf{p})$ is $C^r (r \geqslant 1)$-continuous in $U_k(\mathbf{x}_k^*)$ with Eq.(4.28). The linearized system is $\mathbf{y}_{k+j+1} = D\mathbf{f}(\mathbf{x}_k^*, \mathbf{p})\mathbf{y}_{k+j}$ ($\mathbf{y}_{k+j} = \mathbf{x}_{k+j} - \mathbf{x}_k^*$) in $U_k(\mathbf{x}_k^*)$. Consider a pair of complex eigenvalues $\alpha_i \pm \mathbf{i}\beta_i$ ($i \in N = \{1, 2, \cdots, n\}, \mathbf{i} = \sqrt{-1}$) of matrix $D\mathbf{f}(\mathbf{x}^*, \mathbf{p})$ with a pair of eigenvectors $\mathbf{u}_i \pm \mathbf{i}\mathbf{v}_i$. On the invariant plane of $(\mathbf{u}_i, \mathbf{v}_i)$, consider $\mathbf{r}_k^{(i)} = \mathbf{y}_k^{(i)} = \mathbf{y}_{k+}^{(i)} + \mathbf{y}_{k-}^{(i)}$ with Eqs.(4.73) and (4.75). For any arbitrarily small $\varepsilon > 0$, the stability of the fixed point $\mathbf{x}_k^*$ on the invariant plane of $(\mathbf{u}_i, \mathbf{v}_i)$ can be determined.

(i) $\mathbf{x}_k^{(i)}$ at the fixed point $\mathbf{x}_k^*$ on the plane of $(\mathbf{u}_i, \mathbf{v}_i)$ is spirally stable if

$$r_{k+1}^{(i)} - r_k^{(i)} < 0. \quad (4.99)$$

(ii) $\mathbf{x}_k^{(i)}$ at the fixed point $\mathbf{x}_k^*$ on the plane of $(\mathbf{u}_i, \mathbf{v}_i)$ is spirally unstable if

$$r_{k+1}^{(i)} - r_k^{(i)} > 0. \quad (4.100)$$

(iii) $\mathbf{x}_k^{(i)}$ at the fixed point $\mathbf{x}_k^*$ on the plane of $(\mathbf{u}_i, \mathbf{v}_i)$ is spirally stable with the m_kth-order singularity if for $\theta_k^{(i)} \in [0, 2\pi]$

$$\rho_i = \sqrt{\alpha_i^2 + \beta_i^2} = 1,$$
$$G_{r_{k+1}^{(i)}}^{(s_k^{(i)})}(\theta_k) = 0 \text{ for } s_k^{(i)} = 1, 2, \cdots, m_i - 1, \quad (4.101)$$
$$r_{k+1}^{(i)} - r_k^{(i)} < 0.$$

(iv) $\mathbf{x}_k^{(i)}$ at the fixed point $\mathbf{x}_k^*$ on the plane of $(\mathbf{u}_i, \mathbf{v}_i)$ is spirally unstable with the m_kth-order singularity if for $\theta_k^{(i)} \in [0, 2\pi]$

$$\rho_i = \sqrt{\alpha_i^2 + \beta_i^2} = 1,$$

$$G^{(s_k^{(i)})}_{r_{k+1}^{(i)}}(\theta_k) = 0 \text{ for } s_k^{(i)} = 1, 2, \cdots, m_i - 1 \qquad (4.102)$$

$$r_{k+1}^{(i)} - r_k^{(i)} > 0.$$

(v) $\mathbf{x}_k^{(i)}$ at the fixed point $\mathbf{x}_k^*$ on the plane of $(\mathbf{u}_i, \mathbf{v}_i)$ is circular if for $\theta_k^{(i)} \in [0, 2\pi]$

$$r_{k+1}^{(i)} - r_k^{(i)} = 0. \qquad (4.103)$$

(vi) $\mathbf{x}_k^{(i)}$ at the fixed point $\mathbf{x}_k^*$ on the plane of $(\mathbf{u}_i, \mathbf{v}_i)$ is degenerate in the direction of $\mathbf{u}_i$ if

$$\beta_i = 0 \quad \text{and} \quad \theta_{k+1}^{(i)} - \theta_k^{(i)} = 0. \qquad (4.104)$$

Theorem 4.7 *Consider a discrete, nonlinear dynamical system* $\mathbf{x}_{k+1} = \mathbf{f}(\mathbf{x}_k, \mathbf{p}) \in \mathscr{R}^n$ *in Eq.(4.4) with a fixed point* $\mathbf{x}_k^*$. *The corresponding solution is given by* $\mathbf{x}_{k+j} = \mathbf{f}(\mathbf{x}_{k+j-1}, \mathbf{p})$ *with* $j \in \mathbb{Z}$. *Suppose there is a neighborhood of the fixed point* $\mathbf{x}_k^*$ *(i.e.,* $U_k(\mathbf{x}_k^*) \subset \Omega_\alpha$*), and* $\mathbf{f}(\mathbf{x}_k, \mathbf{p})$ *is* $C^r (r \geqslant 1)$-*continuous in* $U_k(\mathbf{x}_k^*)$ *with Eq.(4.28). The linearized system is* $\mathbf{y}_{k+j+1} = D\mathbf{f}(\mathbf{x}_k^*, \mathbf{p})\mathbf{y}_{k+j}$ *(* $\mathbf{y}_{k+j} = \mathbf{x}_{k+j} - \mathbf{x}_k^*$ *) in* $U_k(\mathbf{x}_k^*)$. *Consider a pair of complex eigenvalues* $\alpha_i \pm i\beta_i (i \in N = \{1, 2, \cdots, n\}, i = \sqrt{-1})$ *of matrix* $D\mathbf{f}(\mathbf{x}^*, \mathbf{p})$ *with a pair of eigenvectors* $\mathbf{u}_i \pm i\mathbf{v}_i$. *On the invariant plane of* $(\mathbf{u}_i, \mathbf{v}_i)$, *consider* $\mathbf{r}_k^{(i)} = \mathbf{y}_k^{(i)} = \mathbf{y}_{k+}^{(i)} + \mathbf{y}_{k-}^{(i)}$ *with Eqs.(4.73) and (4.75). For any arbitrarily small* $\varepsilon > 0$, *the stability of the equilibrium* $\mathbf{x}_k^*$ *on the invariant plane of* $(\mathbf{u}_i, \mathbf{v}_i)$ *can be determined.*

(i) $\mathbf{x}_k^{(i)}$ *at the fixed point* $\mathbf{x}_k^*$ *on the plane of* $(\mathbf{u}_k, \mathbf{v}_k)$ *is spirally stable if and only if*

$$\rho_i < 1. \qquad (4.105)$$

(ii) $\mathbf{x}_k^{(i)}$ *at the fixed point* $\mathbf{x}_k^*$ *on the plane of* $(\mathbf{u}_i, \mathbf{v}_i)$ *is spirally unstable if and only if*

$$\rho_i > 1. \qquad (4.106)$$

(iii) $\mathbf{x}_k^{(i)}$ *at the fixed point* $\mathbf{x}_k^*$ *on the plane of* $(\mathbf{u}_i, \mathbf{v}_i)$ *is stable with the* m_ith-*order singularity if and only if for* $\theta_k^{(i)} \in [0, 2\pi]$

$$\rho_i = \sqrt{\alpha_i^2 + \beta_i^2} = 1,$$

$$G^{(s_k^{(i)})}_{r_k^{(i)}}(\theta_k^{(i)}) = 0 \text{ for } s_k = 1, 2, \cdots, m_i - 1 \qquad (4.107)$$

$$G^{(m_i)}_{r_k^{(i)}}(\theta_k^{(i)}) < 0.$$

(iv) $\mathbf{x}_k^{(i)}$ *at the fixed point* $\mathbf{x}_k^*$ *on the plane of* $(\mathbf{u}_i, \mathbf{v}_i)$ *is spirally unstable with the* m_ith-*order singularity if and only if for* $\theta_k^{(i)} \in [0, 2\pi]$

$$\rho_i = \sqrt{\alpha_i^2 + \beta_i^2} = 1,$$

$$G_{r_k^{(i)}}^{(s_k^{(i)})}(\theta_k^{(i)}) = 0 \ for \ s_k^{(i)} = 0, 1, 2, \cdots, m_i - 1 \qquad (4.108)$$

$$G_{r_k^{(i)}}^{(m_i)}(\theta_k^{(i)}) > 0.$$

(v) $\mathbf{x}_k^{(i)}$ at the fixed point $\mathbf{x}_k^*$ on the plane of $(\mathbf{u}_i, \mathbf{v}_i)$ is circular if and only if for $\theta_k^{(i)} \in [0, 2\pi]$

$$\rho_i = \sqrt{\alpha_i^2 + \beta_i^2} = 1,$$

$$G_{r_k^{(i)}}^{(s_k^{(i)})}(\theta_k^{(i)}) = 0 \ for \ s_k^{(i)} = 0, 1, 2, \cdots. \qquad (4.109)$$

Proof. The proof can be referred to Luo (2011). ∎

4.4 Bifurcation theory

Definition 4.23 Consider a discrete, nonlinear dynamical system $\mathbf{x}_{k+1} = \mathbf{f}(\mathbf{x}_k, \mathbf{p}) \in \mathscr{R}^n$ in Eq.(4.4) with a fixed point $\mathbf{x}_k^*$. The corresponding solution is given by $\mathbf{x}_{k+j} = \mathbf{f}(\mathbf{x}_{k+j-1}, \mathbf{p})$ with $j \in \mathbb{Z}$. Suppose there is a neighborhood of the fixed point $\mathbf{x}_k^*$ (i.e., $U_k(\mathbf{x}_k^*) \subset \Omega_\alpha$), and $\mathbf{f}(\mathbf{x}_k, \mathbf{p})$ is $C^r (r \geqslant 1)$-continuous in $U_k(\mathbf{x}_k^*)$ with Eq.(4.28). The linearized system is $\mathbf{y}_{k+j+1} = D\mathbf{f}(\mathbf{x}_k^*, \mathbf{p})\mathbf{y}_{k+j}$ $(\mathbf{y}_{k+j} = \mathbf{x}_{k+j} - \mathbf{x}_k^*)$ in $U_k(\mathbf{x}_k^*)$ and there are n linearly independent vectors $\mathbf{v}_i$ $(i = 1, 2, \cdots, n)$. For a perturbation of fixed point $\mathbf{y}_k = \mathbf{x}_k - \mathbf{x}_k^*$, let $\mathbf{y}_k^{(i)} = c_k^{(i)}\mathbf{v}_i$ and $\mathbf{y}_{k+1}^{(i)} = c_{k+1}^{(i)}\mathbf{v}_i$.

$$s_k^{(i)} = \mathbf{v}_i^{\mathrm{T}} \cdot \mathbf{y}_k = \mathbf{v}_i^{\mathrm{T}} \cdot (\mathbf{x}_k - \mathbf{x}_k^*) \qquad (4.110)$$

where $s_k^{(i)} = c_k^{(i)}\|\mathbf{v}_i\|^2$.

$$s_{k+1}^{(i)} = \mathbf{v}_i^{\mathrm{T}} \cdot \mathbf{y}_{k+1} = \mathbf{v}_i^{\mathrm{T}} \cdot [\mathbf{f}(\mathbf{x}_k, \mathbf{p}) - \mathbf{x}_k^*]. \qquad (4.111)$$

In the vicinity of point $(\mathbf{x}_{k(0)}^*, \mathbf{p}_0)$, $\mathbf{v}_i^{\mathrm{T}} \cdot \mathbf{f}(\mathbf{x}_k, \mathbf{p})$ can be expended for $(0 < \theta < 1)$ as

$$\mathbf{v}_i^{\mathrm{T}} \cdot [\mathbf{f}(\mathbf{x}_k, \mathbf{p}) - \mathbf{x}_{k(0)}^*] = a_i(s_k^{(i)} - s_{k(0)}^{(i)*}) + \mathbf{b}_i^{\mathrm{T}} \cdot (\mathbf{p} - \mathbf{p}_0)$$

$$+ \sum_{q=2}^{m} \sum_{r=0}^{q} \frac{1}{q!} C_q^r \mathbf{a}_i^{(q-r,r)}(s_k^{(i)} - s_{k(0)}^{(i)*})^{q-r}(\mathbf{p} - \mathbf{p}_0)^r$$

$$+ \frac{1}{(m+1)!}[(s_k^{(i)} - s_{k(0)}^{(i)*})\partial_{s^{(i)}(\cdot)} + (\mathbf{p} - \mathbf{p}_0)\partial_{\mathbf{p}}^{(\cdot)}]^{m+1}$$

$$\times (\mathbf{v}_i^{\mathrm{T}} \cdot \mathbf{f}(\mathbf{x}_{k(0)}^* + \theta\Delta\mathbf{x}_k, \mathbf{p}_0 + \theta\Delta\mathbf{p})) \qquad (4.112)$$

where

$$a_i = \mathbf{v}_i^{\mathrm{T}} \cdot \partial_{s_k^{(i)}} \mathbf{f}(\mathbf{x}_k, \mathbf{p})\Big|_{(\mathbf{x}_{k(0)}^*, \mathbf{p}_0)}, \quad \mathbf{b}_i^{\mathrm{T}} = \mathbf{v}_i^{\mathrm{T}} \cdot \partial_{\mathbf{p}} \mathbf{f}(\mathbf{x}_k, \mathbf{p})\Big|_{(\mathbf{x}_{k(0)}^*, \mathbf{p}_0)},$$

$$\mathbf{a}_i^{(r,s)} = \mathbf{v}_i^{\mathrm{T}} \cdot \partial_{s_k^{(i)}}^{(r)} \partial_{\mathbf{p}}^{(s)} \mathbf{f}(\mathbf{x}_k, \mathbf{p})\Big|_{(\mathbf{x}_{k(0)}^*, \mathbf{p}_0)}. \tag{4.113}$$

If $a_i = 1$ and $\mathbf{p} = \mathbf{p}_0$, the stability of fixed point $\mathbf{x}_k^*$ on an eigenvector $\mathbf{v}_i$ changes from stable to unstable state (or from unstable to stable state). The bifurcation manifold on the direction of $\mathbf{v}_i$ is determined by

$$\mathbf{b}_i^{\mathrm{T}} \cdot (\mathbf{p} - \mathbf{p}_0) + \sum_{q=2}^{m} \sum_{r=0}^{q} \frac{1}{q!} C_q^r \mathbf{a}_i^{(q-r,r)} (s_k^{(i)} - s_{k(0)}^{(i)*})^{q-r} (\mathbf{p} - \mathbf{p}_0)^r = 0. \tag{4.114}$$

In the neighborhood of $(\mathbf{x}_{k(0)}^*, \mathbf{p}_0)$, when other components of fixed point $\mathbf{x}_k^*$ on the eigenvector of $\mathbf{v}_j$ for all $j \neq i$, $(i, j \in N)$ do not change their stability states, equation (4.114) possesses l-branch solutions of equilibrium $s_k^{(i)*}$ $(0 < l \leqslant m)$ with l_1-stable and l_2-unstable solutions $(l_1, l_2 \in \{0, 1, 2, \cdots, l\})$. Such l-branch solutions are called the bifurcation solutions of fixed point $\mathbf{x}_k^*$ on the eigenvector of $\mathbf{v}_i$ in the neighborhood of $(\mathbf{x}_{k(0)}^*, \mathbf{p}_0)$. Such a bifurcation at point $(\mathbf{x}_{k(0)}^*, \mathbf{p}_0)$ is called the hyperbolic bifurcation of mth-order on the eigenvector of $\mathbf{v}_i$. Consider two special cases herein.

(i) If

$$\mathbf{a}_i^{(1,1)} = \mathbf{0} \quad \text{and} \quad \mathbf{b}_i^{\mathrm{T}} \cdot (\mathbf{p} - \mathbf{p}_0) + \frac{1}{2!} a_i^{(2,0)} (s_k^{(i)*} - s_{k0}^{(i)*})^2 = 0 \tag{4.115}$$

where

$$a_i^{(2,0)} = \mathbf{v}_i^{\mathrm{T}} \cdot \partial_{s_k^{(i)}}^{(2)} \partial_{\mathbf{p}}^{(0)} \mathbf{f}(\mathbf{x}_k, \mathbf{p})\Big|_{(\mathbf{x}_{k(0)}^*, \mathbf{p}_0)} = \mathbf{v}_i^{\mathrm{T}} \cdot \partial_{s_k^{(i)}}^{(2)} \mathbf{f}(\mathbf{x}_k, \mathbf{p})\Big|_{(\mathbf{x}_{k(0)}^*, \mathbf{p}_0)}$$

$$= \mathbf{v}_i^{\mathrm{T}} \cdot \partial_{\mathbf{x}}^{(2)} \mathbf{f}(\mathbf{x}_k, \mathbf{p})(\mathbf{v}_k \mathbf{v}_k)\Big|_{(\mathbf{x}_{k(0)}^*, \mathbf{p}_0)} = G_{s_k^{(i)}}^{(2)} (\mathbf{x}_{k(0)}^*, \mathbf{p}_0) \neq 0, \tag{4.116}$$

$$\mathbf{b}_i^{\mathrm{T}} = \mathbf{v}_i^{\mathrm{T}} \cdot \partial_{\mathbf{p}} \mathbf{f}(\mathbf{x}_k, \mathbf{p})\Big|_{(\mathbf{x}_{k(0)}^*, \mathbf{p}_0)} \neq \mathbf{0},$$

$$a_i^{(2,0)} \times [\mathbf{b}_i^{\mathrm{T}} \cdot (\mathbf{p} - \mathbf{p}_0)] < 0, \tag{4.117}$$

such a bifurcation at point $(\mathbf{x}_0^*, \mathbf{p}_0)$ is called the *saddle-node* bifurcation on the eigenvector of $\mathbf{v}_i$.

(ii) If

$$\mathbf{b}_i^{\mathrm{T}} \cdot (\mathbf{p} - \mathbf{p}_0) = 0 \quad \text{and}$$

$$\mathbf{a}_i^{(1,1)} \cdot (\mathbf{p} - \mathbf{p}_0)(s_k^{(i)*} - s_{k(0)}^{(i)*}) + \frac{1}{2!} a_i^{(2,0)} (s_k^{(i)*} - s_{k(0)}^{(i)*})^2 = 0 \tag{4.118}$$

where

$$a_i^{(2,0)} = \mathbf{v}_i^{\mathrm{T}} \cdot \partial_{s_k^{(i)}}^{(2)} \partial_{\mathbf{p}}^{(0)} \mathbf{f}(\mathbf{x}_k,\mathbf{p}) \Big|_{(\mathbf{x}_{k(0)}^*,\mathbf{p}_0)} = \mathbf{v}_i^{\mathrm{T}} \cdot \partial_{s_k^{(i)}}^{(2)} \mathbf{f}(\mathbf{x}_k,\mathbf{p}) \Big|_{(\mathbf{x}_0^*,\mathbf{p}_0)}$$

$$= \mathbf{v}_i^{\mathrm{T}} \cdot \partial_{\mathbf{x}_k}^{(2)} \mathbf{f}(\mathbf{x}_k,\mathbf{p})(\mathbf{v}_i \mathbf{v}_i) \Big|_{(\mathbf{x}_{k(0)}^*,\mathbf{p}_0)} = G_{s_k^{(i)}}^{(2)}(\mathbf{x}_{k(0)}^*,\mathbf{p}_0) \neq 0, \quad (4.119)$$

$$\mathbf{a}_i^{(1,1)} = \mathbf{v}_i^{\mathrm{T}} \cdot \partial_{s_k^{(i)}}^{(1)} \partial_{\mathbf{p}}^{(1)} \mathbf{f}(\mathbf{x}_k,\mathbf{p}) \Big|_{(\mathbf{x}_{k(0)}^*,\mathbf{p}_0)} = \mathbf{v}_i^{\mathrm{T}} \cdot \partial_{s_k^{(i)}} \partial_{\mathbf{p}} \mathbf{f}(\mathbf{x}_k,\mathbf{p}) \Big|_{(\mathbf{x}_{k(0)}^*,\mathbf{p}_0)}$$

$$= \mathbf{v}_i^{\mathrm{T}} \cdot \partial_{\mathbf{x}_k} \partial_{\mathbf{p}} \mathbf{f}(\mathbf{x}_k,\mathbf{p}) \mathbf{v}_i \Big|_{(\mathbf{x}_{k(0)}^*,\mathbf{p}_0)} \neq \mathbf{0}, \quad (4.120)$$

$$a_i^{(2,0)} \times [\mathbf{a}_i^{(1,1)} \cdot (\mathbf{p} - \mathbf{p}_0)] < 0,$$

such a bifurcation at point $(\mathbf{x}_{k(0)}^*, \mathbf{p}_0)$ is called the *transcritical* bifurcation on the eigenvector of $\mathbf{v}_i$.

Definition 4.24 Consider a discrete, nonlinear dynamical system $\mathbf{x}_{k+1} = \mathbf{f}(\mathbf{x}_k,\mathbf{p}) \in \mathscr{R}^n$ in Eq.(4.4) with a fixed point $\mathbf{x}_k^*$. The corresponding solution is given by $\mathbf{x}_{k+j} = \mathbf{f}(\mathbf{x}_{k+j-1},\mathbf{p})$ with $j \in \mathbb{Z}$. Suppose there is a neighborhood of the fixed point $\mathbf{x}_k^*$ (i.e., $U_k(\mathbf{x}_k^*) \subset \Omega_\alpha$), and $\mathbf{f}(\mathbf{x}_k,\mathbf{p})$ is $C^r (r \geqslant 1)$-continuous in $U_k(\mathbf{x}_k^*)$ with Eq.(4.28). The linearized system is $\mathbf{y}_{k+j+1} = D\mathbf{f}(\mathbf{x}_k^*,\mathbf{p})\mathbf{y}_{k+j}$ ($\mathbf{y}_{k+j} = \mathbf{x}_{k+j} - \mathbf{x}_k^*$) in $U_k(\mathbf{x}_k^*)$ and there are n linearly independent vectors $\mathbf{v}_i$ ($i = 1,2,\cdots,n$). For a perturbation of fixed point $\mathbf{y}_k = \mathbf{x}_k - \mathbf{x}_k^*$, let $\mathbf{y}_k^{(i)} = c_k^{(i)} \mathbf{v}_i$ and $\mathbf{y}_{k+1}^{(i)} = c_{k+1}^{(i)} \mathbf{v}_i$. Equations (4.110), (4.111) and (4.113) hold. In the vicinity of point $(\mathbf{x}_{k0}^*,\mathbf{p}_0)$, $\mathbf{v}_i^{\mathrm{T}} \cdot \mathbf{f}(\mathbf{x}_k,\mathbf{p})$ can be expended for $\theta(0 < \theta < 1)$ as

$$\mathbf{v}_i^{\mathrm{T}} \cdot [\mathbf{f}(\mathbf{x}_k,\mathbf{p}) - \mathbf{x}_{k+1(0)}^*] = a_i(s_k^{(i)} - s_{k(0)}^{(i)*}) + \mathbf{b}_i^{\mathrm{T}} \cdot (\mathbf{p} - \mathbf{p}_0)$$

$$+ \sum_{q=2}^{m} \sum_{r=0}^{q} \frac{1}{q!} C_q^r a_i^{(q-r,r)} (s_k^{(i)} - s_{k(0)}^{(i)*})^{q-r} (\mathbf{p} - \mathbf{p}_0)^r$$

$$+ \frac{1}{(m+1)!} [(s_k^{(i)} - s_{k(0)}^{(i)*}) \partial_{s_k^{(i)}(\cdot)} + (\mathbf{p} - \mathbf{p}_0) \partial_{\mathbf{p}}^{(\cdot)}]^{m+1}$$

$$\times (\mathbf{v}_k^{\mathrm{T}} \cdot \mathbf{f}(\mathbf{x}_{k0}^* + \theta \Delta \mathbf{x}_k, \mathbf{p}_0 + \theta \Delta \mathbf{p})) \quad (4.121)$$

and

$$\mathbf{v}_i^{\mathrm{T}} \cdot [\mathbf{f}(\mathbf{x}_{k+1},\mathbf{p}) - \mathbf{x}_{k(0)}^*] = a_i(s_{k+1}^{(i)} - s_{k+1(0)}^*) + \mathbf{b}_i^{\mathrm{T}} \cdot (\mathbf{p} - \mathbf{p}_0)$$

$$+ \sum_{q=2}^{m} \sum_{r=0}^{q} \frac{1}{q!} C_q^r a_i^{(q-r,r)} (s_{k+1}^{(i)} - s_{k+1(0)}^{(i)*})^{q-r} (\mathbf{p} - \mathbf{p}_0)^r$$

$$+ \frac{1}{(m+1)!} [(s_{k+1}^{(i)} - s_{k+1(0)}^*) \partial_{s_{k+1}^{(i)}(\cdot)} + (\mathbf{p} - \mathbf{p}_0) \partial_{\mathbf{p}}^{(\cdot)}]^{m+1}$$

$$\times (\mathbf{v}_i^{\mathrm{T}} \cdot \mathbf{f}(\mathbf{x}_{k+1(0)}^* + \theta \Delta \mathbf{x}_{k+1}, \mathbf{p}_0 + \theta \Delta \mathbf{p})) \quad (4.122)$$

If $a_i = -1$ and $\mathbf{p} = \mathbf{p}_0$, the stability of current equilibrium $\mathbf{x}_k^*$ on an eigen-vector $\mathbf{v}_i$ changes from stable to unstable state (or from unstable to stable state). The bifurcation manifold in the direction of $\mathbf{v}_i$ is determined by

$$\mathbf{b}_i^{\mathrm{T}} \cdot (\mathbf{p}-\mathbf{p}_0) + a_i(s_k^{(i)*}-s_{k(0)}^{(i)*}) + \sum_{q=2}^{m}\sum_{r=0}^{q}\frac{1}{q!}C_q^r \mathbf{a}_i^{(q-r,r)}(s_k^{(i)}-s_{k(0)}^{(i)*})^{q-r}(\mathbf{p}-\mathbf{p}_0)^r$$
$$= (s_{k+1}^{(i)*} - s_{k+1(0)}^{(i)*});$$

$$\mathbf{b}_i^{\mathrm{T}} \cdot (\mathbf{p}-\mathbf{p}_0) + a_i(s_{k+1}^{(i)*}-s_{k+1(0)}^{(i)*}) + \sum_{q=2}^{m}\sum_{r=0}^{q}\frac{1}{q!}C_q^r \mathbf{a}_i^{(q-r,r)}(s_{k+1}^{(i)}-s_{k+1(0)}^{(i)*})^{q-r}(\mathbf{p}-\mathbf{p}_0)^r$$
$$= (s_k^{(i)*} - s_{k(0)}^{(i)*}).$$

$$(4.123)$$

In the neighborhood of $(\mathbf{x}_{k(0)}^*, \mathbf{p}_0)$, when other components of fixed point $\mathbf{x}_{k(0)}^*$ on the eigenvector of $\mathbf{v}_j$ for all $j \neq i$, $(j, i \in N)$ do not change their stability states, equation (4.123) possesses l-branch solutions of equilibrium $s_k^{(i)*}(0 < l \leqslant m)$ with l_1-stable and l_2-unstable solutions $(l_1, l_2 \in \{0, 1, 2, \cdots, l\})$. Such l-branch solutions are called the bifurcation solutions of fixed point $\mathbf{x}_k^*$ on the eigenvector of $\mathbf{v}_i$ in the neighborhood of $(\mathbf{x}_{k(0)}^*, \mathbf{p}_0)$. Such a bifurcation at point $(\mathbf{x}_{k(0)}^*, \mathbf{p}_0)$ is called the *hyperbolic bifurcation* of mth-order with doubling iterations on the eigenvector of $\mathbf{v}_i$. Consider a special case. If

$$\mathbf{b}_i^{\mathrm{T}} \cdot (\mathbf{p}-\mathbf{p}_0) = 0, \quad a_i = -1, \quad a_i^{(2,0)} = 0, \quad \mathbf{a}_i^{(2,1)} = 0, \quad \mathbf{a}_i^{(1,2)} = 0,$$

$$[\mathbf{a}^{(1,1)} \cdot (\mathbf{p}-\mathbf{p}_0) + a_i](s_k^{(i)*} - s_{k(0)}^{(i)*}) + \frac{1}{3!}a_i^{(3,0)}(s_k^* - s_{k(0)}^*)^3 = (s_{k+1}^{(i)*} - s_{k+1(0)}^{(i)*}),$$

$$[\mathbf{a}^{(1,1)} \cdot (\mathbf{p}-\mathbf{p}_0) + a_i](s_{k+1}^{(i)*} - s_{k+1(0)}^{(i)*}) + \frac{1}{3!}a_i^{(3,0)}(s_{k+1}^* - s_{k+1(0)}^*)^3 = (s_k^{(i)*} - s_{k(0)}^{(i)*})$$

$$(4.124)$$

where

$$a_i^{(3,0)} = \mathbf{v}_i^{\mathrm{T}} \cdot \partial_{s_k^{(i)}}^{(3)} \partial_{\mathbf{p}}^{(0)} \mathbf{f}(\mathbf{x}_k, \mathbf{p})\Big|_{(\mathbf{x}_{k(0)}^*, \mathbf{p}_0)} = \mathbf{v}_i^{\mathrm{T}} \cdot \partial_{s_k^{(i)}}^{(3)} \mathbf{f}(\mathbf{x}_k, \mathbf{p})\Big|_{(\mathbf{x}_{k(0)}^*, \mathbf{p}_0)}$$

$$= \mathbf{v}_i^{\mathrm{T}} \cdot \partial_{\mathbf{x}_k}^{(3)} \mathbf{f}(\mathbf{x}_k, \mathbf{p})(\mathbf{v}_i \mathbf{v}_i \mathbf{v}_i)\Big|_{(\mathbf{x}_{k(0)}^*, \mathbf{p}_0)} = G_i^{(3)}(\mathbf{x}_{k(0)}^*, \mathbf{p}_0) \neq 0,$$

$$\mathbf{a}_i^{(1,1)} = \mathbf{v}_i^{\mathrm{T}} \cdot \partial_{s_k^{(i)}}^{(1)} \partial_{\mathbf{p}}^{(1)} \mathbf{f}(\mathbf{x}_k, \mathbf{p})\Big|_{(\mathbf{x}_{k(0)}^*, \mathbf{p}_0)} = \mathbf{v}_i^{\mathrm{T}} \cdot \partial_{s_k^{(i)}} \partial_{\mathbf{p}} \mathbf{f}(\mathbf{x}_k, \mathbf{p})\Big|_{(\mathbf{x}_{k(0)}^*, \mathbf{p}_0)}$$

$$= \mathbf{v}_i^{\mathrm{T}} \cdot \partial_{\mathbf{x}_k} \partial_{\mathbf{p}} \mathbf{f}(\mathbf{x}_k, \mathbf{p}) \mathbf{v}_i\Big|_{(\mathbf{x}_{k(0)}^*, \mathbf{p}_0)} \neq \mathbf{0},$$

$$(4.125)$$

$$a_i^{(3,0)} \times [\mathbf{a}_i^{(1,1)} \cdot (\mathbf{p}-\mathbf{p}_0)] < 0, \qquad (4.126)$$

such a bifurcation at point $(\mathbf{x}_{k(0)}^*, \mathbf{p}_0)$ is called the *pitchfork* bifurcation (or period-doubling bifurcation) on the eigenvector of $\mathbf{v}_i$.

From the proceeding analysis, the bifurcation points possess the higher-order singularity of the flow in *discrete* dynamical system. For the saddle-node bifurcation of the first kind, the $(2m)$th order singularity of the flow at the bifurcation point exists as a saddle of the $(2m)$th order. For the transcritical bifurcation, the $(2m)$th order singularity of the flow at the bifurcation point exists as a saddle of the $(2m)$th order. However, for the stable pitchfork bifurcation (or saddle-node bifurcation of the second kind, or period-doubling bifurcation), the $(2m + 1)$th order singularity of the flow at the bifurcation point exists as an oscillatory sink of the $(2m + 1)$th order. For the unstable pitchfork bifurcation (or the unstable saddle-node bifurcation of the second kind, or unstable period-doubling bifurcation), the $(2m + 1)$th order singularity of the flow at the bifurcation point exists as an oscillatory source of the $(2m + 1)$th order.

Definition 4.25 Consider a discrete, nonlinear dynamical system $\mathbf{x}_{k+1} = \mathbf{f}(\mathbf{x}_k, \mathbf{p}) \in \mathscr{R}^n$ in Eq.(4.4) with a fixed point $\mathbf{x}_k^*$. The corresponding solution is given by $\mathbf{x}_{k+j} = \mathbf{f}(\mathbf{x}_{k+j-1}, \mathbf{p})$ with $j \in \mathbb{Z}$. Suppose there is a neighborhood of the fixed point $\mathbf{x}_k^*$ (i.e., $U_k(\mathbf{x}_k^*) \subset \Omega_\alpha$), and $\mathbf{f}(\mathbf{x}_k, \mathbf{p})$ is C^r ($r \geqslant 1$)-continuous in $U_k(\mathbf{x}_k^*)$ with Eq.(4.28). The linearized system is $\mathbf{y}_{k+j+1} = D\mathbf{f}(\mathbf{x}_k^*, \mathbf{p})\mathbf{y}_{k+j}$ ($\mathbf{y}_{k+j} = \mathbf{x}_{k+j} - \mathbf{x}_k^*$) in $U_k(\mathbf{x}_k^*)$. Consider a pair of complex eigenvalues $\alpha_i \pm i\beta_i$ ($i \in N = \{1, 2, \cdots, n\}, i = \sqrt{-1}$) of matrix $D\mathbf{f}(\mathbf{x}^*, \mathbf{p})$ with a pair of eigenvectors $\mathbf{u}_i \pm i\mathbf{v}_i$. On the invariant plane of $(\mathbf{u}_i, \mathbf{v}_i)$, consider $\mathbf{r}_k^{(i)} = \mathbf{y}_k^{(i)} = \mathbf{y}_{k+}^{(i)} + \mathbf{y}_{k-}^{(i)}$ with

$$\mathbf{r}_k^{(i)} = c_k^{(i)}\mathbf{u}_i + d_k^{(i)}\mathbf{v}_i \quad \text{and} \quad \mathbf{r}_{k+1}^{(i)} = c_{k+1}^{(i)}\mathbf{u}_i + d_{k+1}^{(i)}\mathbf{v}_i. \tag{4.127}$$

and

$$c_k^{(i)} = \frac{1}{\Delta}[\Delta_2(\mathbf{u}_i^{\mathrm{T}} \cdot \mathbf{y}_k) - \Delta_{12}(\mathbf{v}_i^{\mathrm{T}} \cdot \mathbf{y}_k)],$$

$$d_k^{(i)} = \frac{1}{\Delta}[\Delta_1(\mathbf{v}_i^{\mathrm{T}} \cdot \mathbf{y}_k) - \Delta_{12}(\mathbf{u}_i^{\mathrm{T}} \cdot \mathbf{y}_k)]; \tag{4.128}$$

$$\Delta_1 = \|\mathbf{u}_i\|^2, \quad \Delta_2 = \|\mathbf{v}_i\|^2, \quad \Delta_{12} = \mathbf{u}_i^{\mathrm{T}} \cdot \mathbf{v}_i;$$

$$\Delta = \Delta_1\Delta_2 - \Delta_{12}^2.$$

Consider a polar coordinate of (r_k, θ_k) defined by

$$c_k^{(i)} = r_k^{(i)} \cos \theta_k^{(i)}, \quad \text{and} \quad d_k^{(i)} = r_k^{(i)} \sin \theta_k^{(i)};$$

$$r_k^{(i)} = \sqrt{(c_k^{(i)})^2 + (d_k^{(i)})^2}, \quad \text{and} \quad \theta_k^{(i)} = \arctan(d_k^{(i)}/c_k^{(i)}). \tag{4.129}$$

Thus

$$c_{k+1}^{(i)} = \frac{1}{\Delta}[\Delta_2 G_{c_k^{(i)}}(\mathbf{x}_k, \mathbf{p}) - \Delta_{12} G_{d_k^{(i)}}(\mathbf{x}_k, \mathbf{p})],$$

$$d_{k+1}^{(i)} = \frac{1}{\Delta}[\Delta_1 G_{d_k^{(i)}}(\mathbf{x}_k, \mathbf{p}) - \Delta_{12} G_{c_k^{(i)}}(\mathbf{x}_k, \mathbf{p})], \tag{4.130}$$

where

$$
\begin{aligned}
G_{c_k^{(i)}}(\mathbf{x}_k, \mathbf{p}) &= \mathbf{u}_i^{\mathrm{T}} \cdot [\mathbf{f}(\mathbf{x}_k, \mathbf{p}) - \mathbf{x}_{k(0)}^*] \\
&= \mathbf{a}_i^{\mathrm{T}} \cdot (\mathbf{p} - \mathbf{p}_0) + a_{i11}(c_k^{(i)} - c_{k(0)}^{(i)*}) + a_{i12}(d_k^{(i)} - d_{k(0)}^{(i)*}) \\
&\quad + \sum_{q=2}^{m_i} \sum_{r_i=0}^{q} \frac{1}{q!} C_q^{r_i} \mathbf{G}_{c_k^{(i)}}^{(q-r_i, r_i)}(\mathbf{x}_k^*, \mathbf{p}_0)(\mathbf{p} - \mathbf{p}_0)^{r_i} (r_k^{(i)})^{q-r_i} \\
&\quad + \frac{1}{(m_i+1)!}[(c_k^{(i)} - c_{k(0)}^{(i)*})\partial_{c_k^{(i)}} + (d_k^{(i)} - d_{k(0)}^{(i)*})\partial_{d_k^{(i)}} + (\mathbf{p} - \mathbf{p}_0)\partial_{\mathbf{p}}]^{m_i+1} \\
&\quad \times (\mathbf{u}_i^{\mathrm{T}} \cdot \mathbf{f}(\mathbf{x}_{k0}^* + \theta \Delta \mathbf{x}_k, \mathbf{p}_0 + \theta \Delta \mathbf{p})), \\
G_{d_k^{(i)}}(\mathbf{x}_k, \mathbf{p}) &= \mathbf{v}_i^{\mathrm{T}} \cdot [\mathbf{f}(\mathbf{x}_k, \mathbf{p}) - \mathbf{x}_{k(0)}^*] \\
&= \mathbf{b}_i^{\mathrm{T}} \cdot (\mathbf{p} - \mathbf{p}_0) + a_{i21}(c_k^{(i)} - c_{k(0)}^{(i)*}) + a_{i22}(d_k^{(i)} - d_{k(0)}^{(i)*}) \\
&\quad + \sum_{q=2}^{m_i} \sum_{r_i=0}^{q} \frac{1}{q!} C_q^{r_i} \mathbf{G}_{d_k^{(i)}}^{(q-r_i, r_i)}(\mathbf{x}_k^*, \mathbf{p}_0)(\mathbf{p} - \mathbf{p}_0)^{r_i} r_k^{q-r_i} \\
&\quad + \frac{1}{(m_i+1)!}[(c_k^{(i)} - c_{k(0)}^{(i)*})\partial_{c_k^{(i)}} + (d_k^{(i)} - d_{k(0)}^{(i)*})\partial_{d_k^{(i)}} + (\mathbf{p} - \mathbf{p}_0)\partial_{\mathbf{p}}]^{m_i+1} \\
&\quad \times (\mathbf{v}_i^{\mathrm{T}} \cdot \mathbf{f}(\mathbf{x}_{k(0)}^* + \theta \Delta \mathbf{x}, \mathbf{p}_0 + \theta \Delta \mathbf{p}));
\end{aligned}
\tag{4.131}
$$

and

$$
\begin{aligned}
&\mathbf{G}_{c_k^{(i)}}^{(s,r)}(\mathbf{x}_{k(0)}^*, \mathbf{p}_0) \\
&= \mathbf{u}_i^{\mathrm{T}} \cdot [\partial_{\mathbf{x}_k}()\mathbf{u}_i \cos \theta_k^{(i)} + \partial_{\mathbf{x}_k}()\mathbf{v}_i \sin \theta_k^{(i)}]^s \partial_{\mathbf{p}}^{(r)} \mathbf{f}(\mathbf{x}_k, \mathbf{p}) \Big|_{(\mathbf{x}_{k(0)}^*, \mathbf{p}_0)}, \\
&\mathbf{G}_{d_k^{(i)}}^{(s,r)}(\mathbf{x}_{k(0)}^*, \mathbf{p}_0) \\
&= \mathbf{v}_i^{\mathrm{T}} \cdot [\partial_{\mathbf{x}_k}()\mathbf{u}_i \cos \theta_k^{(i)} + \partial_{\mathbf{x}_k}()\mathbf{v}_i \sin \theta_k^{(i)}]^s \partial_{\mathbf{p}}^{(r)} \mathbf{f}(\mathbf{x}_k, \mathbf{p}) \Big|_{(\mathbf{x}_{k(0)}^*, \mathbf{p}_0)};
\end{aligned}
\tag{4.132}
$$

$$
\begin{aligned}
\mathbf{a}_i^{\mathrm{T}} &= \mathbf{u}_i^{\mathrm{T}} \cdot \partial_{\mathbf{p}} \mathbf{f}(\mathbf{x}_k, \mathbf{p}), \quad \mathbf{b}_i^{\mathrm{T}} = \mathbf{v}_i^{\mathrm{T}} \cdot \partial_{\mathbf{p}} \mathbf{f}(\mathbf{x}_k, \mathbf{p}); \\
a_{i11} &= \mathbf{u}_i^{\mathrm{T}} \cdot \partial_{\mathbf{x}_k} \mathbf{f}(\mathbf{x}_k, \mathbf{p})\mathbf{u}_i, \quad a_{i12} = \mathbf{u}_i^{\mathrm{T}} \cdot \partial_{\mathbf{x}_k} \mathbf{f}(\mathbf{x}_k, \mathbf{p})\mathbf{v}_i; \\
a_{i21} &= \mathbf{v}_i^{\mathrm{T}} \cdot \partial_{\mathbf{x}_k} \mathbf{f}(\mathbf{x}_k, \mathbf{p})\mathbf{u}_i, \quad a_{i22} = \mathbf{v}_i^{\mathrm{T}} \cdot \partial_{\mathbf{x}_k} \mathbf{f}(\mathbf{x}_k, \mathbf{p})\mathbf{v}_i.
\end{aligned}
\tag{4.133}
$$

Suppose

$$
\mathbf{a}_i = \mathbf{0} \quad \text{and} \quad \mathbf{b}_i = \mathbf{0}
\tag{4.134}
$$

then

$$
\begin{aligned}
r_{k+1}^{(i)} &= \sqrt{(c_{k+1}^{(i)})^2 + (d_{k+1}^{(i)})^2} = \sqrt{\sum_{m=2}^{\infty} (r_k^{(i)})^m G_{r_{k+1}^{(i)}}^{(m)}} \\
&= \sqrt{G_{r_{k+1}^{(i)}}^{(2,0)}} r_k^{(i)} \sqrt{1 + \lambda^{(i)} + \sum_{m=3}^{\infty} \lambda_m^{(i)} (r_k^{(i)})^{m-2}} \\
\theta_{k+1}^{(i)} &= \arctan(d_{k+1}^{(i)} / c_{k+1}^{(i)})
\end{aligned}
\tag{4.135}
$$

where

$$G^{(2)}_{r^{(i)}_{k+1}} = G^{(2,0)}_{r^{(i)}_{k+1}} + G^{(1,1)}_{r^{(i)}_{k+1}} \quad \text{and} \quad \lambda^{(i)} = G^{(1,1)}_{r^{(i)}_{k+1}} / G^{(2,0)}_{r^{(i)}_{k+1}} \quad \text{with}$$

$$G^{(2,0)}_{r^{(i)}_{k+1}} = [G^{(1,0)}_{c^{(i)}_{k+1}}(\theta_k^{(i)}, \mathbf{p}_0)]^2 + [G^{(1,0)}_{d^{(i)}_{k+1}}(\theta_k^{(i)}, \mathbf{p}_0)]^2,$$

$$G^{(1,1)}_{r^{(i)}_{k+1}} = \sum_{r=2}^{M} \sum_{s=2}^{M} \frac{1}{(s+r-2)!} C^{r-1}_{r+s-2}$$

$$\{[G^{(1,r-1)}_{c^{(i)}_{k+1}}(\theta_k^{(i)}, \mathbf{p}_0) \cdot (\mathbf{p} - \mathbf{p}_0)^{r-1}][G^{(1,s-1)}_{c^{(i)}_{k+1}}(\theta_k^{(i)}, \mathbf{p}_0) \cdot (\mathbf{p} - \mathbf{p}_0)^{s-1}]$$

$$+[G^{(1,r-1)}_{d^{(i)}_{k+1}}(\theta_k^{(i)}, \mathbf{p}_0) \cdot (\mathbf{p} - \mathbf{p}_0)^{r-1}][G^{(1,s-1)}_{d^{(i)}_{k+1}}(\theta_k^{(i)}, \mathbf{p}_0) \cdot (\mathbf{p} - \mathbf{p}_0)^{s-1}]\};$$

$$(4.136)$$

and

$$\lambda_m^{(i)} = G^{(m)}_{r^{(i)}_{k+1}} / G^{(2,0)}_{r^{(i)}_{k+1}} \quad \text{with}$$

$$G^{(m)}_{r^{(i)}_{k+1}} = \sum_{m_i=0}^{M} \sum_{m_j=0}^{M} \frac{1}{m_i!} \frac{1}{m_j!} [\mathbf{G}^{(m_i-r_i,r_i)}_{c^{(i)}_{k+1}}(\theta_k^{(i)}, \mathbf{p}_0) \cdot (\mathbf{p} - \mathbf{p}_0)^{m_i-r_i}$$

$$\times \mathbf{G}^{(m_j-s_j,s_j)}_{c^{(i)}_{k+1}}(\theta_k^{(i)}, \mathbf{p}_0) \cdot (\mathbf{p} - \mathbf{p}_0)^{m_j-s_j}$$

$$+\mathbf{G}^{(m_i-r_i,r_i)}_{d^{(i)}_{k+1}}(\theta_k^{(i)}, \mathbf{p}_0) \cdot (\mathbf{p} - \mathbf{p}_0)^{m_i-r_i}$$

$$\times \mathbf{G}^{(m_j-s_j,s_j)}_{d^{(i)}_{k+1}}(\theta_k^{(i)}, \mathbf{p}_0) \cdot (\mathbf{p} - \mathbf{p}_0)^{m_j-s_j}] \delta_m^{r_i+s_j}$$

$$= \frac{1}{m!} \sum_{q=1}^{m-1} C_m^q \sum_{r=1}^{M} \sum_{s=1}^{M} \frac{1}{(s+r-m)!} C^{r-q}_{r+s-m}$$

$$\{[G^{(q,r-q)}_{c^{(i)}_{k+1}}(\theta_k^{(i)}, \mathbf{p}_0) \cdot (\mathbf{p}-\mathbf{p}_0)^{r-q}][G^{(m-q,s-m+q)}_{c^{(i)}_{k+1}}(\theta_k^{(i)}, \mathbf{p}_0)(\mathbf{p}-\mathbf{p}_0)^{s-m+q}]$$

$$+[G^{(q,r-q)}_{d^{(i)}_{k+1}}(\theta_k^{(i)}, \mathbf{p}_0) \cdot (\mathbf{p}-\mathbf{p}_0)^{r-q}][G^{(m-q,s-m+q)}_{d^{(i)}_{k+1}}(\theta_k^{(i)}, \mathbf{p}_0)] \cdot (\mathbf{p}-\mathbf{p}_0)^{s-m+q}\}.$$

$$(4.137a)$$

$$\mathbf{G}^{(m-r,r)}_{c^{(i)}_{k+1}}(\theta_k, \mathbf{p}_0) = \frac{1}{\Delta}[\Delta_2 \mathbf{G}^{(m-r,r)}_{c^{(i)}_k}(\mathbf{x}^*_{k(0)}, \mathbf{p}_0) - \Delta_{12} \mathbf{G}^{(m-r,r)}_{d^{(i)}_k}(\mathbf{x}^*_{k(0)}, \mathbf{p}_0)],$$

$$\mathbf{G}^{(m-r,r)}_{d^{(i)}_{k+1}}(\theta_k, \mathbf{p}_0) = \frac{1}{\Delta}[\Delta_1 \mathbf{G}^{(m-r,r)}_{d^{(i)}_k}(\mathbf{x}^*_{k(0)}, \mathbf{p}_0) - \Delta_{12} \mathbf{G}^{(m-r,r)}_{c^{(i)}_k}(\mathbf{x}^*_{k(0)}, \mathbf{p}_0)].$$

$$(4.137b)$$

If $G^{(2,0)}_{r^{(i)}_{k+1}} = 1$ and $\mathbf{p} = \mathbf{p}_0$, the stability of current fixed point $\mathbf{x}_k^*$ on an eigenvector plane of $(\mathbf{u}_i, \mathbf{v}_i)$ changes from stable to unstable state (or from unstable to stable state). The bifurcation manifold in the direction of $\mathbf{v}_i$ is determined by

$$\lambda^{(i)} + \sum_{m=3}^{\infty} \lambda_m^{(i)} (r_k^{(i)})^{m-2} = 0. \tag{4.138}$$

Such a bifurcation at the fixed point $(\mathbf{x}_{k(0)}^*, \mathbf{p}_0)$ is called the generalized Neimark bifurcation on the eigenvector plane of $(\mathbf{u}_i, \mathbf{v}_i)$.

For a special case, if

$$\lambda^{(i)} + \lambda_4^{(i)} (r_k^{(i)})^2 = 0, \text{ for } \lambda^{(i)} \times \lambda_4^{(i)} < 0 \text{ and } \lambda_3^{(i)} = 0 \tag{4.139}$$

such a bifurcation at point $(\mathbf{x}_0^*, \mathbf{p}_0)$ is called the Neimark bifurcation on the eigenvector plane of $(\mathbf{u}_i, \mathbf{v}_i)$.

For the repeating eigenvalues of $DP(\mathbf{x}_k^*, \mathbf{p})$, the bifurcation of fixed point $\mathbf{x}_k^*$ can be similarly discussed in the foregoing Theorems 4.5 and 4.6. Herein, such a procedure will not be repeated. From the foregoing analysis of the Neimark bifurcation, the Neimark bifurcation points possess the higher-order singularity of the flow in discrete dynamical system in the radial direction. For the stable Neimark bifurcation, the mth order singularity of the flow at the bifurcation point exists as a sink of the mth order in the radial direction. For the unstable Neimark bifurcation, the mth order singularity of the flow at the bifurcation point exists as a source of the mth order in the radial direction.

Consider a 2D map

$$P : \mathbf{x}_k \to \mathbf{x}_{k+1} \text{ with } \mathbf{x}_{k+1} = \mathbf{f}(\mathbf{x}_k, \mathbf{p}) \tag{4.140}$$

where $\mathbf{x}_k = (x_k, y_k)^{\mathrm{T}}$ and $\mathbf{f} = (f_1, f_2)^{\mathrm{T}}$ with a parameter vector $\mathbf{p}$. The period-n fixed point for Eq.(4.140) is $(\mathbf{x}_k^*, \mathbf{p})$, i.e., $P^{(n)} \mathbf{x}_k^* = \mathbf{x}_{k+n}^*$ where $P^{(n)} = P \circ P^{(n-1)}$ and $P^{(0)} = 1$, and its stability and bifurcation conditions are given as follows.

(i) Period-doubling (flip or pitchfork) bifurcation

$$\mathrm{tr}(DP^{(n)}) + \det(DP^{(n)}) + 1 = 0. \tag{4.141}$$

(ii) Saddle-node bifurcation

$$\det(DP^{(n)}) + 1 = \mathrm{tr}(DP^{(n)}). \tag{4.142}$$

(iii) Neimark bifurcation

$$\det(DP^{(n)}) = 1. \tag{4.143}$$

The bifurcation and stability conditions for the solution of period-n for Eq.(4.140) are summarized in Fig.4.8 with $\det(DP^{(n)}) = \det(DP^{(n)}(\mathbf{x}_{k(0)}^*, \mathbf{p}_0))$ and $\mathrm{tr}(DP^{(n)}) = \mathrm{tr}(DP^{(n)}(\mathbf{x}_{k(0)}^*, \mathbf{p}_0))$. The thick dashed lines are bifurcation lines. The stability of fixed point is given by the eigenvalues in complex plane. The stability of fixed point for higher dimensional systems can be identified by using a naming of stability for linear dynamical systems in Luo (2011,

2012). The saddle-node bifurcation possesses stable saddle-node bifurcation (critical) and unstable saddle-node bifurcation (degenerate).

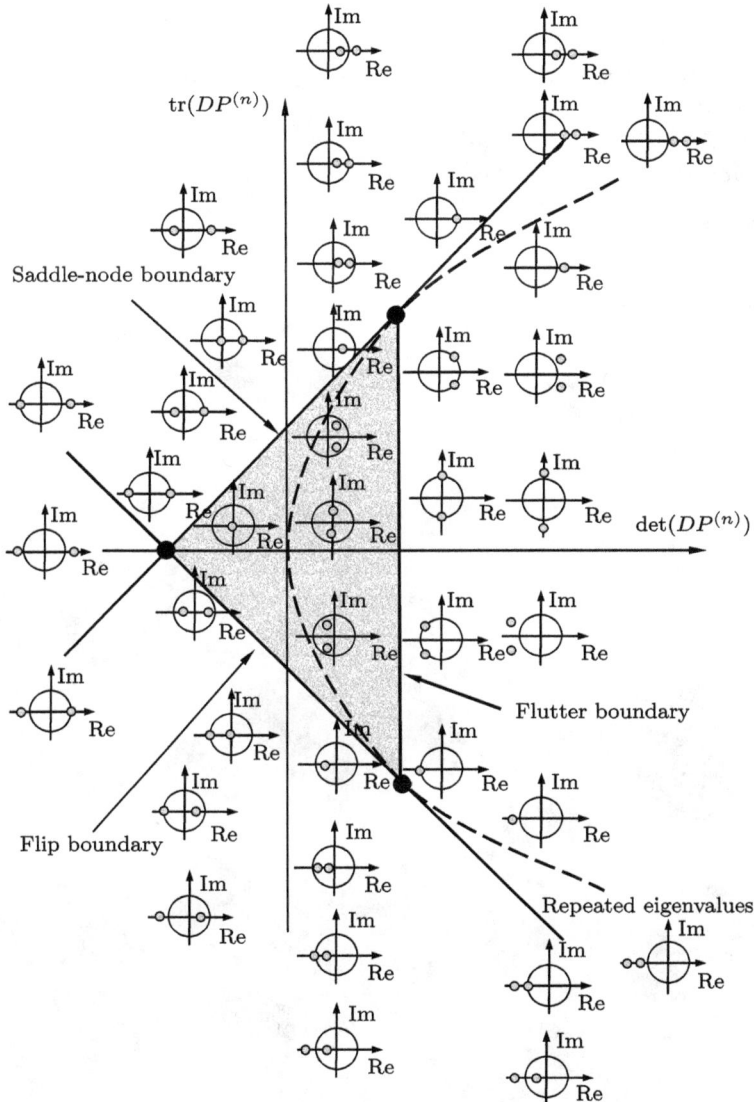

Fig. 4.8 Stability and bifurcation diagrams through the complex plane of eigenvalues for 2D- discrete dynamical systems.

References

Guckenhiemer, J. and Holmes, P., 1990, *Nonlinear Oscillations, Dynamical Systems, and Bifurcations of Vector Fields*, Springer-Verlag, New-York.

Luo, A.C.J., 2011, *Regularity and Complexity in Dynamical Systems*, Springer, New York.

Luo, A.C.J., 2012, *Discrete and Switching Dynamical Systems*, HEP-L&H Scientific, Beijing and Glen Carbon.

Nitecki, Z., 1971, *Differentiable Dynamics: An Introduction to the Orbit Structures of Diffeomorphisms*, MIT Press, Cambridge, MA.

Chapter 5
Periodic Flows in Continuous Systems

This chapter will present periodic flows in nonlinear dynamical systems through the discrete implicit mappings. The period-1 flows in nonlinear dynamical systems will be discussed first by the one-step discrete maps, and then the period-m flows in nonlinear dynamical systems will also be discussed through the one-step discrete maps. Multi-step, implicit discrete maps will be used to discuss the period-1 and period-m motions in nonlinear dynamical systems. Through the discrete nodes in periodic flows, the periodic flows will be approximated by the discrete Fourier series and the frequency space of the periodic flows can be determined through amplitude spectrums.

5.1 Discretization-based methods

As in Luo (2014), periodic flows in continuous dynamical systems will be presented from the discretization-based method. If a nonlinear system has a periodic flow with a period of $T = 2\pi/\Omega$, then such a periodic flow can be determined by discrete points through discrete mappings of the continuous system. The method is stated as follows.

Theorem 5.1 *Consider a nonlinear dynamical system as*

$$\dot{\mathbf{x}} = \mathbf{f}(\mathbf{x}, t, \mathbf{p}) \in \mathscr{R}^n \tag{5.1}$$

where $\mathbf{f}(\mathbf{x}, t, \mathbf{p})$ is a C^r-continuous nonlinear vector function $(r \geqslant 1)$. If such a system has a periodic flow $\mathbf{x}(t)$ with finite norm $||\mathbf{x}||$ and period $T = 2\pi/\Omega$, there is a set of discrete time $t_k (k = 0, 1, \cdots, N)$ with $(N \to \infty)$ during one period T, and the corresponding solution $\mathbf{x}(t_k)$ and the vector field $\mathbf{f}(\mathbf{x}(t_k), t_k, \mathbf{p})$ are exact. Suppose a discrete node $\mathbf{x}_k$ is on the approximate solution of the periodic flow under $||\mathbf{x}(t_k) - \mathbf{x}_k|| \leqslant \varepsilon_k$ with a small $\varepsilon_k \geqslant 0$ and

$$||\mathbf{f}(\mathbf{x}(t_k), t_k, \mathbf{p}) - \mathbf{f}(\mathbf{x}_k, t_k, \mathbf{p})|| \leqslant \delta_k \tag{5.2}$$

with a small $\delta_k \geqslant 0$. During a time interval $t \in [t_k, t_{k+1}]$, there is a mapping $P_k : \mathbf{x}_{k-1} \to \mathbf{x}_k$ $(k = 1, 2, \cdots, N)$, i.e.,

$$\mathbf{x}_k = P_k \mathbf{x}_{k-1} \quad \text{with} \quad \mathbf{g}_k(\mathbf{x}_{k-1}, \mathbf{x}_k, \mathbf{p}) = \mathbf{0}, k = 1, 2, \cdots, N \tag{5.3}$$

where $\mathbf{g}_k$ is an implicit vector function. Consider a mapping structure as

$$
\begin{aligned}
P &= P_N \circ P_{N-1} \circ \cdots \circ P_2 \circ P_1 : \mathbf{x}_0 \to \mathbf{x}_N; \\
\text{with} \quad & P_k : \mathbf{x}_{k-1} \to \mathbf{x}_k (k = 1, 2, \cdots, N).
\end{aligned}
\tag{5.4}
$$

For $\mathbf{x}_N = P\mathbf{x}_0$, if there is a set of points $\mathbf{x}_k^*(k = 0, 1, \cdots, N)$ computed by

$$
\begin{aligned}
\mathbf{g}_k(\mathbf{x}_{k-1}^*, \mathbf{x}_k^*, \mathbf{p}) &= \mathbf{0}, (k = 1, 2, \cdots, N) \\
\mathbf{x}_0^* &= \mathbf{x}_N^*,
\end{aligned}
\tag{5.5}
$$

then the points $\mathbf{x}_k^*(k = 0, 1, \cdots, N)$ are approximations of points $\mathbf{x}(t_k)$ of the periodic solution. In the neighborhood of $\mathbf{x}_k^*$, with $\mathbf{x}_k = \mathbf{x}_k^* + \Delta\mathbf{x}_k$, the linearized equation is given by

$$
\begin{aligned}
\Delta\mathbf{x}_k &= DP_k \cdot \Delta\mathbf{x}_{k-1} \\
\text{with} \quad & \mathbf{g}_k(\mathbf{x}_{k-1}^* + \Delta\mathbf{x}_{k-1}, \mathbf{x}_k^* + \Delta\mathbf{x}_k, \mathbf{p}) = \mathbf{0} \\
& (k = 1, 2, \cdots, N).
\end{aligned}
\tag{5.6}
$$

The resultant Jacobian matrices of the periodic flow are

$$
\begin{aligned}
DP_{k(k-1)\cdots 1} &= DP_k \cdot DP_{k-1} \cdot \cdots \cdot DP_1, (k = 1, 2, \cdots, N) \\
DP &\equiv DP_{N(N-1)\cdots 1} = DP_N \cdot DP_{N-1} \cdot \cdots \cdot DP_1
\end{aligned}
\tag{5.7}
$$

where

$$
\begin{aligned}
DP_k &= \left[\frac{\partial\mathbf{x}_k}{\partial\mathbf{x}_{k-1}}\right]_{(\mathbf{x}_{k-1}^*, \mathbf{x}_k^*)} = -\left[\frac{\partial\mathbf{g}_k}{\partial\mathbf{x}_k}\right]_{(\mathbf{x}_{k-1}^*, \mathbf{x}_k^*)}^{-1} \left[\frac{\partial\mathbf{g}_k}{\partial\mathbf{x}_{k-1}}\right]_{(\mathbf{x}_{k-1}^*, \mathbf{x}_k^*)} \\
& (k = 1, 2, \cdots, N).
\end{aligned}
\tag{5.8}
$$

The eigenvalues of DP and $DP_{k(k-1)\cdots 1}$ for such a periodic flow are determined by

$$
\begin{aligned}
|DP_{k(k-1)\cdots 1} - \overline{\lambda}\mathbf{I}_{n \times n}| &= 0, (k = 1, 2, \cdots, N); \\
|DP - \lambda\mathbf{I}_{n \times n}| &= 0.
\end{aligned}
\tag{5.9}
$$

Thus, the eigenvalues of $DP_{k(k-1)\cdots 1}$ give the properties of $\mathbf{x}_k$ varying with $\mathbf{x}_0$ for the periodic flow. The stability and bifurcation of the periodic flow can be classified by the eigenvalues of $DP(\mathbf{x}_0^*)$ with

$$([n_1^m, n_1^o] : [n_2^m, n_2^o] : [n_3, \kappa_3] : [n_4, \kappa_4]|n_5 : n_6 : [n_7, l, \kappa_7]) \tag{5.10}$$

where n_1 is the total number of real eigenvalues with magnitudes less than one $(n_1 = n_1^m + n_1^o)$, n_2 is the total number of real eigenvalues with magnitude greater than one $(n_2 = n_2^m + n_2^o)$, n_3 is the total number of real eigenvalues equal to $+1$; n_4 is the total number of real eigenvalues equal to -1; n_5 is the total pair number of complex eigenvalues with magnitudes less than one, n_6 is the total pair number of complex eigenvalues with magnitudes greater than one, n_7 is the total pair number of complex eigenvalues with magnitudes equal to one.

(i) *If the magnitudes of all eigenvalues of DP are less than one (i.e., $|\lambda_i| < 1, i = 1, 2, \cdots, n$), the approximate periodic solution is stable.*

(ii) *If at least the magnitude of one eigenvalue of DP is greater than one (i.e., $|\lambda_i| > 1, i \in \{1, 2, \cdots, n\}$), the approximate periodic solution is unstable.*

(iii) *The boundaries between stable and unstable periodic flow with higher order singularity give bifurcation and stability conditions.*

Proof: The proof can be referred to Luo (2015). ∎

To explain how to approximate the periodic flow in an n-dimensional nonlinear dynamical system, consider an $n_1 \times n_2$ plane $(n_1 + n_2 = n)$, as shown in Fig.5.1. N-nodes of the periodic flow are chosen for an approximate solution with a certain accuracy $\|\mathbf{x}(t_k) - \mathbf{x}_k\| \leq \varepsilon_k$ $(\varepsilon_k > 0)$ and $\|\mathbf{f}(\mathbf{x}(t_k), t_k, \mathbf{p}) - \mathbf{f}(\mathbf{x}_k, t_k, \mathbf{p})\| \leq \delta_k$ $(\delta_k > 0)$. Letting $\delta = \max\{\delta_k\}_{k \in \{1,2,\cdots,N\}}$ and $\varepsilon = \max\{\varepsilon_k\}_{k \in \{1,2,\cdots,N\}}$ be small positive quantities prescribed, the periodic flow can be approximately described by a set of mappings P_k with $\mathbf{g}_k(\mathbf{x}_{k-1}, \mathbf{x}_k, \mathbf{p}) = \mathbf{0}$ $(k = 1, 2, \cdots, N)$ with periodicity condition $\mathbf{x}_N = \mathbf{x}_0$. Based on the approximate mapping functions, the nodes of periodic motions are computed approximately, which is depicted by a dashed curve. The exact solution of the periodic flow is described by a solid curve. The exact node points on the periodic flows are depicted with short lines. The red symbols are node points on the approximated solution of the periodic flow. The discrete mapping P_k is developed from the differential equation. With the control of computational accuracy, the nodes of the periodic flow can be obtained with a good approximation.

From the aforementioned method, a set of nonlinear discrete mappings P_k with $\mathbf{g}_k(\mathbf{x}_{k-1}, \mathbf{x}_k, \mathbf{p}) = \mathbf{0}$ $(k = 1, 2, \cdots, N)$ are developed for periodic flows. Such a mapping can be used for numerical simulations. For given $\mathbf{x}_{k-1}$, one can compute $\mathbf{x}_k$ through $\mathbf{g}_k(\mathbf{x}_{k-1}, \mathbf{x}_k, \mathbf{p}) = \mathbf{0}$. For the explicit form, the mapping is directly used for computation of $\mathbf{x}_k$. For the implicit form, the mapping iteration or Newton-Raphson method can be adopted to compute $\mathbf{x}_k$. In addition to a one-step mapping of P_k with $\mathbf{g}_k(\mathbf{x}_{k-1}, \mathbf{x}_k, \mathbf{p}) = \mathbf{0}$, one can develop a multi-step (or l-steps) mapping of P_k with

$$\begin{aligned} \mathbf{g}_k(\mathbf{x}_{k-l}, \cdots, \mathbf{x}_{k-1}, \mathbf{x}_k, \mathbf{p}) = \mathbf{0}, (k = 1, 2, \cdots, N), \\ l \in \{1, 2, \cdots, k\}. \end{aligned} \tag{5.11}$$

(i) If $l = 1$, the one-step mapping is recovered from the multi-step mapping.

(ii) If $l = 2$, the two-step mapping is obtained from the multi-step mapping as

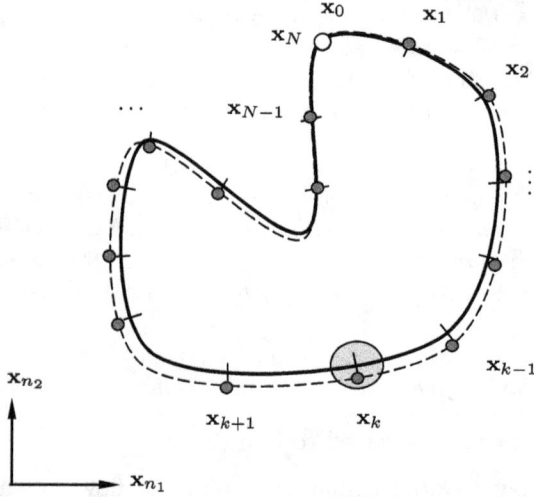

Fig. 5.1 Period-1 flow with N-nodes with short lines. Solid curve: expected exact results. and dashed curve: numerical results. The local shaded area is a small neighborhood of the exact solution at the kth node. The red symbols are for node points on the numerical solution of the periodic flow.

$$\mathbf{g}_k(\mathbf{x}_{k-2}, \mathbf{x}_{k-1}, \mathbf{x}_k, \mathbf{p}) = \mathbf{0}, (k = 1, 2, \cdots, N) \qquad (5.12)$$

which can be expanded as

$$\mathbf{g}_1(\mathbf{x}_0, \mathbf{x}_1, \mathbf{p}) = \mathbf{0},$$
$$\vdots \qquad\qquad\qquad\qquad\qquad (5.13)$$
$$\mathbf{g}_k(\mathbf{x}_{k-2}, \mathbf{x}_{k-1}, \mathbf{x}_k, \mathbf{p}) = \mathbf{0}, (k = 1, 2, \cdots, N).$$

(iii) If $l = k$, the k-steps mapping is obtained, i.e.,

$$\mathbf{g}_k(\mathbf{x}_0, \mathbf{x}_1, \cdots, \mathbf{x}_{k-1}, \mathbf{x}_k, \mathbf{p}) = \mathbf{0}, (k = 1, 2, \cdots, N) \qquad (5.14)$$

and the foregoing equation can be expanded as

$$\mathbf{g}_1(\mathbf{x}_0, \mathbf{x}_1, \mathbf{p}) = \mathbf{0},$$
$$\vdots \qquad\qquad\qquad\qquad\qquad (5.15)$$
$$\mathbf{g}_k(\mathbf{x}_0, \mathbf{x}_1, \cdots, \mathbf{x}_{k-1}, \mathbf{x}_k, \mathbf{p}) = \mathbf{0}, (k = 1, 2, \cdots, N).$$

From the multi-step (or l-steps) mapping of P_k, with the periodicity condition ($\mathbf{x}_0 = \mathbf{x}_N$), the periodic flow can be obtained via

$$\mathbf{g}_k(\mathbf{x}_{k-l}, \cdots, \mathbf{x}_{k-1}, \mathbf{x}_k, \mathbf{p}) = \mathbf{0},$$
$$(k = 1, 2, \cdots, N; l \in \{1, 2, \cdots, k\}) \qquad (5.16)$$
$$\mathbf{x}_0 = \mathbf{x}_N.$$

Suppose node points $\mathbf{x}_k^*$ $(k = 0, 1, 2, \cdots, N)$ of periodic flows are obtained, the stability and bifurcation can be in the vicinity of $\mathbf{x}_k^*$ with $\mathbf{x}_k = \mathbf{x}_k^* + \Delta\mathbf{x}_k$, i.e.,

$$\frac{\partial \mathbf{g}_k}{\partial \mathbf{x}_{k-l}} \frac{\partial \mathbf{x}_{k-l}}{\partial \mathbf{x}_0} + \cdots + \frac{\partial \mathbf{g}_k}{\partial \mathbf{x}_{k-1}} \frac{\partial \mathbf{x}_{k-1}}{\partial \mathbf{x}_0} + \frac{\partial \mathbf{g}_k}{\partial \mathbf{x}_k} \frac{\partial \mathbf{x}_k}{\partial \mathbf{x}_0} = \mathbf{0}_{n\times n} \tag{5.17}$$
$$(k = 1, 2, \cdots, N; l \in \{1, 2, \cdots, k\}).$$

In other words, we have

$$
\begin{bmatrix} \dfrac{\partial \mathbf{x}_1}{\partial \mathbf{x}_0} \\[2mm] \dfrac{\partial \mathbf{x}_2}{\partial \mathbf{x}_0} \\[2mm] \vdots \\[2mm] \dfrac{\partial \mathbf{x}_l}{\partial \mathbf{x}_0} \\[2mm] \vdots \\[2mm] \dfrac{\partial \mathbf{x}_{N-l}}{\partial \mathbf{x}_0} \\[2mm] \vdots \\[2mm] \dfrac{\partial \mathbf{x}_N}{\partial \mathbf{x}_0} \end{bmatrix} = -
\begin{bmatrix}
\dfrac{\partial \mathbf{g}_1}{\partial \mathbf{x}_1} & \mathbf{0}_{n\times n} & \cdots & \mathbf{0}_{n\times n} & \cdots & \mathbf{0}_{n\times n} & \cdots & \mathbf{0}_{n\times n} \\[2mm]
\dfrac{\partial \mathbf{g}_2}{\partial \mathbf{x}_1} & \dfrac{\partial \mathbf{g}_2}{\partial \mathbf{x}_2} & \cdots & \mathbf{0}_{n\times n} & \cdots & \mathbf{0}_{n\times n} & \cdots & \mathbf{0}_{n\times n} \\[2mm]
\vdots & \vdots & & \vdots & & \vdots & & \vdots \\[2mm]
\dfrac{\partial \mathbf{g}_l}{\partial \mathbf{x}_1} & \dfrac{\partial \mathbf{g}_l}{\partial \mathbf{x}_2} & \cdots & \dfrac{\partial \mathbf{g}_l}{\partial \mathbf{x}_l} & \cdots & \mathbf{0}_{n\times n} & \cdots & \mathbf{0}_{n\times n} \\[2mm]
\vdots & \vdots & & \vdots & & \vdots & & \vdots \\[2mm]
\mathbf{0}_{n\times n} & \mathbf{0}_{n\times n} & \cdots & \mathbf{0}_{n\times n} & \cdots & \dfrac{\partial \mathbf{g}_{N-l}}{\partial \mathbf{x}_{N-l}} & \cdots & \mathbf{0}_{n\times n} \\[2mm]
\vdots & \vdots & & \vdots & & \vdots & & \vdots \\[2mm]
\mathbf{0}_{n\times n} & \mathbf{0}_{n\times n} & \cdots & \mathbf{0}_{n\times n} & \cdots & \dfrac{\partial \mathbf{g}_N}{\partial \mathbf{x}_{N-l}} & \cdots & \dfrac{\partial \mathbf{g}_N}{\partial \mathbf{x}_N}
\end{bmatrix}^{-1}
\begin{bmatrix} \dfrac{\partial \mathbf{g}_1}{\partial \mathbf{x}_0} \\[2mm] \dfrac{\partial \mathbf{g}_2}{\partial \mathbf{x}_0} \\[2mm] \vdots \\[2mm] \dfrac{\partial \mathbf{g}_l}{\partial \mathbf{x}_0} \\[2mm] \vdots \\[2mm] \mathbf{0}_{n\times n} \\[2mm] \vdots \\[2mm] \mathbf{0}_{n\times n} \end{bmatrix}
\tag{5.18}
$$

From the mapping structure, we have

$$\Delta\mathbf{x}_N = DP \cdot \Delta\mathbf{x}_0 \quad \text{and} \quad DP = \left[\frac{\partial \mathbf{x}_N}{\partial \mathbf{x}_0}\right]. \tag{5.19}$$

Letting $\Delta\mathbf{x}_N = \lambda\Delta\mathbf{x}_0$, we have

$$(DP - \lambda\mathbf{I}_{n\times n})\Delta\mathbf{x}_0 = \mathbf{0}. \tag{5.20}$$

The eigenvalue of DP is given by $|DP - \lambda\mathbf{I}_{n\times n}| = 0$. In addition, we have

$$\Delta\mathbf{x}_k = DP_{k(k-1)\cdots 1} \cdot \Delta\mathbf{x}_0 \quad \text{and} \quad DP_{k(k-1)\cdots 1} = \left[\frac{\partial \mathbf{x}_k}{\partial \mathbf{x}_0}\right] \tag{5.21}$$
$$(k = 1, 2, \cdots, N).$$

Letting $\Delta\mathbf{x}_k = \lambda\Delta\mathbf{x}_0$, we have

$$(DP_{k(k-1)\cdots 1} - \lambda\mathbf{I}_{n\times n})\Delta\mathbf{x}_0 = \mathbf{0}. \tag{5.22}$$

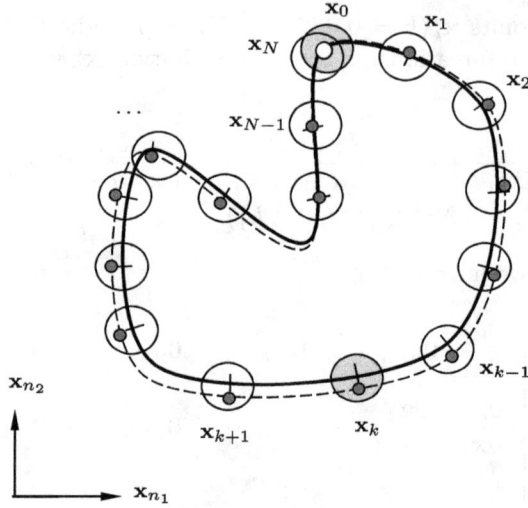

Fig. 5.2 Neighborhoods of N-nodes with short lines for a period-1 flow. Solid curve: expected exact results. and dashed curve: numerical results, The local shaded area is a small neighborhood of the exact solution at the kth node. The red symbols are node points on the approximate solution of the periodic flow.

The eigenvalues of $DP_{k(k-1)\cdots 1}$ are given by $|DP_{k(k-1)\cdots 1} - \lambda \mathbf{I}_{n \times n}| = 0$. Such eigenvalues tell effects of variation of $\mathbf{x}_0$ on node $\mathbf{x}_k$ in its vicinity. The neighborhood of $\mathbf{x}_k^*$ (i.e., $U_k(\mathbf{x}_k^*)$) is presented in Fig.5.2 through large circles. In such a neighborhood, the eigenvalues are used to measure the effects $\Delta \mathbf{x}_k$ of $\mathbf{x}_k^*$ varying with $\Delta \mathbf{x}_0$ at $\mathbf{x}_0^*$.

(i) If $l = 1$, equation (5.17) becomes

$$\frac{\partial \mathbf{g}_k}{\partial \mathbf{x}_{k-1}} \frac{\partial \mathbf{x}_{k-1}}{\partial \mathbf{x}_0} + \frac{\partial \mathbf{g}_k}{\partial \mathbf{x}_k} \frac{\partial \mathbf{x}_k}{\partial \mathbf{x}_0} = \mathbf{0} (k = 1, 2, \cdots, N). \tag{5.23}$$

The deformation of the foregoing equation yields

$$\frac{\partial \mathbf{g}_k}{\partial \mathbf{x}_{k-1}} + \frac{\partial \mathbf{g}_k}{\partial \mathbf{x}_k} \frac{\partial \mathbf{x}_k}{\partial \mathbf{x}_{k-1}} = \mathbf{0} (k = 1, 2, \cdots, N). \tag{5.24}$$

That is,

$$\frac{\partial \mathbf{x}_k}{\partial \mathbf{x}_{k-1}} = - \left[\frac{\partial \mathbf{g}_k}{\partial \mathbf{x}_k} \right]^{-1} \cdot \frac{\partial \mathbf{g}_k}{\partial \mathbf{x}_{k-1}} (k = 1, 2, \cdots, N). \tag{5.25}$$

From Eq.(5.23), the following matrix form can be formed.

$$
\begin{bmatrix}
\dfrac{\partial \mathbf{g}_1}{\partial \mathbf{x}_1} & \mathbf{0}_{n\times n} & \cdots & \mathbf{0}_{n\times n} & \mathbf{0}_{n\times n} & \cdots & \mathbf{0}_{n\times n} & \mathbf{0}_{n\times n} \\[2mm]
\dfrac{\partial \mathbf{g}_2}{\partial \mathbf{x}_1} & \dfrac{\partial \mathbf{g}_2}{\partial \mathbf{x}_2} & \cdots & \mathbf{0}_{n\times n} & \mathbf{0}_{n\times n} & \cdots & \mathbf{0}_{n\times n} & \mathbf{0}_{n\times n} \\[2mm]
\vdots & \vdots & & \mathbf{0}_{n\times n} & \mathbf{0}_{n\times n} & & \vdots & \vdots \\[2mm]
\mathbf{0}_{n\times n} & \mathbf{0}_{n\times n} & \cdots & \dfrac{\partial \mathbf{g}_{k-1}}{\partial \mathbf{x}_{k-1}} & \mathbf{0}_{n\times n} & \cdots & \mathbf{0}_{n\times n} & \mathbf{0}_{n\times n} \\[2mm]
\mathbf{0}_{n\times n} & \vdots & \cdots & \dfrac{\partial \mathbf{g}_k}{\partial \mathbf{x}_{k-1}} & \dfrac{\partial \mathbf{g}_k}{\partial \mathbf{x}_k} & \cdots & \mathbf{0}_{n\times n} & \mathbf{0}_{n\times n} \\[2mm]
\vdots & \vdots & \vdots & \vdots & & \vdots & \vdots \\[2mm]
\mathbf{0}_{n\times n} & \mathbf{0}_{n\times n} & \cdots & \mathbf{0}_{n\times n} & \mathbf{0}_{n\times n} & \cdots & \dfrac{\partial \mathbf{g}_{N-1}}{\partial \mathbf{x}_{N-1}} & \mathbf{0}_{n\times n} \\[2mm]
\mathbf{0}_{n\times n} & \mathbf{0}_{n\times n} & \cdots & \mathbf{0}_{n\times n} & \mathbf{0}_{n\times n} & \cdots & \dfrac{\partial \mathbf{g}_N}{\partial \mathbf{x}_{N-1}} & \dfrac{\partial \mathbf{g}_N}{\partial \mathbf{x}_N}
\end{bmatrix}
\begin{bmatrix}
\dfrac{\partial \mathbf{x}_1}{\partial \mathbf{x}_0} \\[2mm]
\dfrac{\partial \mathbf{x}_2}{\partial \mathbf{x}_0} \\[2mm]
\vdots \\[2mm]
\dfrac{\partial \mathbf{x}_{k-1}}{\partial \mathbf{x}_0} \\[2mm]
\dfrac{\partial \mathbf{x}_k}{\partial \mathbf{x}_0} \\[2mm]
\vdots \\[2mm]
\dfrac{\partial \mathbf{x}_{N-1}}{\partial \mathbf{x}_0} \\[2mm]
\dfrac{\partial \mathbf{x}_N}{\partial \mathbf{x}_0}
\end{bmatrix}
= -
\begin{bmatrix}
\dfrac{\partial \mathbf{g}_1}{\partial \mathbf{x}_0} \\[2mm]
\mathbf{0}_{n\times n} \\[2mm]
\vdots \\[2mm]
\mathbf{0}_{n\times n} \\[2mm]
\mathbf{0}_{n\times n} \\[2mm]
\vdots \\[2mm]
\mathbf{0}_{n\times n} \\[2mm]
\mathbf{0}_{n\times n}
\end{bmatrix}
\tag{5.26}
$$

So we have

$$
DP = \left[\frac{\partial \mathbf{x}_N}{\partial \mathbf{x}_0} \right] = \left[\frac{\partial \mathbf{x}_N}{\partial \mathbf{x}_{N-1}} \right] \cdot \cdots \cdot \left[\frac{\partial \mathbf{x}_1}{\partial \mathbf{x}_0} \right]. \tag{5.27}
$$

(ii) For $l = k$, equation (5.16) with periodicity condition ($\mathbf{x}_0 = \mathbf{x}_N$) gives node points $\mathbf{x}_k^*$ ($k = 0, 1, 2, \cdots, N$). The stability and bifurcation can be analyzed in the vicinity of $\mathbf{x}_k^*$ with $\mathbf{x}_k = \mathbf{x}_k^* + \Delta \mathbf{x}_k$. Equation (5.17) becomes

$$
\frac{\partial \mathbf{g}_k}{\partial \mathbf{x}_0} + \frac{\partial \mathbf{g}_k}{\partial \mathbf{x}_1} \frac{\partial \mathbf{x}_1}{\partial \mathbf{x}_0} + \cdots + \frac{\partial \mathbf{g}_k}{\partial \mathbf{x}_{k-1}} \frac{\partial \mathbf{x}_{k-1}}{\partial \mathbf{x}_0} + \frac{\partial \mathbf{g}_k}{\partial \mathbf{x}_k} \frac{\partial \mathbf{x}_k}{\partial \mathbf{x}_0} = \mathbf{0}_{n\times n}. \tag{5.28}
$$
$$
(k = 1, 2, \cdots, N)
$$

In other words,

$$
\begin{bmatrix}
\dfrac{\partial \mathbf{g}_1}{\partial \mathbf{x}_1} & \cdots & \mathbf{0}_{n\times n} & \mathbf{0}_{n\times n} \\[2mm]
\vdots & & \vdots & \vdots \\[2mm]
\dfrac{\partial \mathbf{g}_{N-1}}{\partial \mathbf{x}_1} & \cdots & \dfrac{\partial \mathbf{g}_{N-1}}{\partial \mathbf{x}_{N-1}} & \mathbf{0}_{n\times n} \\[2mm]
\dfrac{\partial \mathbf{g}_N}{\partial \mathbf{x}_1} & \cdots & \dfrac{\partial \mathbf{g}_N}{\partial \mathbf{x}_{N-1}} & \dfrac{\partial \mathbf{g}_N}{\partial \mathbf{x}_N}
\end{bmatrix}
\begin{bmatrix}
\dfrac{\partial \mathbf{x}_1}{\partial \mathbf{x}_0} \\[2mm]
\vdots \\[2mm]
\dfrac{\partial \mathbf{x}_{N-1}}{\partial \mathbf{x}_0} \\[2mm]
\dfrac{\partial \mathbf{x}_N}{\partial \mathbf{x}_0}
\end{bmatrix}
= -
\begin{bmatrix}
\dfrac{\partial \mathbf{g}_1}{\partial \mathbf{x}_0} \\[2mm]
\vdots \\[2mm]
\dfrac{\partial \mathbf{g}_{N-1}}{\partial \mathbf{x}_0} \\[2mm]
\dfrac{\partial \mathbf{g}_N}{\partial \mathbf{x}_0}
\end{bmatrix}
\tag{5.29}
$$

and

$$
\begin{bmatrix}
\dfrac{\partial \mathbf{x}_1}{\partial \mathbf{x}_0} \\[2mm]
\vdots \\[2mm]
\dfrac{\partial \mathbf{x}_{N-1}}{\partial \mathbf{x}_0} \\[2mm]
\dfrac{\partial \mathbf{x}_N}{\partial \mathbf{x}_0}
\end{bmatrix}
= -
\begin{bmatrix}
\dfrac{\partial \mathbf{g}_1}{\partial \mathbf{x}_1} & \cdots & \mathbf{0}_{n\times n} & \mathbf{0}_{n\times n} \\[2mm]
\vdots & & \vdots & \vdots \\[2mm]
\dfrac{\partial \mathbf{g}_{N-1}}{\partial \mathbf{x}_1} & \cdots & \dfrac{\partial \mathbf{g}_{N-1}}{\partial \mathbf{x}_{N-1}} & \mathbf{0}_{n\times n} \\[2mm]
\dfrac{\partial \mathbf{g}_N}{\partial \mathbf{x}_1} & \cdots & \dfrac{\partial \mathbf{g}_N}{\partial \mathbf{x}_{N-1}} & \dfrac{\partial \mathbf{g}_N}{\partial \mathbf{x}_N}
\end{bmatrix}^{-1}
\begin{bmatrix}
\dfrac{\partial \mathbf{g}_1}{\partial \mathbf{x}_0} \\[2mm]
\vdots \\[2mm]
\dfrac{\partial \mathbf{g}_{N-1}}{\partial \mathbf{x}_0} \\[2mm]
\dfrac{\partial \mathbf{g}_N}{\partial \mathbf{x}_0}
\end{bmatrix}.
\tag{5.30}
$$

Using $\partial \mathbf{x}_k / \partial \mathbf{x}_0$, the eigenvalues are determined by

$$
|DP_{k(k-1)\cdots 1} - \bar{\lambda}\mathbf{I}_{n\times n}| = 0 \quad \text{with} \quad DP_{k(k-1)\cdots 1} = \left[\frac{\partial \mathbf{x}_k}{\partial \mathbf{x}_0}\right].
\tag{5.31}
$$

which is used to measure the properties of node points on the periodic flow.

The multi-step mappings are developed from the previously determined nodes of periodic motion. During time interval $t \in [t_0, t_0 + T]$, the periodic flow can be determined by

$$
\mathbf{x}(t) = \mathbf{x}(t_l) + \int_{t_l}^{t} \mathbf{f}(\mathbf{x}, t, \mathbf{p}) dt, l \in \{0, 1, \cdots, k-1\}.
\tag{5.32}
$$

For such a periodic flow, at most, all of N-nodes during the time interval $t \in [t_0, t_0 + T]$ are selected, and the corresponding points $\mathbf{x}(t_k)$ $(k = 1, 2, \cdots, N)$. Under $||\mathbf{x}(t_k) - \mathbf{x}_k|| \leqslant \varepsilon_k$ with $\varepsilon_k \geqslant 0$,

$$
||\mathbf{f}(\mathbf{x}(t_k), t_k, \mathbf{p}) - \mathbf{f}(\mathbf{x}_k, t_k, \mathbf{p})|| \leqslant \delta_k.
\tag{5.33}
$$

Suppose that $\mathbf{x}_0, \cdots, \mathbf{x}_N$ are given, $\mathbf{f}(\mathbf{x}_k, t_k, \mathbf{p})$ $(k = 0, 1, \cdots, N)$ can be determined. An interpolation polynomial $\mathbf{P}(t, \mathbf{x}_0, \cdots, \mathbf{x}_N, t_0, \cdots, t_N, \mathbf{p})$ is determined, which can be used to approximate $\mathbf{f}(\mathbf{x}, t, \mathbf{p})$. That is,

$$
\mathbf{f}(\mathbf{x}, t, \mathbf{p}) \approx \mathbf{P}(t, \mathbf{x}_0, \cdots, \mathbf{x}_N, t_0, \cdots, t_N, \mathbf{p})
\tag{5.34}
$$

and $\mathbf{x}(t_k) \approx \mathbf{x}_k$ can be computed by

$$
\mathbf{x}_k = \mathbf{x}_l + \int_{t_l}^{t_k} \mathbf{P}(t, \mathbf{x}_0, \cdots, \mathbf{x}_N, t_0, \cdots, t_N, \mathbf{p}) dt
\tag{5.35}
$$
$$
(l \in \{0, 1, \cdots, k-1\}).
$$

Therefore, we have

$$
\mathbf{x}_k = \mathbf{x}_l + \bar{\mathbf{g}}_k(\mathbf{x}_0, \cdots, \mathbf{x}_N, \mathbf{p})(l \in \{0, 1, \cdots, k-1\}).
\tag{5.36}
$$

The mapping P_k $(k \in \{1, 2, \cdots, N\})$ for a specific l is

$$
\mathbf{g}_k(\mathbf{x}_0, \cdots, \mathbf{x}_N, \mathbf{p}) = \mathbf{0}.
\tag{5.37}
$$

The periodic motion is determined by mapping P_k $(k = 1, 2, \cdots, N)$ and periodicity conditions

$$\mathbf{g}_k(\mathbf{x}_0^*, \cdots, \mathbf{x}_N^*, \mathbf{p}) = \mathbf{0} \quad \text{for} \quad k = 1, 2, \cdots, N$$
$$\mathbf{x}_0^* = \mathbf{x}_N^*. \tag{5.38}$$

From the foregoing equation, node points $\mathbf{x}_k^*$ $(k = 0, 1, 2, \cdots, N)$ can be determined. The corresponding stability and bifurcation can be discussed in the neighborhood of $\mathbf{x}_k^*$ with $\mathbf{x}_k = \mathbf{x}_k^* + \Delta\mathbf{x}_k$. The derivative of Eq.(5.38) with respect to $\mathbf{x}_0$ gives

$$\frac{\partial \mathbf{g}_k}{\partial \mathbf{x}_0} + \frac{\partial \mathbf{g}_k}{\partial \mathbf{x}_1}\frac{\partial \mathbf{x}_1}{\partial \mathbf{x}_0} + \cdots + \frac{\partial \mathbf{g}_k}{\partial \mathbf{x}_{k-1}}\frac{\partial \mathbf{x}_{k-1}}{\partial \mathbf{x}_0} + \frac{\partial \mathbf{g}_k}{\partial \mathbf{x}_k}\frac{\partial \mathbf{x}_k}{\partial \mathbf{x}_0} = \mathbf{0}_{n \times n} \tag{5.39}$$
$$(k = 1, 2, \cdots, N).$$

In other words, we have

$$
\begin{bmatrix}
\dfrac{\partial \mathbf{g}_1}{\partial \mathbf{x}_1} & \cdots & \dfrac{\partial \mathbf{g}_1}{\partial \mathbf{x}_{N-1}} & \dfrac{\partial \mathbf{g}_1}{\partial \mathbf{x}_N} \\
\vdots & & \vdots & \vdots \\
\dfrac{\partial \mathbf{g}_{N-1}}{\partial \mathbf{x}_1} & \cdots & \dfrac{\partial \mathbf{g}_{N-1}}{\partial \mathbf{x}_{N-1}} & \dfrac{\partial \mathbf{g}_{N-1}}{\partial \mathbf{x}_N} \\
\dfrac{\partial \mathbf{g}_N}{\partial \mathbf{x}_1} & \cdots & \dfrac{\partial \mathbf{g}_N}{\partial \mathbf{x}_{N-1}} & \dfrac{\partial \mathbf{g}_N}{\partial \mathbf{x}_N}
\end{bmatrix}
\begin{bmatrix}
\dfrac{\partial \mathbf{x}_1}{\partial \mathbf{x}_0} \\
\vdots \\
\dfrac{\partial \mathbf{x}_{N-1}}{\partial \mathbf{x}_0} \\
\dfrac{\partial \mathbf{x}_N}{\partial \mathbf{x}_0}
\end{bmatrix}
= -
\begin{bmatrix}
\dfrac{\partial \mathbf{g}_1}{\partial \mathbf{x}_0} \\
\vdots \\
\dfrac{\partial \mathbf{g}_{N-1}}{\partial \mathbf{x}_0} \\
\dfrac{\partial \mathbf{g}_N}{\partial \mathbf{x}_0}
\end{bmatrix}, \tag{5.40}
$$

and

$$
\begin{bmatrix}
\dfrac{\partial \mathbf{x}_1}{\partial \mathbf{x}_0} \\
\vdots \\
\dfrac{\partial \mathbf{x}_{N-1}}{\partial \mathbf{x}_0} \\
\dfrac{\partial \mathbf{x}_N}{\partial \mathbf{x}_0}
\end{bmatrix}
= -
\begin{bmatrix}
\dfrac{\partial \mathbf{g}_1}{\partial \mathbf{x}_1} & \cdots & \dfrac{\partial \mathbf{g}_1}{\partial \mathbf{x}_{N-1}} & \dfrac{\partial \mathbf{g}_1}{\partial \mathbf{x}_N} \\
\vdots & & \vdots & \vdots \\
\dfrac{\partial \mathbf{g}_{N-1}}{\partial \mathbf{x}_1} & \cdots & \dfrac{\partial \mathbf{g}_{N-1}}{\partial \mathbf{x}_{N-1}} & \dfrac{\partial \mathbf{g}_{N-1}}{\partial \mathbf{x}_N} \\
\dfrac{\partial \mathbf{g}_N}{\partial \mathbf{x}_1} & \cdots & \dfrac{\partial \mathbf{g}_N}{\partial \mathbf{x}_{N-1}} & \dfrac{\partial \mathbf{g}_N}{\partial \mathbf{x}_N}
\end{bmatrix}^{-1}
\begin{bmatrix}
\dfrac{\partial \mathbf{g}_1}{\partial \mathbf{x}_0} \\
\vdots \\
\dfrac{\partial \mathbf{g}_{N-1}}{\partial \mathbf{x}_0} \\
\dfrac{\partial \mathbf{g}_N}{\partial \mathbf{x}_0}
\end{bmatrix}. \tag{5.41}
$$

From the above discussion, the discrete mapping can be developed through many forward and backward nodes. The periodic flow in a nonlinear dynamical system can be determined through the following Theorem.

Theorem 5.2 *Consider a nonlinear dynamical system in Eq.(5.1). If such a dynamical system has a periodic flow $\mathbf{x}(t)$ with finite norm $\|\mathbf{x}\|$ and period $T = 2\pi/\Omega$, there is a set of discrete time t_k $(k = 0, 1, \cdots, N)$ with $N \to \infty$ during one period T, and the corresponding solution $\mathbf{x}(t_k)$ and vector field $\mathbf{f}(\mathbf{x}(t_k), t_k, \mathbf{p})$ are exact. Suppose a discrete node $\mathbf{x}_k$ is on the approximate*

solutions of the periodic flow under $||\mathbf{x}(t_k) - \mathbf{x}_k|| \leqslant \varepsilon_k$ *with a small* $\varepsilon_k \geqslant 0$
and

$$||\mathbf{f}(\mathbf{x}(t_k), t_k, \mathbf{p}) - \mathbf{f}(\mathbf{x}_k, t_k, \mathbf{p})|| \leqslant \delta_k \qquad (5.42)$$

with a small $\delta_k \geqslant 0$. *During a time interval* $t \in [t_{k-1}, t_k]$, *there is a mapping* $P_k : \mathbf{x}_{k-1} \to \mathbf{x}_k \ (k = 1, 2, \cdots, N)$ *as*

$$\mathbf{x}_k = P_k \mathbf{x}_{k-1} \quad \text{with} \quad \mathbf{g}_k(\mathbf{x}_{s_{kl_1}}, \cdots, \mathbf{x}_{s_{k1}}, \mathbf{x}_{s_{k0}}, \mathbf{x}_{s_{k(-1)}}, \cdots, \mathbf{x}_{s_{k(-l_2)}}, \mathbf{p}) = \mathbf{0},$$

$$s_{kj} = \text{mod}(k - j + N, N), j = -l_2, -l_2 + 1, \cdots, l_1 - 1, l_1;$$

$$l_1, l_2 \in \{0, 1, 2, \cdots, N\}, 1 \leqslant l_1 + l_2 \leqslant N; l_1 \geqslant 1; (k = 1, 2, \cdots, N)$$
$$(5.43)$$

where $\mathbf{g}_k$ *is an implicit vector function. Consider a mapping structure as*

$$P = P_N \circ \cdots \circ P_2 \circ P_1 : \mathbf{x}_0 \to \mathbf{x}_N;$$
$$\text{with} \quad P_k : \mathbf{x}_{k-1} \to \mathbf{x}_k (k = 1, 2, \cdots, N). \qquad (5.44)$$

For $\mathbf{x}_N = P\mathbf{x}_0$, *if there is a set of points* $\mathbf{x}_k^* \ (k = 0, 1, \cdots, N)$ *computed by*

$$\mathbf{g}_k(\mathbf{x}_{s_{kl_1}}^*, \cdots, \mathbf{x}_{s_{k1}}^*, \mathbf{x}_{s_{k0}}^*, \mathbf{x}_{s_{k(-1)}}^*, \cdots, \mathbf{x}_{s_{k(-l_2)}}^*, \mathbf{p}) = \mathbf{0},$$

$$(k = 1, 2, \cdots, N) \qquad (5.45)$$

$$\mathbf{x}_0^* = \mathbf{x}_N^*,$$

then the points $\mathbf{x}_k^* \ (k = 0, 1, \cdots, N)$ *are approximations of points* $\mathbf{x}(t_k)$ *of the periodic solution. In the neighborhood of* $\mathbf{x}_k^*$, *with* $\mathbf{x}_k = \mathbf{x}_k^* + \Delta\mathbf{x}_k$, *the linearized equation is given by*

$$\frac{\partial \mathbf{g}_k}{\partial \mathbf{x}_0} + \cdots + \frac{\partial \mathbf{g}_k}{\partial \mathbf{x}_{k-1}} \frac{\partial \mathbf{x}_{k-1}}{\partial \mathbf{x}_0} + \frac{\partial \mathbf{g}_k}{\partial \mathbf{x}_k} \frac{\partial \mathbf{x}_k}{\partial \mathbf{x}_0} + \frac{\partial \mathbf{g}_k}{\partial \mathbf{x}_{k+1}} \frac{\partial \mathbf{x}_{k+1}}{\partial \mathbf{x}_0} + \cdots + \frac{\partial \mathbf{g}_k}{\partial \mathbf{x}_N} \frac{\partial \mathbf{x}_N}{\partial \mathbf{x}_0} = \mathbf{0}$$

$$\text{with} \quad \frac{\partial \mathbf{g}_k}{\partial \mathbf{x}_\alpha} = \mathbf{0}(\alpha \neq s_{kj}), j = -l_2, -l_2 + 1, \cdots, l_1 - 1, l_1; (k = 1, 2, \cdots, N).$$
$$(5.46)$$

The resultant Jacobian matrices of the periodic flow are

$$DP_{k(k-1)\cdots1} = \left[\frac{\partial \mathbf{x}_k}{\partial \mathbf{x}_0}\right]_{(\mathbf{x}_0^*, \mathbf{x}_1^*, \cdots, \mathbf{x}_N^*)} \quad (k = 1, 2, \cdots, N),$$

$$(5.47)$$

$$\text{and} \quad DP = DP_{N(N-1)\cdots1} = \left[\frac{\partial \mathbf{x}_N}{\partial \mathbf{x}_0}\right]_{(\mathbf{x}_0^*, \mathbf{x}_1^*, \cdots, \mathbf{x}_N^*)}$$

where

$$
\begin{bmatrix}
\dfrac{\partial \mathbf{x}_1}{\partial \mathbf{x}_0} \\[2ex]
\vdots \\[2ex]
\dfrac{\partial \mathbf{x}_{N-1}}{\partial \mathbf{x}_0} \\[2ex]
\dfrac{\partial \mathbf{x}_N}{\partial \mathbf{x}_0}
\end{bmatrix}
= -
\begin{bmatrix}
\dfrac{\partial \mathbf{g}_1}{\partial \mathbf{x}_1} & \cdots & \dfrac{\partial \mathbf{g}_1}{\partial \mathbf{x}_{N-1}} & \dfrac{\partial \mathbf{g}_1}{\partial \mathbf{x}_N} \\[2ex]
\vdots & & \vdots & \vdots \\[2ex]
\dfrac{\partial \mathbf{g}_{N-1}}{\partial \mathbf{x}_1} & \cdots & \dfrac{\partial \mathbf{g}_{N-1}}{\partial \mathbf{x}_{N-1}} & \dfrac{\partial \mathbf{g}_{N-1}}{\partial \mathbf{x}_N} \\[2ex]
\dfrac{\partial \mathbf{g}_N}{\partial \mathbf{x}_1} & \cdots & \dfrac{\partial \mathbf{g}_N}{\partial \mathbf{x}_{N-1}} & \dfrac{\partial \mathbf{g}_N}{\partial \mathbf{x}_N}
\end{bmatrix}^{-1}
\begin{bmatrix}
\dfrac{\partial \mathbf{g}_1}{\partial \mathbf{x}_0} \\[2ex]
\vdots \\[2ex]
\dfrac{\partial \mathbf{g}_{N-1}}{\partial \mathbf{x}_0} \\[2ex]
\dfrac{\partial \mathbf{g}_N}{\partial \mathbf{x}_0}
\end{bmatrix}.
\tag{5.48}
$$

The properties of discrete points $\mathbf{x}_k$ $(k = 1, 2, \cdots, N)$ can be estimated by the eigenvalues of $DP_{k(k-1)\cdots 1}$ as

$$
|DP_{k(k-1)\cdots 1} - \lambda \mathbf{I}_{n\times n}| = 0 \ (k = 1, 2, \cdots, N).
\tag{5.49}
$$

The eigenvalues of DP for such a periodic flow are determined by

$$
|DP - \lambda \mathbf{I}_{n\times n}| = 0.
\tag{5.50}
$$

Thus, the stability and bifurcation of the periodic flow can be classified by the eigenvalues of $DP(\mathbf{x}_0^*)$ with

$$
([n_1^m, n_1^o] : [n_2^m, n_2^o] : [n_3, \kappa_3] : [n_4, \kappa_4] | n_5 : n_6 : [n_7, l, \boldsymbol{\kappa}_7]).
\tag{5.51}
$$

(i) If the magnitudes of all eigenvalues of $DP(|\lambda_i| < 1, i = 1, 2, \cdots, n)$ are less than one, the approximate periodic solution is stable.

(ii) If at least the magnitude of one eigenvalue of $DP(|\lambda_i| > 1, i \in \{1, 2, \cdots, n\})$ is greater than one, the approximate periodic solution is unstable.

(iii) The boundaries between stable and unstable periodic flow with higher order singularity give bifurcation and stability conditions.

Proof. The proof can be referred to Luo (2015). ∎

From the stability and bifurcation analysis, the period-1 flow under period $T = 2\pi/\Omega$, based on the set of discrete mapping P_k with $\mathbf{g}_k(\mathbf{x}_{k-1}, \mathbf{x}_k, \mathbf{p}) = 0$ $(k = 1, 2, \cdots, N)$, is stable or unstable. If the period-doubling bifurcation occurs, the periodic flow will become a periodic flow under period $T' = 2T$, and such a periodic flow is called the period-2 flow. Due to the period-doubling, $2N$ nodes of the period-2 flow will be employed to describe the period-2 flow. Thus, consider a mapping structure of the period-2 flow with $2N$ mappings as

$$
\begin{aligned}
&P = P_{2N} \circ P_{2N-1} \circ \cdots \circ P_2 \circ P_1 : \mathbf{x}_0 \to \mathbf{x}_{2N}; \\
&\text{with} \quad P_k : \mathbf{x}_{k-1} \to \mathbf{x}_k (k = 1, 2, \cdots, 2N).
\end{aligned}
\tag{5.52}
$$

For $\mathbf{x}_{2N} = P\mathbf{x}_0$, there is a set of points $\mathbf{x}_k^*$ $(k = 0, 1, \cdots, 2N)$ computed by

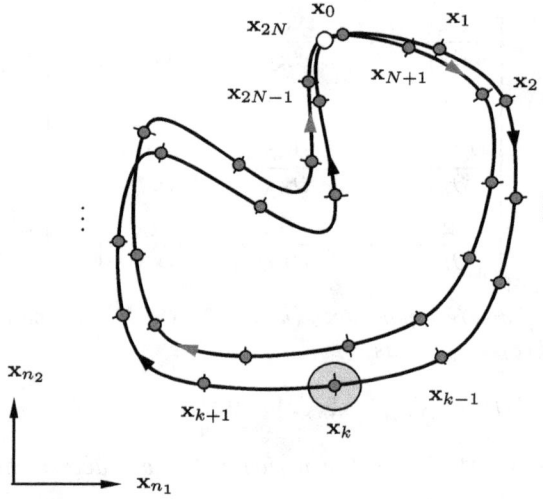

Fig. 5.3 Period-2 flow with $2N$-nodes with short lines. Solid curve: numerical results. The symbols are node points on the periodic flow.

$$\mathbf{g}_k(\mathbf{x}_{k-1}^*, \mathbf{x}_k^*, \mathbf{p}) = \mathbf{0}, (k = 1, 2, \cdots, 2N)$$
$$\mathbf{x}_0^* = \mathbf{x}_{2N}^*. \tag{5.53}$$

After period-doubling, the period-1 flow becomes period-2 flow. The nodes points increase to $2N$ points during two periods $(2T)$. The period-2 flow can be sketched in Fig.5.3. The node points are determined through the discrete mapping with a mathematical relation in Eq.(5.52). On the other hand,

$$T' = 2T = \frac{2(2\pi)}{\Omega} = \frac{2\pi}{\omega} \Rightarrow \omega = \frac{\Omega}{2}. \tag{5.54}$$

During the period of T', there is a periodic flow, which can be described by node points $\mathbf{x}_k$ $(k = 1, 2, \cdots, N')$. Since the period-1 flow is described by node points $\mathbf{x}_k$ $(k = 1, 2, \cdots, N)$ during the period T, due to $T' = 2T$, the period-2 flow can be described by $N' \geqslant 2N$ nodes. Thus, the corresponding mapping P_k is defined as

$$P_k : \mathbf{x}_{k-1}^{(2)} \rightarrow \mathbf{x}_k^{(2)} \ (k = 1, 2, \cdots, 2N) \tag{5.55}$$

and

$$\mathbf{g}_k(\mathbf{x}_{k-1}^{(2)*}, \mathbf{x}_k^{(2)*}, \mathbf{p}) = \mathbf{0} \ (k = 1, 2, \cdots, 2N)$$
$$\mathbf{x}_0^{(2)*} = \mathbf{x}_{2N}^{(2)*}. \tag{5.56}$$

In general, for period $T' = mT$, there is a period-m flow which can be described by $N' \geqslant mN$. The corresponding mapping P_k is given by

$$P_k : \mathbf{x}_{k-1}^{(m)} \to \mathbf{x}_k^{(m)} \ (k = 1, 2, \cdots, mN) \tag{5.57}$$

and

$$\mathbf{g}_k^{(m)}(\mathbf{x}_{k-1}^{(m)*}, \mathbf{x}_k^{(m)*}, \mathbf{p}) = \mathbf{0} \ (k = 1, 2, \cdots, mN)$$
$$\mathbf{x}_0^{(m)*} = \mathbf{x}_{mN}^{(m)*}. \tag{5.58}$$

From the above discussion, the period-m flow in a nonlinear system can be described through mN nodes for period mT, as stated in the following theorem.

Theorem 5.3 *Consider a nonlinear dynamical system in Eq.(5.1). If such a dynamical system has a period-m flow $\mathbf{x}^{(m)}(t)$ with finite norm $||\mathbf{x}^{(m)}||$ and period $mT(T = 2\pi/\Omega)$, there is a set of discrete time $t_k(k = 0, 1, \cdots, mN)$ with $N \to \infty$ during m-periods (mT), and the corresponding solution $\mathbf{x}^{(m)}(t_k)$ and the vector field $\mathbf{f}(\mathbf{x}^{(m)}(t_k), t_k, \mathbf{p})$ are exact. Suppose a discrete node $\mathbf{x}_k^{(m)}$ is on the approximate solutions of the periodic flow under $||\mathbf{x}^{(m)}(t_k) - \mathbf{x}_k^{(m)}|| \leqslant \varepsilon_k$ with a small $\varepsilon_k \geqslant 0$ and*

$$||\mathbf{f}(\mathbf{x}^{(m)}(t_k), t_k, \mathbf{p}) - \mathbf{f}(\mathbf{x}_k^{(m)}, t_k, \mathbf{p})|| \leqslant \delta_k \tag{5.59}$$

with a small $\delta_k \geqslant 0$. During a time interval $t \in [t_{k-1}, t_k]$, there is a mapping $P_k : \mathbf{x}_{k-1}^{(m)} \to \mathbf{x}_k^{(m)}(k = 1, 2, \cdots, mN)$, i.e.,

$$\mathbf{x}_k^{(m)} = P_k \mathbf{x}_{k-1}^{(m)} \quad \text{with} \quad \mathbf{g}_k(\mathbf{x}_{k-1}^{(m)}, \mathbf{x}_k^{(m)}, \mathbf{p}) = \mathbf{0}, k = 1, 2, \cdots, mN \tag{5.60}$$

where $\mathbf{g}_k$ is an implicit vector function. Consider a mapping structure as

$$P = P_{mN} \circ P_{mN-1} \circ \cdots \circ P_1 : \mathbf{x}_0^{(m)} \to \mathbf{x}_{mN}^{(m)};$$
$$\text{with} \quad P_k : \mathbf{x}_{k-1}^{(m)} \to \mathbf{x}_k^{(m)} \ (k = 1, 2, \cdots, mN). \tag{5.61}$$

For $\mathbf{x}_{mN}^{(m)} = P\mathbf{x}_0^{(m)}$, if there is a set of points $\mathbf{x}_k^{(m)*}(k = 0, 1, \cdots, mN)$ computed by

$$\mathbf{g}_k(\mathbf{x}_{k-1}^{(m)*}, \mathbf{x}_k^{(m)*}, \mathbf{p}) = \mathbf{0} \ (k = 1, 2, \cdots, mN)$$
$$\mathbf{x}_0^{(m)*} = \mathbf{x}_{mN}^{(m)*}, \tag{5.62}$$

then the points $\mathbf{x}_k^{(m)*}(k = 0, 1, \cdots, mN)$ are approximations of points $\mathbf{x}^{(m)}(t_k)$ of the periodic solution. In the neighborhood of $\mathbf{x}_k^{(m)*}$, with $\mathbf{x}_k^{(m)} = \mathbf{x}_k^{(m)*} + \Delta\mathbf{x}_k^{(m)}$, the linearized equation is given by

$$\Delta\mathbf{x}_k^{(m)} = DP_k \cdot \Delta\mathbf{x}_{k-1}^{(m)} \quad \text{with} \quad \mathbf{g}_k(\mathbf{x}_{k-1}^{(m)*} + \Delta\mathbf{x}_{k-1}^{(m)}, \mathbf{x}_k^{(m)*} + \Delta\mathbf{x}_k^{(m)}, \mathbf{p}) = \mathbf{0}$$
$$(k = 1, 2, \cdots, mN). \tag{5.63}$$

The resultant Jacobian matrices of the periodic flow are

$$DP_{k(k-1)\cdots 1} = DP_k \cdot DP_{k-1} \cdot \cdots \cdot DP_1, (k = 1, 2, \cdots, mN);$$

$$DP \equiv DP_{mN(mN-1)\cdots 1} = DP_{mN} \cdot DP_{mN-1} \cdot \cdots \cdot DP_1 \tag{5.64}$$

where

$$DP_k = \left[\frac{\partial \mathbf{x}_k^{(m)}}{\partial \mathbf{x}_{k-1}^{(m)}} \right]_{(\mathbf{x}_{k-1}^{(m)*}, \mathbf{x}_k^{(m)*})} = -\left[\frac{\partial \mathbf{g}_k}{\partial \mathbf{x}_k^{(m)}} \right]^{-1} \left[\frac{\partial \mathbf{g}_k}{\partial \mathbf{x}_{k-1}^{(m)}} \right] \Bigg|_{(\mathbf{x}_{k-1}^{(m)*}, \mathbf{x}_k^{(m)*})}. \tag{5.65}$$

The eigenvalues of $DP(\mathbf{x}_0^{(m)*})$ and $DP_{k(k-1)\cdots 1}$ for such a periodic flow are determined by

$$|DP_{k(k-1)\cdots 1} - \overline{\lambda} \mathbf{I}_{n \times n}| = 0, (k = 1, 2, \cdots, mN);$$

$$|DP - \lambda \mathbf{I}_{n \times n}| = 0. \tag{5.66}$$

Thus, the eigenvalues of $DP_{k(k-1)\cdots 1}$ give the properties of $\mathbf{x}_k$ varying with $\mathbf{x}_0$. The stability and bifurcation of the periodic flow can be classified by the eigenvalues of $DP(\mathbf{x}_0^{(m)*})$ with

$$([n_1^{\mathrm{m}}, n_1^{\mathrm{o}}] : [n_2^{\mathrm{m}}, n_2^{\mathrm{o}}] : [n_3, \kappa_3] : [n_4, \kappa_4] | n_5 : n_6 : [n_7, l, \boldsymbol{\kappa}_7]). \tag{5.67}$$

(i) *If the magnitudes of all eigenvalues of $DP^{(m)}$ are less than one (i.e., $|\lambda_i| < 1, i = 1, 2, \cdots, n$), the approximate period-m solution is stable.*

(ii) *If at least the magnitude of one eigenvalue of $DP^{(m)}$ is greater than one (i.e., $|\lambda_i| > 1, i \in \{1, 2, \cdots, n\}$), the approximate period-m solution is unstable.*

(iii) *The boundaries between stable and unstable period-m flow with higher order singularity give bifurcation and stability conditions.*

Proof. The proof can be referred to Luo (2015).   ∎

The discrete mapping for a period-m flow with multiple steps can be developed by using many forward and backward nodes. The period-m flow in a nonlinear dynamical system can be determined through the following Theorem.

Theorem 5.4 *Consider a nonlinear dynamical system in Eq.(5.1). If such a dynamical system has a period-m flow $\mathbf{x}^{(m)}(t)$ with finite norm $||\mathbf{x}^{(m)}||$ and m-periods $mT(T = 2\pi/\Omega)$, there is a set of discrete time $t_k(k = 0, 1, \cdots, mN)$ with $(N \to \infty)$ during m-periods mT, and the corresponding solution $\mathbf{x}^{(m)}(t_k)$ and the vector field $\mathbf{f}(\mathbf{x}^{(m)}(t_k), t_k, \mathbf{p})$ are exact. Suppose a discrete node $\mathbf{x}_k^{(m)}$ is on the approximate solution of the period-m flow under $||\mathbf{x}^{(m)}(t_k) - \mathbf{x}_k^{(m)}|| \leqslant \varepsilon_k$ with a small $\varepsilon_k \geqslant 0$ and*

$$||\mathbf{f}(\mathbf{x}^{(m)}(t_k), t_k, \mathbf{p}) - \mathbf{f}(\mathbf{x}_k^{(m)}, t_k, \mathbf{p})|| \leqslant \delta_k \tag{5.68}$$

with a small $\delta_k \geqslant 0$. During a time interval $t \in [t_{k-1}, t_k]$, there is a mapping $P_k : \mathbf{x}_{k-1}^{(m)} \to \mathbf{x}_k^{(m)} (k = 1, 2, \cdots, mN)$, i.e.,

$$\mathbf{x}_k^{(m)} = P_k \mathbf{x}_{k-1}^{(m)}$$

$$\text{with} \quad \mathbf{g}_k(\mathbf{x}_{s_{kl_1}}^{(m)}, \cdots, \mathbf{x}_{s_{k1}}^{(m)}, \mathbf{x}_{s_{k0}}^{(m)}, \mathbf{x}_{s_{k(-1)}}^{(m)}, \cdots, \mathbf{x}_{s_{k(-l_2)}}^{(m)}, \mathbf{p}) = \mathbf{0},$$

$$s_{kj} = \text{mod}(k - j + mN, mN), j = -l_2, -l_2 + 1, \cdots, l_1 - 1, l_1;$$

$$l_1, l_2 \in \{0, 1, 2, \cdots, mN\}, 1 \leqslant l_1 + l_2 \leqslant mN, l_1 \geqslant 1; (k = 1, 2, \cdots, mN)$$

$$(5.69)$$

where $\mathbf{g}_k$ is an implicit vector function. Consider a mapping structure as

$$P = P_{mN} \circ P_{mN-1} \circ \cdots \circ P_2 \circ P_1 : \mathbf{x}_0^{(m)} \to \mathbf{x}_{mN}^{(m)};$$

$$\text{with} \quad P_k : \mathbf{x}_{k-1}^{(m)} \to \mathbf{x}_k^{(m)} (k = 1, 2, \cdots, mN).$$

$$(5.70)$$

For $\mathbf{x}_{mN}^{(m)} = P\mathbf{x}_0^{(m)}$, if there is a set of points $\mathbf{x}_k^{(m)*} (k = 0, 1, \cdots, mN)$ computed by

$$\mathbf{g}_k(\mathbf{x}_{s_{kl_1}}^{(m)*}, \cdots, \mathbf{x}_{s_{k1}}^{(m)*}, \mathbf{x}_{s_{k0}}^{(m)*}, \mathbf{x}_{s_{k(-1)}}^{(m)*}, \cdots, \mathbf{x}_{s_{k(-l_2)}}^{(m)*}, \mathbf{p}) = \mathbf{0},$$

$$(k = 1, 2, \cdots, mN)$$

$$\mathbf{x}_0^{(m)*} = \mathbf{x}_{mN}^{(m)*},$$

$$(5.71)$$

then the points $\mathbf{x}_k^{(m)*} (k = 0, 1, \cdots, mN)$ are approximations of points $\mathbf{x}^{(m)}(t_k)$ of the periodic solution. In the neighborhood of $\mathbf{x}_k^{(m)*}$, with $\mathbf{x}_k^{(m)} = \mathbf{x}_k^{(m)*} + \Delta\mathbf{x}_k^{(m)}$, the linearized equation is given by

$$\frac{\partial \mathbf{g}_k}{\partial \mathbf{x}_0^{(m)}} + \cdots + \frac{\partial \mathbf{g}_k}{\partial \mathbf{x}_{k-1}^{(m)}} \frac{\partial \mathbf{x}_{k-1}^{(m)}}{\partial \mathbf{x}_0^{(m)}} + \frac{\partial \mathbf{g}_k}{\partial \mathbf{x}_k^{(m)}} \frac{\partial \mathbf{x}_k^{(m)}}{\partial \mathbf{x}_0^{(m)}} + \frac{\partial \mathbf{g}_k}{\partial \mathbf{x}_{k+1}^{(m)}} \frac{\partial \mathbf{x}_{k+1}^{(m)}}{\partial \mathbf{x}_0^{(m)}}$$

$$+ \cdots + \frac{\partial \mathbf{g}_k}{\partial \mathbf{x}_{mN}^{(m)}} \frac{\partial \mathbf{x}_{mN}^{(m)}}{\partial \mathbf{x}_0^{(m)}} = \mathbf{0},$$

$$\frac{\partial \mathbf{g}_k}{\partial \mathbf{x}_\alpha^{(m)}} = \mathbf{0}(\alpha \neq s_{kj}), j = -l_2, -l_2 + 1, \cdots, l_1 - 1, l_1;$$

$$(k = 1, 2, \cdots, mN).$$

$$(5.72)$$

The resultant Jacobian matrices of the periodic flow are

$$DP_{k(k-1)\cdots 1} = \left[\frac{\partial \mathbf{x}_k^{(m)}}{\partial \mathbf{x}_0^{(m)}} \right]_{(\mathbf{x}_0^{(m)*}, \mathbf{x}_1^{(m)*}, \cdots, \mathbf{x}_{mN}^{(m)*})} \quad (k = 1, 2, \cdots, mN),$$

$$\text{and} \quad DP = DP_{(mN)(mN-1)\cdots 1} = \left[\frac{\partial \mathbf{x}_N^{(m)}}{\partial \mathbf{x}_0^{(m)}} \right]_{(\mathbf{x}_0^{(m)*}, \mathbf{x}_1^{(m)*}, \cdots, \mathbf{x}_{mN}^{(m)*})}$$

$$(5.73)$$

where

$$
\begin{bmatrix}
\dfrac{\partial \mathbf{x}_1^{(m)}}{\partial \mathbf{x}_0^{(m)}} \\
\vdots \\
\dfrac{\partial \mathbf{x}_{mN-1}^{(m)}}{\partial \mathbf{x}_0^{(m)}} \\
\dfrac{\partial \mathbf{x}_{mN}^{(m)}}{\partial \mathbf{x}_0^{(m)}}
\end{bmatrix}
=
\begin{bmatrix}
\dfrac{\partial \mathbf{g}_1}{\partial \mathbf{x}_1^{(m)}} & \cdots & \dfrac{\partial \mathbf{g}_1}{\partial \mathbf{x}_{mN-1}^{(m)}} & \dfrac{\partial \mathbf{g}_1}{\partial \mathbf{x}_{mN}^{(m)}} \\
\vdots & & \vdots & \vdots \\
\dfrac{\partial \mathbf{g}_{mN-1}}{\partial \mathbf{x}_1^{(m)}} & \cdots & \dfrac{\partial \mathbf{g}_{mN-1}}{\partial \mathbf{x}_{mN-1}^{(m)}} & \dfrac{\partial \mathbf{g}_{mN-1}}{\partial \mathbf{x}_{mN}^{(m)}} \\
\dfrac{\partial \mathbf{g}_{mN}}{\partial \mathbf{x}_1^{(m)}} & \cdots & \dfrac{\partial \mathbf{g}_{mN}}{\partial \mathbf{x}_{mN-1}^{(m)}} & \dfrac{\partial \mathbf{g}_{mN}}{\partial \mathbf{x}_{mN}^{(m)}}
\end{bmatrix}^{-1}
\begin{bmatrix}
\dfrac{\partial \mathbf{g}_1}{\partial \mathbf{x}_0^{(m)}} \\
\vdots \\
\dfrac{\partial \mathbf{g}_{mN-1}}{\partial \mathbf{x}_0^{(m)}} \\
\dfrac{\partial \mathbf{g}_{mN}}{\partial \mathbf{x}_0^{(m)}}
\end{bmatrix}.
$$

$$(5.74)$$

The properties of discrete points $\mathbf{x}_k (k = 1, 2, \cdots, mN)$ *can be estimated by the eigenvalues of* $DP_{k(k-1)\cdots 1}$ *as*

$$
|DP_{k(k-1)\cdots 1} - \overline{\lambda} \mathbf{I}_{n \times n}| = 0 (k = 1, 2, \cdots, mN). \tag{5.75}
$$

The eigenvalues of DP *for such a periodic flow are determined by*

$$
|DP - \lambda \mathbf{I}_{n \times n}| = 0. \tag{5.76}
$$

Thus, the stability and bifurcation of the period-m flow can be classified by the eigenvalues of $DP(\mathbf{x}_0^{(m)*})$ *with*

$$
([n_1^m, n_1^o] : [n_2^m, n_2^o] : [n_3, \kappa_3] : [n_4, \kappa_4] | n_5 : n_6 : [n_7, l, \kappa_7]). \tag{5.77}
$$

(i) *If the magnitudes of all eigenvalues of* $DP(\mathbf{x}_0^{(m)*})$ *are less than one (i.e.,* $|\lambda_i| < 1, i = 1, 2, \cdots, n$ *), the approximate period-m solution is stable.*

(ii) *If at least the magnitude of one eigenvalue of* $DP(\mathbf{x}_0^{(m)*})$ *is greater than one (i.e.,* $|\lambda_i| > 1, i \in \{1, 2, \cdots, n\}$ *), the approximate period-m solution is unstable.*

(iii) *The boundaries between stable and unstable periodic flow with higher order singularity give bifurcation and stability conditions.*

Proof. The proof is similar to Theorem 5.1.   ∎

5.2 Discrete Fourier series

Consider a nonlinear dynamical system. If such a dynamical system has a period-m flow $\mathbf{x}^{(m)}(t)$ with finite norm $\|\mathbf{x}^{(m)}\|$ and period $mT (T = 2\pi/\Omega)$, then

$$
\mathbf{x}^{(m)}(t + mT) = \mathbf{x}^{(m)}(t). \tag{5.78}
$$

From the Fourier series theory of periodic function, a definition is introduced.

Definition 5.1 Consider a nonlinear dynamical system and such a dynamical system has a flow $\mathbf{x}(t)$ on the time interval $t \in (0, T)$. Assume there are node points t_j ($j = 0, 1, 2, \cdots, N$) with $t_0 = 0$ and $t_N = T$. If $\mathbf{x}(t_j)$ is finite ($j = 0, 1, 2, \cdots, N$) and $\mathbf{x}(t)$ is continuous for $t \in (t_{i-1}, t_i)(i = 1, 2, \cdots, N)$, such a flow $\mathbf{x}(t)$ is called to be piecewise continuous on the time interval $t \in (0, T)$.

Definition 5.2 Consider a nonlinear dynamical system and such a dynamical system has a period-m flow $\mathbf{x}^{(m)}(t)$ with finite norm $\|\mathbf{x}^{(m)}\|$ and period mT ($T = 2\pi/\Omega$). If $\mathbf{x}^{(m)}(t)$ is a piecewise continuous flow on $t \in (0, mT)$, there is the Fourier series $\mathbf{S}^{(m)}(t) \in \mathscr{R}^n$ for the period-m flow $\mathbf{x}^{(m)}(t) \in \mathscr{R}^n$ as

$$\mathbf{S}^{(m)}(t) = \mathbf{a}_0^{(m)} + \sum_{j=1}^{\infty} \mathbf{b}_{j/m} \cos\left(\frac{j}{m}\Omega t\right) + \mathbf{c}_{j/m} \sin\left(\frac{j}{m}\Omega t\right). \qquad (5.79)$$

If $\mathbf{S}^{(m)}(t) = \mathbf{x}^{(m)}(t)$, the coefficients $\mathbf{a}_0^{(m)}$, $\mathbf{b}_{j/m}$, and $\mathbf{c}_{j/m}$ in Eq.(5.79) are by the Euler's formulas

$$\mathbf{a}_0^{(m)} = \frac{1}{mT} \int_0^{mT} \mathbf{x}^{(m)}(t)dt,$$

$$\mathbf{b}_{j/m} = \frac{2}{mT} \int_0^{mT} \mathbf{x}^{(m)}(t) \cos\left(\frac{j}{m}\Omega t\right) dt (j = 1, 2, \cdots), \qquad (5.80)$$

$$\mathbf{c}_{j/m} = \frac{2}{mT} \int_0^{mT} \mathbf{x}^{(m)}(t) \sin\left(\frac{j}{m}\Omega t\right) dt (j = 1, 2, \cdots)$$

and

$$\mathbf{a}_0^{(m)} = (a_{01}^{(m)}, a_{02}^{(m)}, \cdots, a_{0n}^{(m)})^{\mathrm{T}} \in \mathscr{R}^n;$$

$$\mathbf{b}_{j/m} = (b_{j/m1}, b_{j/m2}, \cdots, b_{j/mn})^{\mathrm{T}} \in \mathscr{R}^n, \qquad (5.81)$$

$$\mathbf{c}_{j/m} = (c_{j/m1}, c_{j/m2}, \cdots, c_{j/mn})^{\mathrm{T}} \in \mathscr{R}^n.$$

Theorem 5.5 *Consider a nonlinear dynamical system and such a dynamical system has a period-m flow $\mathbf{x}^{(m)}(t)$ with finite norm $\|\mathbf{x}^{(m)}\|$ and period $mT(T = 2\pi/\Omega)$. If $D^{(l+1)}\mathbf{x}^{(m)}(t)(l \geqslant 0)$ is a piecewise continuous flow on $t \in (0, mT)$ and has a left-hand derivative and right-hand derivative $D^{(l+1)}\mathbf{x}^{(m)}(t)$ with $\|D^{(l+1)}\mathbf{x}^{(m)}(t)\| < K$ at each point in such time interval, then the Fourier series $\mathbf{S}^{(m)}(t) \in \mathscr{R}^n$ for the period-m flow $\mathbf{x}^{(m)}(t) \in \mathscr{R}^n$ is convergent with order l, and $\mathbf{x}^{(m)}(t)$ is continuous with the lth order differentiation. Thus $\mathbf{S}^{(m)}(t) = \mathbf{x}^{(m)}(t)$, i.e.,*

$$\mathbf{x}^{(m)}(t) = \mathbf{a}_0^{(m)} + \sum_{j=1}^{\infty} \mathbf{b}_{j/m} \cos\left(\frac{j}{m}\Omega t\right) + \mathbf{c}_{j/m} \sin\left(\frac{j}{m}\Omega t\right). \qquad (5.82)$$

If $\mathbf{x}^{(m)}(t)$ is discontinuous at $t = t_i$, then the following equation exists

$$\mathbf{x}^{(m)}(t_i) = \frac{1}{2}[\mathbf{x}^{(m)}(t_i^-) + \mathbf{x}^{(m)}(t_i^+)]. \tag{5.83}$$

where $\mathbf{x}^{(m)}(t_i^-)$ and $\mathbf{x}^{(m)}(t_i^+)$ are the left-hand and right-hand limits, respectively. Thus, the Fourier series of $\mathbf{x}^{(m)}(t)$ can be expressed as in Eq.(5.82).

Proof. The proof can be referred to Luo (2015).   ∎

Remark (i) The piecewise continuous periodic flow in a dynamical system cannot be expressed to the Fourier series expansion. Such a piecewise continuous periodic flow should be investigated through the discontinuous dynamical systems theory (e.g., Filippov, 1988; Luo, 2009, 2011). (ii) If a periodic flow possesses the kth derivatives that are continuous, then the Fourier series expansion of the periodic flow is convergent with $1/j^k$. The detailed discussion of the Fourier series theory for periodic functions can be referred to Churchill (1941).

Definition 5.3 Consider a nonlinear dynamical system and such a dynamical system has a period-m flow $\mathbf{x}^{(m)}(t)$ with finite norm $||\mathbf{x}^{(m)}||$ and period $mT(T = 2\pi/\Omega)$. If $\mathbf{x}^{(m)}(t)$ is a continuous flow on $t \in (0, mT)$, there exists the finite Fourier series $\mathbf{T}_M^{(m)}(t) \in \mathscr{R}^n$ for the period-m flow $\mathbf{x}^{(m)}(t) \in \mathscr{R}^n$ as

$$\mathbf{T}_M^{(m)}(t) = \mathbf{a}_0^{(m)} + \sum_{j=1}^{M} \mathbf{b}_{j/m} \cos\left(\frac{j}{m}\Omega t\right) + \mathbf{c}_{j/m} \sin\left(\frac{j}{m}\Omega t\right) \tag{5.84}$$

which is called a trigonometric polynomial of order M.

From discrete mapping structures, the node points of period-m flows are computed. Consider the node points of period-m flows as $\mathbf{x}_k^{(m)} = (x_{1k}^{(m)}, x_{2k}^{(m)}, \cdots, x_{nk}^{(m)})^{\mathrm{T}}$ for $k = 0, 1, 2, \cdots, mN$ in a nonlinear dynamical system. The approximate expression for period-m flow is determined by the Fourier series as

$$\mathbf{x}^{(m)}(t) \approx \mathbf{a}_0^{(m)} + \sum_{j=1}^{M} \mathbf{b}_{j/m} \cos\left(\frac{j}{m}\Omega t\right) + \mathbf{c}_{j/m} \sin\left(\frac{j}{m}\Omega t\right). \tag{5.85}$$

There are $(2M + 1)$ unknown vector coefficients of $\mathbf{a}_0^{(m)}, \mathbf{b}_{j/m}$ and $\mathbf{c}_{j/m}$. To determine such unknowns, at least we have the given nodes $\mathbf{x}_k^{(m)}(k = 0, 1, 2, \cdots, mN)$ with $mN+1 \geqslant 2M+1$. In other words, we have $M \leqslant mN/2$. The node points $\mathbf{x}_k^{(m)}$ on the period-m flow can be expressed by the finite Fourier series for $t_k \in [0, mT]$

$$\mathbf{x}^{(m)}(t_k) \equiv \mathbf{x}_k^{(m)} = \mathbf{a}_0^{(m)} + \sum_{j=1}^{mN/2} \mathbf{b}_{j/m} \cos\left(\frac{j}{m}\Omega t_k\right) + \mathbf{c}_{j/m} \sin\left(\frac{j}{m}\Omega t_k\right)$$

$$= \mathbf{a}_0^{(m)} + \sum_{j=1}^{mN/2} \mathbf{b}_{j/m} \cos\left(\frac{j}{m}\frac{2k\pi}{N}\right) + \mathbf{c}_{j/m} \sin\left(\frac{j}{m}\frac{2k\pi}{N}\right) \quad (5.86)$$

$$(k = 0, 1, \cdots, mN).$$

Theorem 5.6 *Consider a nonlinear dynamical system and such a dynamical system has a period-m flow* $\mathbf{x}^{(m)}(t)$ *with finite norm* $||\mathbf{x}^{(m)}||$ *and period* $mT(T = 2\pi/\Omega)$. *If the node points of period-m flows in a nonlinear dynamical system are* $\mathbf{x}_k^{(m)} = (x_{1k}^{(m)}, x_{2k}^{(m)}, \cdots, x_{nk}^{(m)})^{\mathrm{T}}$ *for* $k = 0, 1, 2, \cdots, mN$ *with*

$$t_k = k\Delta t = \frac{2k\pi}{\Omega N} \quad \text{with} \quad \Delta t = \frac{T}{N} = \frac{2\pi}{\Omega N}. \quad (5.87)$$

Then there is a trigonometric polynomial $\mathbf{T}_M^{(m)}(t)$, *and* $\mathbf{x}^{(m)}(t)$ *can be approximated by* $\mathbf{T}_M^{(m)}(t)$ *under the minimization of* $\sum_{k=0}^{mN} [(\mathbf{x}^{(m)}(t_k) - \mathbf{T}_{mN/2}^{(m)}(t_k)]^2$ *(i.e.,* $\mathbf{x}^{(m)}(t) \approx \mathbf{T}_{mN/2}^{(m)}(t))$. *That is,*

$$\mathbf{x}^{(m)}(t) \approx \mathbf{a}_0^{(m)} + \sum_{j=1}^{mN/2} \mathbf{b}_{j/m} \cos\left(\frac{j}{m}\Omega t\right) + \mathbf{c}_{j/m} \sin\left(\frac{j}{m}\Omega t\right) \quad (5.88)$$

where

$$\mathbf{a}_0^{(m)} = \frac{1}{mN} \sum_{k=0}^{mN} \mathbf{x}_k^{(m)},$$

$$\mathbf{b}_{j/m} = \frac{2}{mN} \sum_{k=0}^{mN} \mathbf{x}_k^{(m)} \cos\left(k\frac{2j\pi}{mN}\right), \quad (5.89)$$

$$\mathbf{c}_{j/m} = \frac{2}{mN} \sum_{k=0}^{mN} \mathbf{x}_k^{(m)} \sin\left(k\frac{2j\pi}{mN}\right)$$

$$(j = 1, 2, \cdots, mN/2).$$

Proof. The proof can be referred to Luo (2015). ∎

In the above theorem, the coefficients for discrete Fourier series can be computed by direct use of Euler formulas through the discrete nodes. For a period-m flow $\mathbf{x}^{(m)}(t)$ with finite norm $||\mathbf{x}^{(m)}||$ and period $mT(T = 2\pi/\Omega)$, consider the nodes of period-m flows in a nonlinear system are $\mathbf{x}_k^{(m)} = (x_{1k}^{(m)}, x_{2k}^{(m)}, \cdots, x_{nk}^{(m)})^{\mathrm{T}}$ for $k = 0, 1, 2, \cdots, mN$. The integration in the coefficients of the Fourier series is by the interpolation of the discrete nodes. Let $h = \Delta t = T/N$ where $T = 2\pi/\Omega$ and $\mathbf{x}^{(m)}(t_0) = \mathbf{x}^{(m)}(t_{mN})$. For simplicity, let $t_0 = 0$. Application of the trapezoidal rules to the Euler formulas of the Fourier series produces the discrete Euler formulas.

(i) The constant term $\mathbf{a}_0^{(m)}$ is discussed as follows.

$$\mathbf{a}_0^{(m)} = \frac{1}{mT} \int_0^{mT} \mathbf{x}^{(m)}(t) dt$$

$$= \frac{1}{mT} \left[\frac{1}{2} \mathbf{x}^{(m)}(t_0) + \mathbf{x}^{(m)}(t_1) + \cdots + \mathbf{x}^{(m)}(t_{mN-1}) + \frac{1}{2} \mathbf{x}^{(m)}(t_{mN}) \right] h$$

$$- \frac{h^3}{12mT} \sum_{k=1}^{mN} \frac{d^2 \mathbf{x}^{(m)}(t)}{dt^2} \bigg|_{t=t_k^c} \tag{5.90}$$

where $t_k^c \in [t_{k-1}, t_k]$ for $k = 1, 2, \cdots, mN$.

$$\left\| \mathbf{a}_0^{(m)} - \frac{1}{mN} \sum_{k=0}^{mN} \mathbf{x}^{(m)}(t_k) \right\| \leqslant \frac{h^2}{12} L. \tag{5.91}$$

where $\max_k \| d^2 \mathbf{x}^{(m)}(t) / dt^2 |_{t=t_k^c} \| = L$. Thus,

$$\mathbf{a}_0^{(m)} \approx \frac{1}{mN} \sum_{k=0}^{mN} \mathbf{x}^{(m)}(t_k) \approx \frac{1}{mN} \sum_{k=0}^{mN} \mathbf{x}_k^{(m)}. \tag{5.92}$$

(ii) The cosine terms coefficients $\mathbf{b}_{j/m} (j = 1, 2, \cdots, mN/2)$ are discussed.

$$\mathbf{b}_{j/m} = \frac{2}{mT} \int_0^{mT} \mathbf{x}^{(m)}(t) \cos\left(\frac{j}{m}\Omega t\right) dt$$

$$= \frac{2}{mT} \left[\frac{1}{2} \mathbf{x}^{(m)}(t_0) \cos\left(\frac{j}{m}\Omega t_0\right) + \mathbf{x}^{(m)}(t_1) \cos\left(\frac{j}{m}\Omega t_1\right) + \cdots \right.$$

$$\left. + \mathbf{x}^{(m)}(t_{mN-1}) \cos\left(\frac{j}{m}\Omega t_{mN-1}\right) + \frac{1}{2} \mathbf{x}^{(m)}(t_{mN}) \cos\left(\frac{j}{m}\Omega t_{mN}\right) \right] h$$

$$- \frac{h^3}{6mT} \sum_{k=1}^{mN} \frac{d^2}{dt^2} \left[\mathbf{x}^{(m)}(t) \cos\left(\frac{j}{m}\Omega t\right) \right] \bigg|_{t=t_k^c}, \tag{5.93}$$

From the foregoing equation, we have

$$\left\| \mathbf{b}_{j/m} - \frac{2}{mN} \sum_{k=0}^{mN} \mathbf{x}^{(m)}(t_k) \cos\left(\frac{j}{m}\Omega t_k\right) \right\| \leqslant \frac{h^2}{6} L_1. \tag{5.94}$$

where $\max_k \| d^2 [\mathbf{x}^{(m)}(t) \cos(j\Omega t)/m]/dt^2 |_{t=t_k^c} \| = L_1$. Thus, the cosine coefficients in discrete Fourier series is

$$\mathbf{b}_{j/m} \approx \frac{2}{mN} \sum_{k=0}^{mN} \mathbf{x}^{(m)}(t_k) \cos\left(\frac{j}{m}\Omega t_k\right) \approx \frac{2}{mN} \sum_{k=0}^{mN} \mathbf{x}_k^{(m)} \cos\left(\frac{j}{m}\Omega t_k\right).$$

$$(5.95)$$

(iii) The sine terms coefficients $\mathbf{c}_{j/m}(j = 1, 2, \cdots, mN/2)$ can be discussed similarly. That is,

$$\mathbf{c}_{j/m} = \frac{2}{mT} \int_0^{mT} \mathbf{x}^{(m)}(t) \sin\left(\frac{j}{m}\Omega t\right) dt$$

$$= \frac{2}{mT}\left[\frac{1}{2}\mathbf{x}^{(m)}(t_0)\sin\left(\frac{j}{m}\Omega t_0\right) + \mathbf{x}^{(m)}(t_1)\sin\left(\frac{j}{m}\Omega t_1\right) + \cdots\right.$$

$$\left. + \mathbf{x}^{(m)}(t_{mN-1})\sin\left(\frac{j}{m}\Omega t_{mN-1}\right) + \frac{1}{2}\mathbf{x}^{(m)}(t_{mN})\sin\left(\frac{j}{m}\Omega t_{mN}\right)\right]h$$

$$- \frac{h^3}{6mT}\sum_{k=1}^{mN}\frac{d^2}{dt^2}\left[\mathbf{x}^{(m)}(t)\sin\left(\frac{j}{m}\Omega t\right)\right]\Bigg|_{t=t_k^c}. \qquad (5.96)$$

From the foregoing equation, we have

$$\left\|\mathbf{c}_{j/m} - \frac{2}{mN}\sum_{k=0}^{mN}\mathbf{x}^{(m)}(t_k)\sin\left(\frac{j}{m}\Omega t_k\right)\right\| \leqslant \frac{h^2}{6}L_2. \qquad (5.97)$$

where $\max_k \|d^2[\mathbf{x}^{(m)}(t)\sin(j\Omega t)/m]/dt^2|_{t=t_k^c}\| = L_2$. Thus, the sine coefficients in discrete Fourier series is

$$\mathbf{c}_{j/m} \approx \frac{2}{mN}\sum_{k=0}^{mN}\mathbf{x}^{(m)}(t_k)\sin\left(\frac{j}{m}\Omega t_k\right) \approx \frac{2}{mN}\sum_{k=0}^{mN}\mathbf{x}_k^{(m)}\sin\left(\frac{j}{m}\Omega t_k\right).$$

$$(5.98)$$

In fact, other interpolations can be used to obtain the Euler formulas, which is not presented.

The harmonic amplitudes and harmonic phases for period-m motion are

$$A_{j/ms} = \sqrt{b_{j/ms}^2 + c_{j/ms}^2}, \varphi_{j/ms} = \arctan\frac{c_{j/ms}}{b_{j/ms}}, (s = 1, 2, \cdots, n). \quad (5.99)$$

Thus the approximation of period-m motion in Eq.(5.88) is given by

$$\mathbf{x}^{(m)}(t) \approx \mathbf{a}_0^{(m)} + \sum_{j=1}^{mN/2}\mathbf{b}_{j/m}\cos\left(\frac{j}{m}\Omega t\right) + \mathbf{c}_{j/m}\sin\left(\frac{j}{m}\Omega t\right). \qquad (5.100)$$

The foregoing equation can be expressed as

$$x_s^{(m)}(t) = a_{0s}^{(m)} + \sum_{j=1}^{mN/2} A_{j/ms} \cos\left(\frac{j}{m}\Omega t - \varphi_{j/ms}\right) \tag{5.101}$$

$(s = 1, 2, \cdots, n).$

References

Churchill, R.N., 1941, *Fourier Series and Boundary Value Problems*, McGraw-Hill, New York.

Filippov, A.F., 1988, *Differential Equations with Discontinuous Righthand Sides*, Kluwer Academic, Dordrecht.

Kreyszig, E., 1988, *Advanced Engineering Mathematics*, John Wiley & Sons, New York.

Luo, A.C.J., 2009, *Discontinuous Dynamical Systems on Time-varying Domains*, Higher Education Press/Springer, Beijing/Heidelberg.

Luo, A.C.J., 2011, *Discontinuous Dynamical Systems*, Higher Education Press/Springer, Beijing/Heidelberg.

Luo, A.C.J., 2012a, *Discrete and Switching Dynamical Systems*, Higher Education Press/L&H Scientific, Beijing/Glen Carbon.

Luo, A.C.J., 2012b, *Regularity and Complexity in Dynamical Systems*, Springer: New York.

Luo, A.C.J., 2015, Periodic flows to chaos based on implicit mappings of nonlinear dynamical systems, *International Journal of Bifurcation and Chaos*, **25**(3), Article No. 1550044 (62 pages).

Chapter 6
Periodic Motions to Chaos in Pendulum

In this chapter, periodic motions to chaos in a periodically forced pendulum will be predicted analytically by a semi-analytical method. The method is based on discretization of differential equations of the dynamical system to obtain implicit maps. Using the implicit maps, mapping structures for specific periodic motions will be developed, and the corresponding periodic motions can be predicted analytically through such mapping structures. Analytical bifurcation trees of periodic motions to chaos will be obtained, and the corresponding stability and bifurcation analysis of periodic motions to chaos will be carried out by eigenvalue analysis. From the analytical predictions of periodic motions to chaos, the corresponding frequency-amplitude characteristics will be obtained for a better understanding of motions complexity in the periodically forced pendulum. Finally, numerical simulations of selected periodic motions will be completed for illustrated. The non-travelable and travelable periodic motions on the bifurcation trees are discovered. Through this chapter, one can better understand the periodic motions to chaos in the periodically forced pendulums. Based on the perturbation method, one cannot achieve the adequate solutions presented herein for periodic motions of the periodically forced pendulum.

6.1 Periodic motions in pendulum

In this section, the implicit discretization of pendulum will be presented to obtain the corresponding implicit maps. From the discrete nodes and mappings, periodic motions in pendulums can be predicted, and the stability and bifurcation of periodic motions will be discussed through eigenvalues analysis.

6.1.1 Implicit discretization

A periodically driven pendulum system is described by

$$\ddot{x} + \delta\dot{x} + \alpha\sin x = Q_0\cos\Omega t \qquad (6.1)$$

where δ is the damping coefficient, α is the stiffness, Q_0 and Ω are excitation amplitude and frequency, respectively. The system is expressed in state space as

$$\dot{x} = y, \ \dot{y} = Q_0\cos\Omega t - \delta\dot{x} - \alpha\sin x. \qquad (6.2)$$

From Luo (2015a,b), such a nonlinear dynamical system can be discretized. Using a midpoint scheme for the time interval of $t \in [t_{k-1}, t_k]$, an implicit map $P_k(k = 0, 1, 2, \cdots)$ is formed as

$$P_k : (x_{k-1}, y_{k-1}) \to (x_k, y_k) \Rightarrow (x_k, y_k) = P_k(x_{k-1}, y_{k-1}) \qquad (6.3)$$

with the implicit relations

$$x_k = x_{k-1} + \frac{1}{2}h(y_{k-1} + y_k),$$

$$y_k = y_{k-1} + h\left\{Q_0\cos\Omega\left(t_{k-1} + \frac{1}{2}h\right) - \frac{1}{2}\delta(y_{k-1} + y_k)\right. \qquad (6.4)$$

$$\left. -\alpha\sin\left[\frac{1}{2}(x_{k-1} + x_k)\right]\right\}.$$

The foregoing discretization experiences an accuracy of $O(h^3)$ for each step. To keep computational accuracy less than 10^{-9}, $h < 10^{-3}$ needs to be maintained.

6.1.2 Periodic motions

From the above implicit discretization, the discrete implicit map for the pendulum is obtained. Using the implicit map, a period-m motion in such a pendulum system can be expressed by a mapping structure, i.e.,

$$P = \underbrace{P_{mN} \circ P_{mN-1} \circ \cdots \circ P_2 \circ P_1}_{mN-\text{actions}} : (x_0, y_0) \to (x_{mN}, y_{mN}) \qquad (6.5)$$

with

$$P_k : (x_{k-1}^{(m)}, y_{k-1}^{(m)}) \to (x_k^{(m)}, y_k^{(m)})$$

$$\Rightarrow (x_k^{(m)}, y_k^{(m)}) = P_k(x_{k-1}^{(m)}, y_{k-1}^{(m)}) \ (k = 1, 2, \cdots, mN). \qquad (6.6)$$

From Eq.(6.4), the governing algebraic equations for each mapping P_k are

$$x_k^{(m)} = x_{k-1}^{(m)} + \frac{1}{2}h(y_{k-1}^{(m)} + y_k^{(m)}),$$

$$y_k^{(m)} = y_{k-1}^{(m)} + h\left\{Q_0 \cos\Omega\left(t_{k-1}^{(m)} + \frac{1}{2}h\right) - \frac{1}{2}\delta(y_{k-1}^{(m)} + y_k^{(m)})\right.$$

$$\left. -\alpha\sin\left[\frac{1}{2}(x_{k-1}^{(m)} + x_k^{(m)})\right]\right\},$$

$$(k = 1, 2, \cdots, mN)$$

(6.7)

Using the periodicity condition, we have

$$(x_{mN}^{(m)}, y_{mN}^{(m)}) = (x_0^{(m)} + 2l\pi, y_0^{(m)}), \ l = 0, \pm 1, \pm 2, \cdots ; m = 1, 2, \cdots. \quad (6.8)$$

New vector functions are introduced as

$$\mathbf{g}_k = (g_{k1}, g_{k2})^T, \quad \mathbf{x}_k^{(m)} = (x_k^{(m)}, y_k^{(m)})^T. \quad (6.9)$$

The governing equations for P_k in Eq.(6.7) are rewritten by

$$\mathbf{g}_k(\mathbf{x}_{k-1}^{(m)}, \mathbf{x}_k^{(m)}, \mathbf{p}) = \mathbf{0} \quad (6.10)$$

where

$$g_{k1} = x_k^{(m)} - x_{k-1}^{(m)} - \frac{1}{2}h(y_{k-1}^{(m)} + y_{k-1}^{(m)}),$$

$$g_{k2} = y_k^{(m)} - y_{k-1}^{(m)} - h\left\{Q_0 \cos\Omega\left(t_{k-1}^{(m)} + \frac{1}{2}h\right)\right.$$

$$\left. -\frac{1}{2}\delta(y_{k-1}^{(m)} + y_k^{(m)}) - \alpha\sin\left[\frac{1}{2}(x_{k-1}^{(m)} + x_k^{(m)})\right]\right\}. \quad (6.11)$$

For a period-1 motion, the corresponding mapping structure is

$$P = P_N \circ P_{N-1} \circ \cdots \circ P_2 \circ P_1 : \mathbf{x}_0 \to \mathbf{x}_N;$$

$$\text{with} \quad P_k : \mathbf{x}_{k-1} \to \mathbf{x}_k (k = 1, 2, \cdots, N). \quad (6.12)$$

where the implicit mapping P_k with $\mathbf{g}_k(\mathbf{x}_{k-1}, \mathbf{x}_k, \mathbf{p}) = 0 (k = 1, 2, \cdots, N)$ and the periodicity condition is $(x_{mN}^{(m)}, y_{mN}^{(m)}) = (x_0^{(m)} + 2l\pi, y_0^{(m)})$. Such a mapping structure for period-1 motion is presented in Fig.6.1. The node points and the mapping are depicted by circles on the trajectory with arrows. The periodicity guarantees that the initial and final nodes overlap with each other. The mapping structure consists of N nodes and N mappings. The N nodes corresponding to a set of points $\mathbf{x}_k^*(k = 0, 1, \cdots, N)$ are computed by $2(N+1)$ implicit vector functions as

$$\mathbf{g}_k(\mathbf{x}_{k-1}^*, \mathbf{x}_k^*, \mathbf{p}) = 0, (k = 1, 2, \cdots, N)$$

$$\mathbf{x}_0^* = \mathbf{x}_N^*. \quad (6.13)$$

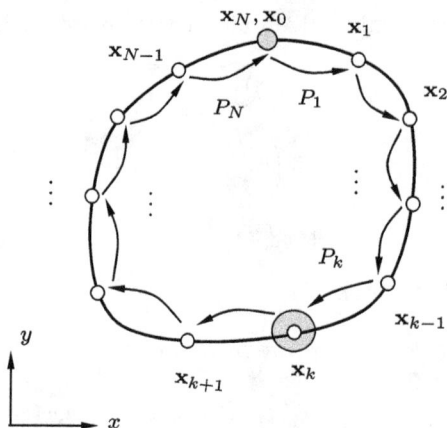

Fig. 6.1 Mapping structure of period-1 motion with N-nodes.

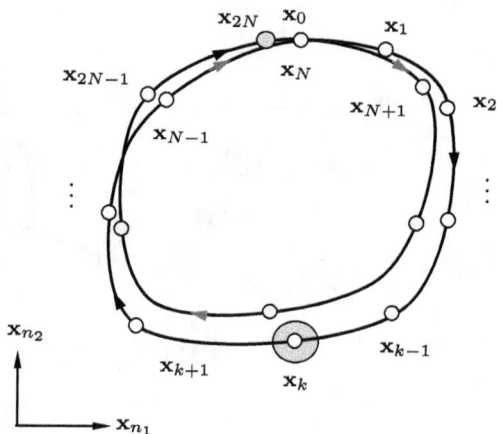

Fig. 6.2 Mapping structure of period-2 motion with $2N$-nodes.

Through a period-doubling of the period-1 motion, a period-2 motion could be obtained, and the corresponding mapping structure is

$$P = P_{2N} \circ P_{2N-1} \circ \cdots \circ P_2 \circ P_1 : \mathbf{x}_0^{(2)} \to \mathbf{x}_{2N}^{(2)};$$
$$\text{with} \quad P_k : \mathbf{x}_{k-1}^{(2)} \to \mathbf{x}_k^{(2)} (k = 1, 2, \cdots, N). \tag{6.14}$$

In a similar fashion, a sketch of the mapping structure of such a period-2 motion is presented in Fig. 6.2. The mapping structure consists of $2N$ nodes and $2N$ mappings. The $2N$ node points $\mathbf{x}_k^*$ ($k = 0, 1, \cdots, 2N$) are computed by the following implicit vector functions.

$$\mathbf{g}_k(\mathbf{x}_{k-1}^{(2)*}, \mathbf{x}_k^{(2)*}, \mathbf{p}) = \mathbf{0}, (k = 1, 2, \cdots, 2N)$$
$$\mathbf{x}_0^{(2)*} = \mathbf{x}_{2N}^{(2)*} \tag{6.15}$$

Similarly, the solution of a period-m motion can be determined through the $2(mN + 1)$ equations. That is,

$$
\begin{aligned}
\mathbf{g}_k(\mathbf{x}_{k-1}^{(m)*}, \mathbf{x}_k^{(m)*}, \mathbf{p}) &= 0, \ (k = 1, 2, \cdots, mN), \\
\mathbf{x}_0^{(m)*} &= \mathbf{x}_{mN}^{(m)*}.
\end{aligned}
\tag{6.16}
$$

If the node points $\mathbf{x}_k^{(m)*}$ $(k = 1, 2, \cdots, mN)$ of the period-m motions are obtained, the stability and bifurcation of such a periodic motion can be studied through the eigenvalue analysis. In vicinity of $\mathbf{x}_k^{(m)*}$, $\mathbf{x}_k^{(m)} = \mathbf{x}_k^{(m)*} + \Delta\mathbf{x}_k^{(m)}$ and one obtains

$$
\Delta\mathbf{x}_{mN} = DP\Delta\mathbf{x}_0^{(m)} = \prod_{k=mN}^{1} DP_k \Delta\mathbf{x}_0^{(m)}.
\tag{6.17}
$$

For each mapping P_k, the linearized equation is

$$
\Delta\mathbf{x}_k^{(m)} = DP_k\Delta\mathbf{x}_{k-1}^{(m)}, (k = 1, 2, \cdots, mN),
\tag{6.18}
$$

where DP_k is the Jacobian matrix of each mapping as

$$
DP_k = \left[\frac{\partial\mathbf{x}_k^{(m)}}{\partial\mathbf{x}_{k-1}^{(m)}} \right]_{(\mathbf{x}_k^{(m)*}, \mathbf{x}_{k-1}^{(m)*})}, (k = 1, 2, \cdots, mN).
\tag{6.19}
$$

To determine the stability and bifurcations of such a period-m motion, the eigenvalues of the resultant Jacobian matrix are computed by

$$
|DP - \lambda\mathbf{I}| = 0,
\tag{6.20}
$$

where the resultant Jacobian matrix is

$$
DP = \prod_{k=mN}^{1} DP_k = \prod_{k=mN}^{1} \left[\frac{\partial\mathbf{x}_k^{(m)}}{\partial\mathbf{x}_{k-1}^{(m)}} \right]_{(\mathbf{x}_k^{(m)*}, \mathbf{x}_{k-1}^{(m)*})}.
\tag{6.21}
$$

For a 2-dimemsional mapping, there are two eigenvalues. From [Luo, 2012, 2015], the stability conditions are given as follows:

(i) If the magnitudes of two eigenvalues are less than one (i.e., $|\lambda_i| < 1, i = 1, 2$), the period-$m$ motion is stable.

(ii) If one of two eigenvalue magnitudes is greater than one (i.e., $|\lambda_i| > 1$, $i \in \{1, 2\}$), the period-m motion is unstable.

For the bifurcation conditions, we have the following:

(i) If $\lambda_i = 1, i \in \{1, 2\}$ and $|\lambda_j| < 1, j \in \{1, 2\}$ but $j \neq i$, the saddle-node bifurcation of the period-m motion occurs.

(ii) If $\lambda_i = -1, i \in \{1,2\}$ and $|\lambda_j| < 1, j \in \{1,2\}$ but $j \neq i$, the period-doubling bifurcation of the period-m motion occurs.

(iii) If $|\lambda_{1,2}| = 1$ with $\lambda_{1,2} = \alpha \pm i\beta$, the Neimark bifurcation of the period-m motion occurs.

6.2 Bifurcation trees to chaos

From Luo and Guo (2016), this section will present the bifurcation trees of period-m motions ($m = 1,3,5\cdots$) to chaos in the periodically forced pendulum. For the periodically excited pendulum discussed above, a set of system parameters is considered as $\alpha = 1.5, \delta = 0.75, Q_0 = 5.0$.

6.2.1 Period-1 motions to chaos

In this section, analytical prediction of bifurcation trees will be presented for period-1, period-2, and period-4 motions to chaos. The corresponding stability and bifurcation analysis will be presented through the eigenvalues of the specific periodic motions.

A global view of the analytical prediction of bifurcation tree from period-1 motions to chaos is presented in Fig.6.3 for such a periodically excited pendulum. The solid and dashed curves depict stable and unstable periodic motions, respectively. The pairs of asymmetric motions are presented with black and red colors, respectively. The acronyms 'SN' and 'PD' represent saddle-node and period-doubling bifurcations, respectively. The displacement and velocity of the periodic nodes $x_{\text{mod }(k,N)}$ and $y_{\text{mod }(k,N)}$ for $\text{mod}(k,N) = 0$ varying with excitation frequency are presented in Figs.6.3 (a) and (b), respectively. The period-1, period-2, and period-4 motions are labeled with P-1, P-2, and P-4, respectively. In order to provide better views, five segments, $\Omega \in (0.05, 0.25), (0.25, 0.70), (0.70, 0.90), (1.1, 1.5)$ and $\Omega \in (1.28, 1.31)$ are zoomed and presented in Fig.6.4, respectively. The stability ranges and bifurcation points are listed in Tables 6.1—6.5.

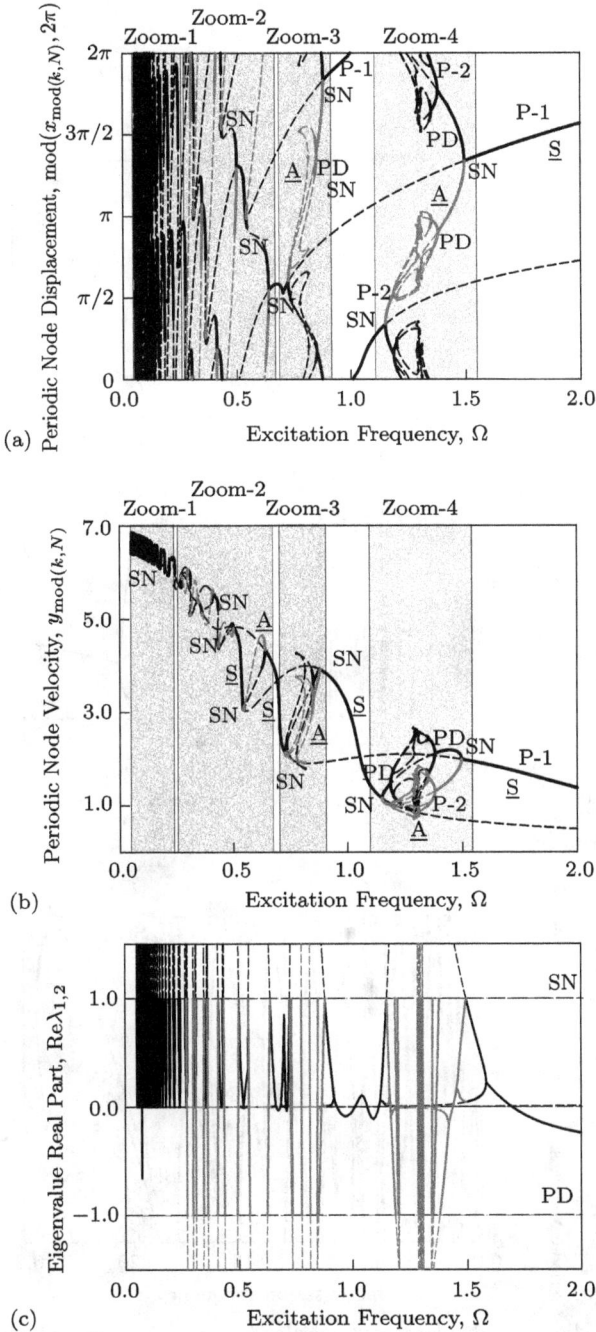

Fig. 6.3 A global view of bifurcation trees for period-1 to chaos varying with excitation frequency $\Omega \in (0.0, 2.0)$: (a) displacement $\mathrm{mod}(x_{\mathrm{mod}(k,N)}, 2\pi)$, (b) velocity $y_{\mathrm{mod}(k,N)}$; Eigenvalues: (c) real part, (d) imaginary part. (e) Magnitude. ($\alpha = 1.5, \delta = 0.75, Q_0 = 5.0$).

(d)

(e)

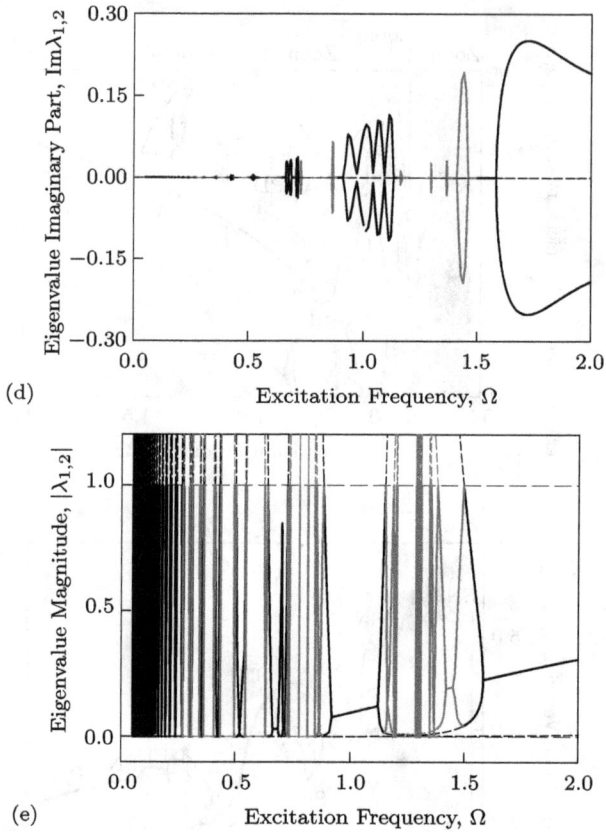

Fig. 6.3 Continued.

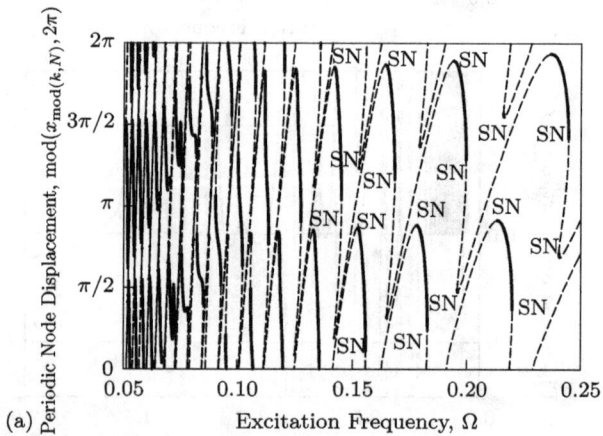

(a)

Fig. 6.4 Periodic node displacement $\mod(x_{\mod(k,N)}, 2\pi)$ and velocity $y_{\mod(k,N)}$ for bifurcation trees of period-1 motions to chaos: (a,b) zoom-1 view ($\Omega \in (0.05, 0.25)$) (c,d) zoom-2 view ($\Omega \in (0.25, 0.70)$), (e,f) zoom-3 view ($\Omega \in (0.70, 0.90)$), (g,h) zoom-4 view ($\Omega \in (1.1, 1.5)$), (i,j) detail view of zoom-4 window ($\Omega \in (1.28, 1.31)$). ($\alpha = 1.5, \delta = 0.75, Q_0 = 5.0$).

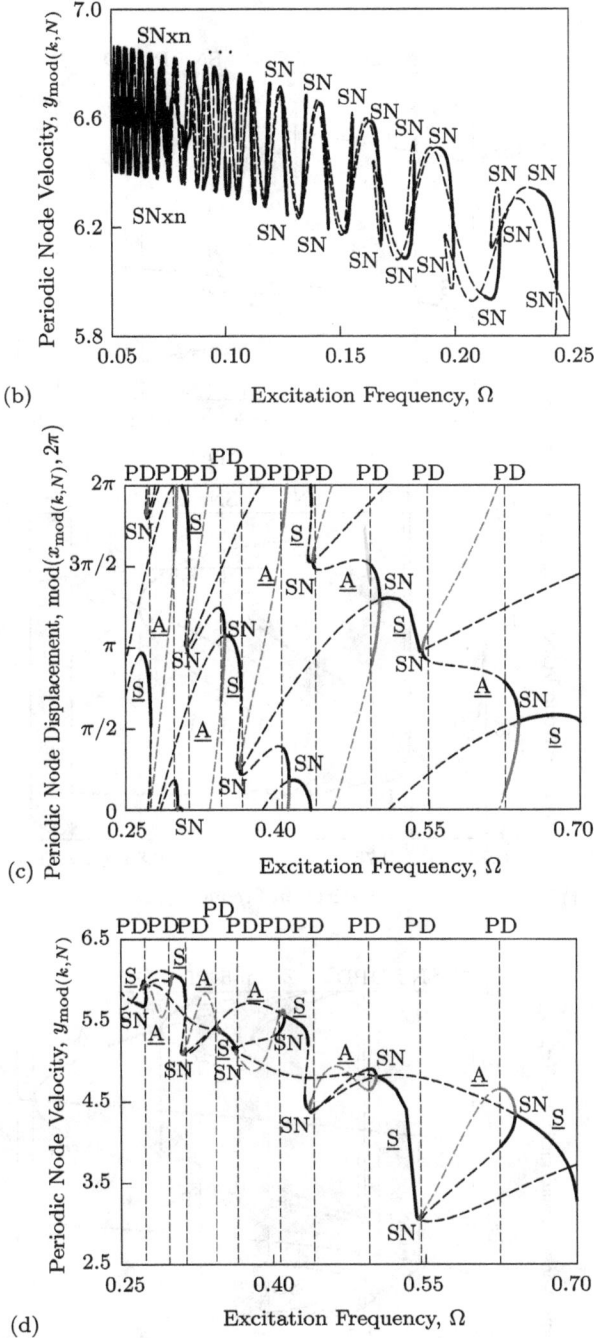

(b)

(c)

(d)

Fig. 6.4 Continued.

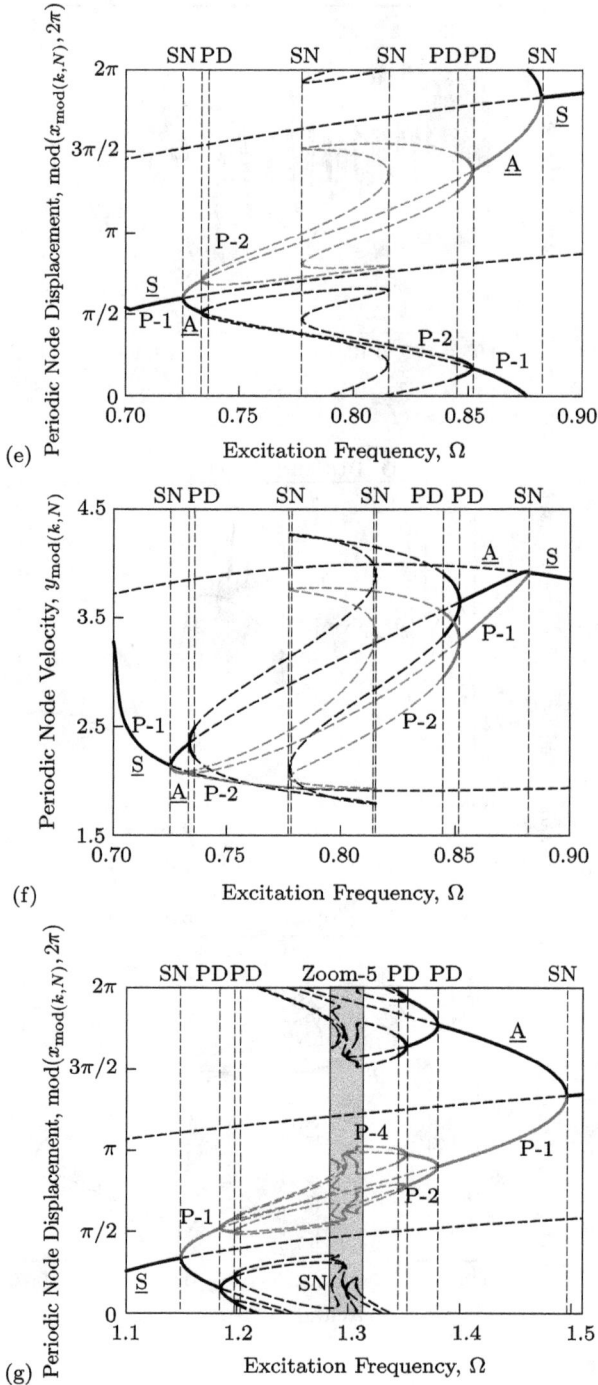

(e)

(f)

(g)

Fig. 6.4 Continued.

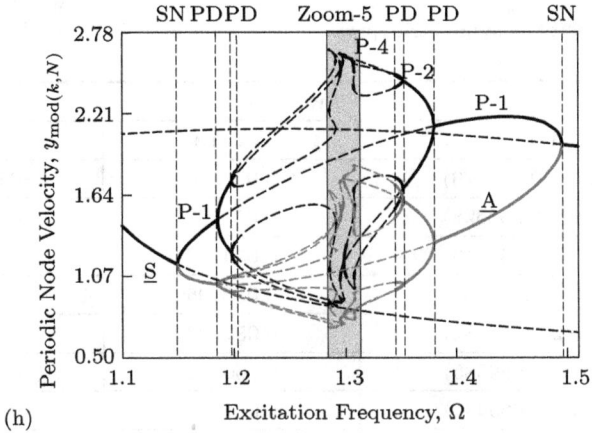

(h)

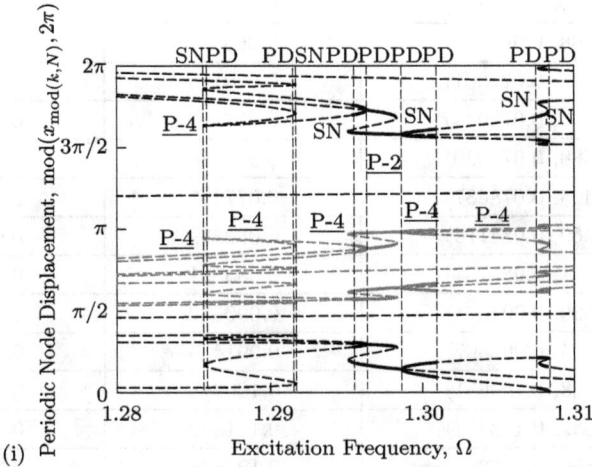

(i)

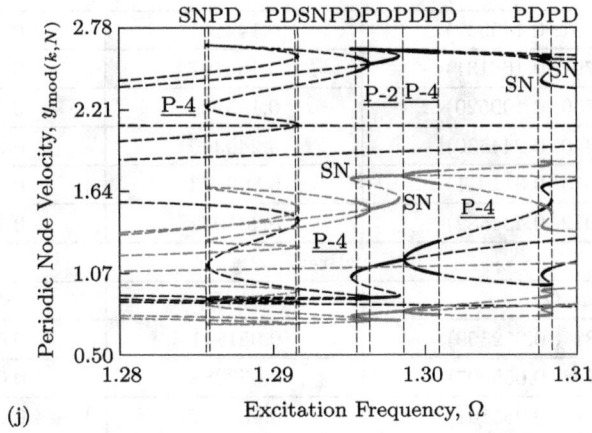

(j)

Fig. 6.4 Continued.

Table 6.1 Unstable symmetric period-1 motions associated with jumping ($\alpha = 1.5, \delta = 0.75, Q_0 = 5.0$)

Branch 1		
Ω	SN (L)	SN (R)
(0.051916, 0.053579)	0.051916	0.053579
(0.053167, 0.053698)	0.053167	0.053698
(0.054737, 0.056513)	0.054737	0.056513
(0.056105, 0.056707)	0.056105	0.056707
(0.057892, 0.059786)	0.057892	0.059786
(0.060072, 0.059385)	0.060072	0.059385
(0.061448, 0.063461)	0.061448	0.063461
(0.063072, 0.063857)	0.063072	0.063857
(0.065498, 0.067619)	0.065498	0.067619
(0.070175, 0.072361)	0.070175	0.072361
(0.072009, 0.073044)	0.072009	0.073044
(0.075688, 0.077820)	0.075688	0.077820
(0.077495, 0.078687)	0.077495	0.078687
(0.082369, 0.084175)	0.082369	0.084175
(0.083881, 0.085257)	0.083881	0.085257
(0.090539, 0.091671)	0.090539	0.091671
(0.091405, 0.092997)	0.091405	0.092997
(0.100268, 0.100661)	0.100268	0.100661
(0.111332, 0.113474)	0.111332	0.113474
(0.124896, 0.127387)	0.124896	0.127387
(0.142151, 0.145042)	0.142151	0.145042
(0.164797, 0.168131)	0.164797	0.168131
(0.195739, 0.199529)	0.195739	0.199529
(0.240340, 0.244530)	0.240340	0.244530
(0.309673, 0.314011)	0.309673	0.314011
(0.431072, 0.434582)	0.431072	0.434582
Branch 2		
Ω	SN (L)	SN (R)
(0.051811, 0.052309)	0.051811	0.052309
(0.053288, 0.055007)	0.053288	0.055007
(0.054597, 0.055161)	0.054597	0.055161
(0.056269, 0.058103)	0.056269	0.058103
(0.057698, 0.058341)	0.057698	0.058341

Continued

Branch 2		
Ω	SN (L)	SN (R)
(0.059614, 0.061569)	0.059614	0.061569
(0.061173, 0.061907)	0.061173	0.061907
(0.063404, 0.065474)	0.063404	0.065474
(0.065092, 0.065933)	0.065092	0.065933
(0.067747, 0.069909)	0.067747	0.069909
(0.069546, 0.070511)	0.069546	0.070511
(0.072809, 0.074991)	0.072809	0.074991
(0.074652, 0.075762)	0.074652	0.075762
(0.078858, 0.080872)	0.078858	0.080872
(0.080563, 0.081843)	0.080563	0.081843
(0.086259, 0.087761)	0.086259	0.087761
(0.087483, 0.088962)	0.087483	0.088962
(0.095205, 0.095952)	0.095205	0.095952
(0.095693, 0.097406)	0.095693	0.097406
(0.105762, 0.105876)	0.105762	0.105876
(0.105586, 0.107574)	0.105586	0.107574
(0.117730, 0.120040)	0.117730	0.120040
(0.132975, 0.135660)	0.132975	0.135660
(0.152656, 0.155765)	0.152656	0.155765
(0.178976, 0.182540)	0.178976	0.182540
(0.215837, 0.219842)	0.215837	0.219842
(0.270817, 0.275135)	0.270817	0.275135
(0.360809, 0.364953)	0.360809	0.364953

Note: L and R denote left and right. SN is for saddle-node bifurcation.

Table 6.2 Excitation frequencies for unstable symmetric to asymmetric period-1 motions ($\alpha = 1.5, \delta = 0.75, Q_0 = 5.0$)

Branch 1		
Ω	SN (L)	SN (R)
(0.050522, 0.053579)	0.050522	0.053579
(0.051917, 0.052190)	0.051917	0.052190
(0.053248, 0.053695)	0.053248	0.053695
(0.053167, 0.056512)	0.053167	0.056512

Continued

Branch 1		
Ω	SN (L)	SN (R)
(0.054742, 0.055002)	0.054742	0.055002
(0.056168, 0.056702)	0.056168	0.056702
(0.056105, 0.059785)	0.056105	0.059785
(0.057908, 0.058110)	0.057908	0.058110
(0.059421, 0.060058)	0.059421	0.060058
(0.059385, 0.063461)	0.059385	0.063461
(0.063071, 0.063824)	0.063071	0.063824
(0.063072, 0.067617)	0.063072	0.067617
(0.067200, 0.067246)	0.067200	0.067246
(0.067246, 0.072357)	0.067246	0.072357
(0.072009, 0.077811)	0.072009	0.077811
(0.077707, 0.078007)	0.077707	0.078007
(0.077496, 0.084155)	0.077496	0.084155
(0.083882, 0.091624)	0.083882	0.091624
(0.091407, 0.100543)	0.091407	0.100543
(0.102243, 0.100399)	0.102243	0.100399
(0.100402, 0.111371)	0.100402	0.111371
(0.111339, 0.124771)	0.111339	0.124771
(0.124914, 0.141750)	0.124914	0.141750
(0.142192, 0.163880)	0.142192	0.163880
(0.164893, 0.193830)	0.164893	0.193830
(0.195970, 0.236450)	0.195970	0.236450
(0.240910, 0.301500)	0.240910	0.301500
(0.311200, 0.411900)	0.311200	0.411900
(0.435780, 0.639000)	0.435780	0.639000
(0.725370, 1.496000)	0.725370	1.496000
Branch 2		
Ω	SN (L)	SN (R)
(0.050614, 0.050884)	0.050614	0.050884
(0.051897, 0.052308)	0.051897	0.052308
(0.051811, 0.055007)	0.051811	0.055007
(0.053291, 0.053561)	0.053291	0.053561
(0.054670, 0.055158)	0.054670	0.055158
(0.054597, 0.058103)	0.054597	0.058103
(0.056277, 0.056518)	0.056277	0.056518
(0.057749, 0.058333)	0.057749	0.058333
(0.057698, 0.061568)	0.057698	0.061568

Continued

Branch 2		
Ω	SN (L)	SN (R)
(0.059650, 0.059779)	0.059650	0.059779
(0.061192, 0.061886)	0.061192	0.061886
(0.061173, 0.065473)	0.061173	0.065473
(0.065069, 0.065881)	0.065069	0.065881
(0.065092, 0.069907)	0.065092	0.069907
(0.069484, 0.070381)	0.069484	0.070381
(0.069546, 0.074985)	0.069546	0.074985
(0.074652, 0.080859)	0.074652	0.080859
(0.080563, 0.087731)	0.080563	0.087731
(0.087484, 0.095877)	0.087484	0.095877
(0.095695, 0.105683)	0.095695	0.105683
(0.105591, 0.117697)	0.105591	0.117697
(0.117741, 0.132730)	0.117741	0.132730
(0.133002, 0.152040)	0.133002	0.152040
(0.152719, 0.177650)	0.152719	0.177650
(0.179125, 0.213120)	0.179125	0.213120
(0.216199, 0.265220)	0.216199	0.265220
(0.271740, 0.348530)	0.271740	0.348530
(0.363400, 0.501500)	0.363400	0.501500
(0.543300, 0.881900)	0.543300	0.881900

Note: L and R denote left and right. SN is for saddle-node bifurcation.

Table 6.3 Excitation frequencies for unstable asymmetric period-1 to period-2 motions ($\alpha = 1.5, \delta = 0.75, Q_0 = 5.0$)

Ω	PD (L)	PD (R)
(0.272110, 0.298880)	0.272110	0.298880
(0.311800, 0.344860)	0.311800	0.344860
(0.364430, 0.406460)	0.364430	0.406460
(0.437580, 0.493120)	0.437580	0.493120
(0.546910, 0.624500)	0.546910	0.624500
(0.733600, 0.851700)	0.733600	0.851700
(1.185900, 1.380900)	1.185900	1.380900

Note: L and R denote left and right. PD is for period-doubling bifurcation.

Table 6.4 Excitation frequencies for unstable period-2 motions ($\alpha = 1.5$, $\delta = 0.75$, $Q_0 = 5.0$)

With Period-4 Motions		
Ω	PD (L)	PD (R)
(0.735946, 0.815082)	0.735946	0.815082
(0.777674, 0.844840)	0.777674	0.844840
(1.198330, 1.296350)	1.198330	1.296350
(1.298600, 1.352000)	1.298600	1.352000
With Jumping Phenomena		
Ω	PD (L)	PD (R)
(0.777632, 0.815116)	0.777632	0.815116
(1.295148, 1.298300)	1.295148	1.298300

Note: L and R denote left and right. SN and PD are for saddle-node and period-doubling bifurcations.

Table 6.5 Excitation frequencies for unstable period-4 motions ($\alpha = 1.5$, $\delta = 0.75$, $Q_0 = 5.0$)

With Period-8 Motions		
Ω	PD (L)	PD (R)
(1.295424, 1.285843)	1.295424	1.285843
(1.201350, 1.291596)	1.201350	1.291596
(1.300908, 1.307618)	1.300908	1.307618
(1.308240, 1.344570)	1.308240	1.344570
With Jumping Phenomena		
Ω	SN (L)	SN (R)
(1.285755, 1.291662)	1.285755	1.291662
(1.307720, 1.308359)	1.307720	1.308359

Note: L and R denote left and right. SN and PD are for saddle-node and period-doubling bifurcations.

Two branches of symmetric period-1 motions exist for the whole range of $\Omega \in (0, +\infty)$. The two branches of symmetric period-1 motions form a double helix structure together. The double helix structure of the symmetric period-1 motions cannot be observed easily from the displacement due to the modulus of displacement into the range of $\mathrm{mod}(x_{\mathrm{mod}\ (k,N)}, 2\pi) \in (0, 2\pi)$. However, from the velocity plot, such a double helix structure of symmetric period-1 motion can be observed as in Fig.6.3(b). The two branches alternatively become stable and unstable. The asymmetric period-1 motion starts from one branch of the symmetric period-1 motions through a saddle-node bifurcation, and will join to the other branch through another saddle-node bifurcation.

As $\Omega \to 0$, periodic motions are very complicated, as shown in Figs.6.4(a) and (b). The two branches of symmetric period-1 motions frequently switch between stable and unstable motions through saddle-node bifurcations. The saddle-node bifurcations also occur between unstable symmetric period-1 motions and stable asymmetric period-1 motions. In addition, jumping phenomena of symmetric and asymmetric period-1 motions also take place frequently. The unstable period-1 motion is enclosed by two jumping phenomena with two saddle-node bifurcations. Cascaded period-doubling bifurcations leading to chaos also happen once the asymmetric period-1 motions appear. However the asymmetric period-1 to period-m motions generated by period-doubling bifurcations are not illustrated herein due to the very tiny stable ranges. As the excitation frequency increases, the change of motion starts to slow down gradually.

From $\Omega = 0.25$ to 0.7, pairs of asymmetric period-1 motions are plotted in Figs.6.4(c) and(d). The asymmetric period-1 motions do not possess any jumping phenomenon. The asymmetric period-1 motions connect the two branches of symmetric period-1 motions together. There are four pairs of asymmetric period-1 motions for $\Omega \in (0.25, 0.70)$. The four branches are for $\Omega \in (0.3112, 0.3485)$, $(0.3634, 0.4118)$, $(0.4358, 0.5015)$, and $(0.5434, 0.6390)$. The corresponding cascaded period doubling bifurcations to chaos in the four branches exist for $\Omega \in (0.3118, 0.3485)$, $(0.3644, 0.4065)$, $(0.4376, 0.4931)$, and $(0.5469, 0.6245)$, respectively. Due to the very tiny stable range of the period-m motions after period-doubling bifurcations, the period-m motions $(m = 2^l, l = 1, 2, \cdots)$ are not illustrated herein.

For $\Omega \in (0.7254, 0.8820)$, a pair of asymmetric period-1 motions is generated through saddle-node bifurcations of the symmetric period-1 motions, as shown in Figs.6.4(e) and (f). This pair of asymmetric period-1 motions possesses period-doubling bifurcations at $\Omega \approx 0.7336$ and 0.8517, and such bifurcation points produce unstable asymmetric period-1 motions and stable period-2 motions. The period-2 motions are in the range of $\Omega \in (0.7336, 0.8517)$. Two pairs of cascade period-doubling bifurcations and one jumping phenomenon exist for such period-2 motions. The saddle-node bifurcations associated with the jumping phenomenon occur at $\Omega \approx 0.7776$ and 0.81512. A pair of the cascade period-doubling bifurcation trees occurs at $\Omega \approx 0.7359$ and 0.81508. The other pair of the bifurcation trees exists at $\Omega \approx 0.7777$ and 0.8448. The cascade period-doubling bifurcation scenarios generate period-4 motions to chaos. The period-4, period-8, and other motions are not illustrated herein due to the very tiny stable ranges.

Finally, for $\Omega \in (1.1500, 1.4960)$, another branch of asymmetric period-1 motions exists in pair, as shown in Figs.6.4(g) and (h). The corresponding saddle-node bifurcations for the asymmetric period-1 motions are at $\Omega \approx 1.15$ and 1.496. The asymmetric period-1 motions possess period-doubling bifurcations at $\Omega \approx 1.1859$ and 1.3809. The period-doubling bifurcations yield unstable asymmetric period-1 motions and stable period-2 motions. The period-2 motions experience two pairs of period-doubling bifurcations and a pair of jumping phenomena. The two jumping phenomena are caused by the saddle-node bifurcations at $\Omega \approx 1.2951$ and 1.2983, as shown in Figs.6.4(i) and (j). The corresponding unstable period-2 motions connect the stable motions from $\Omega \approx 1.2951$ to 1.2983. The first pair of period-doubling bifurcations occurs at $\Omega \approx 1.1983$ and 1.2963, which induce unstable period-2 motions and a branch of stable period-4 motions. This period-4 motion has two pairs of further cascade period-doubling bifurcations at $\Omega \approx 1.2014, 1.2916$ and $\Omega \approx 1.2858, 1.2954$. The two pairs of cascade period-doubling bifurcations lead to period-8, period-16 motions to chaos. In addition a jumping phenomenon occurs in the range of $\Omega \in (1.2857, 1.2917)$, with the corresponding saddle-node bifurcations at $\Omega \approx 1.2857$ and 1.2917. Similarly, the period-2 motions encounter the second pair of period-doubling bifurcation at $\Omega \approx 1.2986$ and 1.3520, which introduce unstable period-2 motions and stable period-4 motions. The period-4 motions encounter two pairs of further cascade period-doubling bifurcations at $\Omega \approx 1.3009, 1.3076$ and $\Omega \approx 1.3082, 1.3446$. Again these cascade period-doubling bifurcations induce period-8, period-16 motions to chaos. Two jumping phenomena also exists for $\Omega \in (1.3077, 1.3084)$. The corresponding saddle-node bifurcations occur at $\Omega \approx 1.3077$ and $\Omega \approx 1.3084$. The period-8, period-16, and other periodic motions are not demonstrated herein due to the very tiny stable ranges.

6.2.2 Period-3 motions to chaos

In this section, analytical prediction of bifurcation trees will be presented for period-3, period-6 motions to chaos. The bifurcation trees of symmetric period-3 motions to chaos will be presented in Fig.6.5 for $\Omega \in (0.741784, 0.802574)$ with two saddle-node bifurcation points at $\Omega \approx 0.741784$ and 0.802574 for catastrophes and motion appearance. In addition, for $\Omega \in (0.762588, 0.763558)$, there are three solutions of symmetric period-3 motions with jumping phenomena at the two saddle-node bifurcation points of $\Omega \approx 0.762588, 0.763558$. For $\Omega \in (0.763920, 0.801850)$, the symmetric period-3 motions become unstable, and at the two saddle-node bifurcations of $\Omega \approx 0.763920$ and 0.801850, the asymmetric period-3 motions will appear. Because the frequency range for the stable asymmetric period-3 motion is very short for such a frequency range, such an asymmetric period-3 motion will not be presented. For $\Omega \in (0.744290, 0.759830)$, the symmetric period-3 motions also become unstable, and at the two saddle-node bifurcations of $\Omega \approx 0.763920$ and 0.801850, the asymmetric period-3 motions will appear and exist in such a frequency range. However, the frequency range for the stable asymmetric period-3 motion is relatively large for such a frequency range, such asymmetric period-3 motion is presented for bi-

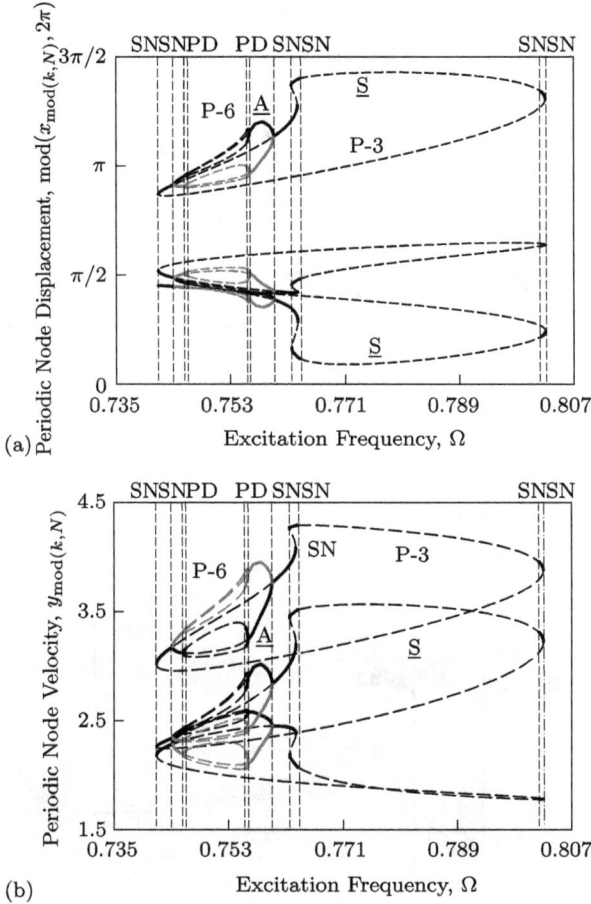

Fig. 6.5 Period-3 motions to chaos for $\Omega \in (0.735, 0.807)$: (a) displacement $\text{mod}(x_{\text{mod}(k,N)}, 2\pi)$; (b) velocity $y_{\text{mod}(k,N)} (\alpha = 1.5, \delta = 0.75, Q_0 = 5.0)$.

furcation tree. The asymmetric period-3 motion has two period-doubling bifurcations at $\Omega \approx 0.745990, 0.756100$ and the period-6 motions appear. The asymmetric period-3 motion becomes unstable and the period-6 motions exist for $\Omega \in (0.745990, 0.756100)$. For $\Omega \in (0.746450, 0.755716)$, the period-6 motions become unstable with two period-doubling bifurcations at $\Omega \approx 0.746450, 0.755716$. With the two period-doubling bifurcations, a period-12 motion appears and exists for $\Omega \in (0.746450, 0.755716)$. Continuously, the bifurcation tree of symmetric period-3 motion to chaos can be obtained. The frequency ranges for unstable period-3 and period-6 motions and bifurcation points are listed in Tables 6.6 and 6.7.

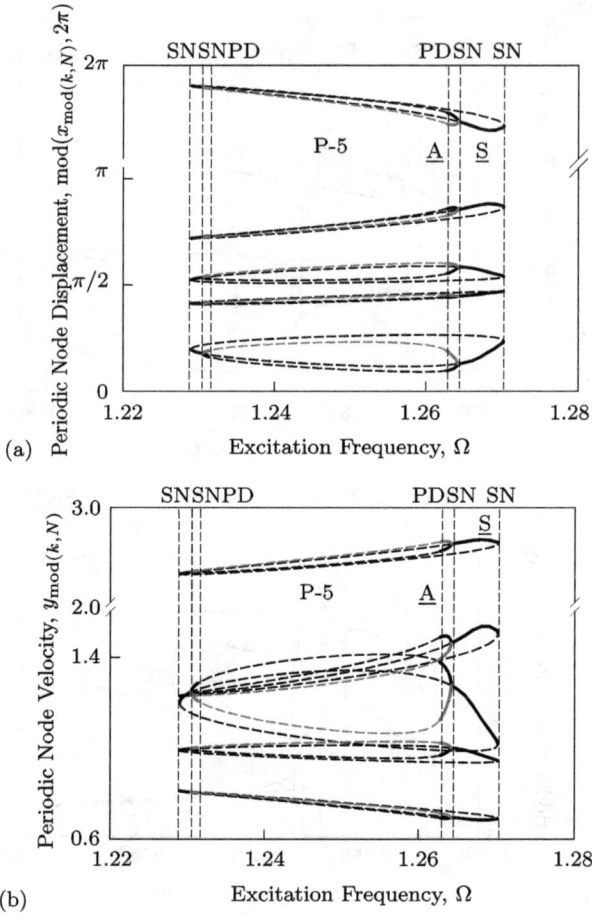

(a)

(b)

Fig. 6.6 Period-5 motions to chaos for $\Omega \in (1.22, 1.28)$: (a) displacement $\text{mod}(x_{\text{mod}(k,N)}, 2\pi)$; (b) velocity $y_{\text{mod}(k,N)}$ ($\alpha = 1.5, \delta = 0.75, Q_0 = 5.0$).

Table 6.6 Unstable symmetric period-3 Motions ($\alpha = 1.5, \delta = 0.75, Q_0 = 5.0$)

With Asymmetric Period-3 motions		
Ω	SN (L)	SN (R)
(0.763920, 0.801850)	0.763920	0.801850
(0.744290, 0.759830)	0.744290	0.759830

Note: L and R denote left and right. SN is for saddle-node bifurcation.

Table 6.7 Unstable asymmetric period-3 and unstable period-6 motions ($\alpha = 1.5, \delta = 0.75, Q_0 = 5.0$)

Unstable Period-3 Motions		
Ω	PD (L)	PD (R)
(0.745990, 0.756100)	0.745990	0.756100
Unstable Period-6 Motions		
Ω	PD (L)	PD (R)
(0.746450, 0.755716)	0.746450	0.755716

Note: L and R denote left and right. PD is for period-doubling bifurcation.

6.2.3 Period-5 motions to chaos

In this section, the analytical prediction of bifurcation trees will be presented for period-5 motions to chaos. The bifurcation trees of symmetric period-5 motions to chaos are presented in Fig.6.5 for $\Omega \in (1.229014, 1.270134)$ with two saddle-node bifurcation points at $\Omega \approx 1.229014$ and 1.270134 for catastrophes and motion appearance. For $\Omega \in (1.230530, 1.264160)$, the symmetric period-5 motions become unstable, and at the two saddle-node bifurcations of $\Omega \approx 1.230530$ and 1.264160, the asymmetric period-5 motions will appear. The asymmetric period-5 motion has two period-doubling bifurcations at $\Omega \approx 1.231350, 1.262742$ and the period-10 motion will appear. The asymmetric period-5 motion becomes unstable, and the period-10 motion exists for $\Omega \in (1.231350, 1.262742)$. Continuously, the bifurcation tree of symmetric period-5 motion to chaos can be obtained. The frequency ranges for unstable period-5 and period-10 motions and bifurcation points are listed in Tables 6.8 and 6.9.

Table 6.8 Excitation frequencies for unstable symmetric period-5 motions ($\alpha = 1.5, \delta = 0.75, Q_0 = 5.0$)

With Asymmetric Period-5 Motions		
Ω	SN (L)	SN (R)
(1.230530, 1.264160)	1.230530	1.264160
With Jumping Phenomena		
Ω	SN (L)	SN (R)
(1.229014, 1.270134)	1.229014	1.270134

Note: L and R denote left and right. SN is for saddle-node bifurcation.

Table 6.9 Excitation frequencies for unstable period-5 to period-10 motions ($\alpha = 1.5, \delta = 0.75, Q_0 = 5.0$)

Ω	PD (L)	PD (R)
(1.231350, 1.262742)	1.231350	1.262742

Note: L and R denote left and right. PD is for period-doubling bifurcation.

6.3 Frequency-amplitude characteristics

Consider the node points of period-m motions as $\mathbf{x}_k^{(m)} = (x_k^{(m)}, y_k^{(m)})^{\mathrm{T}}$ for $k = 0, 1, 2, \cdots, mN$. The approximate expression for period-m motions is determined by the Fourier series as

$$\mathbf{x}^{(m)}(t) \approx \mathbf{a}_0^{(m)} + \sum_{j=1}^{M} \mathbf{b}_{j/m} \cos\left(\frac{j}{m}\Omega t\right) + \mathbf{c}_{j/m} \sin\left(\frac{j}{m}\Omega t\right). \tag{6.22}$$

There are $2M + 1$ unknown vector coefficients of $\mathbf{a}_0^{(m)}, \mathbf{b}_{j/m}$, and $\mathbf{c}_{j/m}$. To determine such unknowns, the given nodes $\mathbf{x}_k^{(m)}$ ($k = 0, 1, 2, \cdots, mN$) with $mN + 1 \geqslant 2M + 1$ are used. In other words, $M \leqslant mN/2$. The node points $\mathbf{x}_k^{(m)}$ on the period-m motions can be expressed by the finite Fourier series as for $t_k \in [0, mT]$

$$\mathbf{x}^{(m)}(t_k) \equiv \mathbf{x}_k^{(m)} = \mathbf{a}_0^{(m)} + \sum_{j=1}^{mN/2} \mathbf{b}_{j/m} \cos\left(\frac{j}{m}\frac{2k\pi}{N}\right) + \mathbf{c}_{j/m} \sin\left(\frac{j}{m}\frac{2k\pi}{N}\right)$$
$$(k = 0, 1, \cdots, mN - 1) \tag{6.23}$$

where

$$T = \frac{2\pi}{\Omega} = N\Delta t; \quad \Omega t_k = \Omega k\Delta t = \frac{2k\pi}{N};$$

$$\mathbf{a}_0^{(m)} = \frac{1}{mN}\sum_{k=0}^{mN} \mathbf{x}_k^{(m)},$$

$$\left.\begin{array}{l} \mathbf{b}_{j/m} = \frac{2}{mN}\sum_{k=1}^{mN} \mathbf{x}_k^{(m)} \cos\left(k\frac{2j\pi}{mN}\right), \\[4mm] \mathbf{c}_{j/m} = \frac{2}{mN}\sum_{k=1}^{mN} \mathbf{x}_k^{(m)} \sin\left(k\frac{2j\pi}{mN}\right) \end{array}\right\} (j = 1, 2, \cdots, mN/2) \tag{6.24}$$

and

$$\mathbf{a}_0^{(m)} = (a_{01}^{(m)}, a_{02}^{(m)})^{\mathrm{T}}, \quad \mathbf{b}_{j/m} = (b_{j/m1}, b_{j/m2})^{\mathrm{T}}, \quad \mathbf{c}_{j/m} = (c_{j/m1}, c_{j/m2})^{\mathrm{T}}. \tag{6.25}$$

The harmonic amplitudes and phases for period-m motion are

$$\begin{aligned}
A_{j/m1} &= \sqrt{b_{j/m1}^2 + c_{j/m1}^2}, \quad \varphi_{j/m1} = \arctan \frac{c_{j/m1}}{b_{j/m1}}, \\
A_{j/m2} &= \sqrt{b_{j/m2}^2 + c_{j/m2}^2}, \quad \varphi_{j/m2} = \arctan \frac{c_{j/m2}}{b_{j/m2}}.
\end{aligned} \tag{6.26}$$

Thus the approximate expression for period-m motions in Eq.(6.22) becomes by

$$\mathbf{x}^{(m)}(t) \approx \mathbf{a}_0^{(m)} + \sum_{j=1}^{mN/2} \mathbf{b}_{j/m} \cos\left(\frac{j}{m}\Omega t\right) + \mathbf{c}_{j/m} \sin\left(\frac{j}{m}\Omega t\right). \tag{6.27}$$

The foregoing equation can be expressed as

$$\begin{bmatrix} x^{(m)}(t) \\ y^{(m)}(t) \end{bmatrix} \equiv \begin{bmatrix} x_1^{(m)}(t) \\ x_2^{(m)}(t) \end{bmatrix} \approx \begin{bmatrix} a_{01}^{(m)} \\ a_{02}^{(m)} \end{bmatrix} + \sum_{j=1}^{mN/2} \begin{bmatrix} A_{j/m1} \cos\left(\dfrac{j}{m}\Omega t - \varphi_{j/m1}\right) \\ A_{j/m2} \cos\left(\dfrac{j}{m}\Omega t - \varphi_{j/m2}\right) \end{bmatrix}. \tag{6.28}$$

For simplicity, only frequency-amplitudes for displacement $x^{(m)}(t)$ are presented. Similarly, frequency-amplitudes for velocity $y^{(m)}(t)$ can also be determined. Thus the displacement can be expressed as

$$x^{(m)}(t) \approx a_0^{(m)} + \sum_{j=1}^{mN/2} b_{j/m} \cos\left(\frac{j}{m}\Omega t\right) + c_{j/m} \sin\left(\frac{j}{m}\Omega t\right) \tag{6.29}$$

and

$$x^{(m)}(t) \approx a_0^{(m)} + \sum_{j=1}^{mN/2} A_{j/m} \cos\left(\frac{j}{m}\Omega t - \varphi_{j/m}\right), \tag{6.30}$$

where

$$A_{j/m} = \sqrt{b_{j/m}^2 + c_{j/m}^2}, \quad \varphi_{j/m} = \arctan \frac{c_{j/m}}{b_{j/m}}. \tag{6.31}$$

The bifurcation trees of period-m motions to chaos can be illustrated by the frequency-amplitude curves. In all plots of frequency-amplitude curves, the acronyms SN and PD are the saddle-node and period-doubling bifurcations, respectively. The unstable and stable solutions of period-m motions are represented by dashed and solid curves, respectively.

6.3.1 Period-1 to period-4 motions

The bifurcation trees of period-1 motion to chaos are presented is Fig.6.7 through the period-1 to period-4 motions, as presented in Fig.6.7. The given parameters are $\alpha = 1.5$, $\delta = 0.75$, $Q_0 = 5.0$. The constant terms $a_0^{(m)}$ ($m = 1, 2, 4$) are presented in Fig.6.7(i) for the solution center at $\mathrm{mod}(a_0^{(m)}, l\pi) = 0$. The bifurcation trees are clearly observed. For the solution center on the left side of $\mathrm{mod}(a_0^{(m)}, l\pi) = 0$, $\mathrm{mod}(l\pi - a_0^{(m)L}, 2\pi) = \mathrm{mod}(a_0^{(m)R} - l\pi, 2\pi)$ ($l = 0, 1, 2, \cdots$). For the symmetric period-m motion, $\mathrm{mod}(a_0^{(m)}, l\pi) = 0$ ($l = 0, 1, 2, \cdots$), labeled by "S". However, for asymmetric period-m motion, $\mathrm{mod}(a_0^{(m)}, l\pi) \neq 0$, labeled by "A". For the symmetric period-1 motions to asymmetric period-1 motions, the saddle-node bifurcation occur, i.e. at $\Omega \approx 0.3112, 0.3485, 0.3634, 0.4118, 0.4358, 0.5015$, etc. For such saddle-node bifurcations, the asymmetric periodic motions appear, and the symmetric motions change from stable to unstable solutions or from unstable to stable solutions. As $\Omega \to 0$, jumping phenomena of the symmetric motions occurs with the saddle-node bifurcations at $\Omega \approx 0.1643, 0.1679, 0.1789, 0.1827, 0.1956, 0.1997$, etc. The symmetric period-1 motion is only from the stable to unstable solution or from the unstable to stable solution. When the asymmetric period-1 motion experiences a period-doubling bifurcations, the period-2 motions will appear and the asymmetric period-1 motions are from the stable to unstable solution. The frequencies of $\Omega \approx 0.7336, 0.8517, 1.1859, 1.3809$, etc are not only for the period-doubling bifurcations of the asymmetric period-1 motions but also for the saddle-node bifurcations of the period-2 motion appearance. The period-2 motions also have jumping phenomenon, with saddle-node bifurcations at $\Omega \approx 0.7776, 0.81512, 1.2951, 1.2983$, etc. When the period-2 motion possesses a period-doubling bifurcation, the period-4 motion appears, and the period-2 motion is from the stable to unstable solution. The frequencies of $\Omega \approx 0.7359, 0.81508, 0.8448, 0.7777, 1.1983, 1.2963, 1.2986, 1.3520$, etc are for the period-doubling bifurcations of period-2 motions and for the saddle-node bifurcations of the period-4 motion appearance. The period-4 motions jumping points with saddle-node bifurcations are at $\Omega \approx 1.2857, 1.2917, 1.3077, 1.3084$. The frequencies of $\Omega \approx 1.2014, 1.2916, 1.2858, 1.2954, 1.3009, 1.3076, 1.3082, 1.3446$ are for the period-doubling bifurcations of period-4 motions and for the saddle-node bifurcation of the period-8 motion appearance. All period-2 and period-4 motions are on the branches of asymmetric period-1 motions, and the centers of the periodic motions are on the right side of the y-axis. In Fig.6.7(ii), the harmonic amplitude $A_{1/4}$ is presented. For period-1 and period-2 motions, $A_{1/4} = 0$. The bifurcation points are clearly observed, and the quantity level of the harmonic amplitude $A_{1/4}$ for period-4 motion is $A_{1/4} \sim 0.5$. In Fig.6.7(iii), the harmonic amplitudes $A_{1/2}$ for period-4 and period-2 motions are presented. For the first branch, only the period-2 motion is presented because the range of stable period-4 motion is very small and more discrete nodes are needed to obtain such a period-4 motion. The quantity level of the harmonic amplitude is $A_{1/2} \sim 2.0$. In Fig.6.7(iv), harmonic amplitude $A_{3/4}$ is presented, which is similar to the harmonic amplitude $A_{1/4}$. The quantity level of such a harmonic amplitude is $A_{3/4} \sim 0.2$. The other harmonic amplitudes $A_{k/4}$ ($k = 4l + 1, 4l + 3,\ l = 1, 2 \cdots$) will not be presented herein for reduction of abundant illustrations. In Fig.6.7(v), the primary harmonic amplitude A_1 versus excitation frequency Ω is presented for the period-1 to period-4 motion. The bifurcation trees are clearly observed. The entire skeleton of frequency-amplitude for the symmetric period-1 motion is presented, and the asymmetric period-1 motions and the corresponding period-2 and period-4 motions are attached to the symmetric period-1 motion. The maximum quantity level of the primary amplitude is $A_1 \sim 130.0$, which varies with the excitation frequency. The bifurcation points are presented as before. In Fig.6.7(vi), the harmonic amplitude $A_{3/2}$ is presented and the bifurcation trees are similar to the harmonic amplitude $A_{1/2}$. The quantity level of the harmonic amplitude $A_{3/2}$ is $A_{3/2} \sim 0.4$. To reduce abundant illustrations, $A_{k/2}$ ($k = 2l + 1,\ l = 2, 3, \cdots$) will not be presented any more. In Figs.6.7(vii) and (ix), harmonic amplitudes A_2 and A_4 are presented which are similar to constant term $a_0^{(m)}$. The bifurcation trees have the similar structures for the different harmonic amplitudes but the corresponding quantity levels of harmonic amplitudes are different. The overall maximum quantity levels are $A_2 \sim 0.3$ and $A_4 \sim 0.3$, respectively. The quantity levels drop to 10^{-2} for $\Omega \in (1.0, 2.0)$ as the order increases, i.e., $A_2 \sim 0.2$, and $A_4 \sim 10^{-2}$. In Fig.6.7(viii), the harmonic amplitudes A_3 are presented, which are similar to the primary harmonic amplitude A_1. For $\Omega \in (1.0, 2.0)$, the quantity level of harmonic amplitude is $A_3 \sim 0.2$. Different are the bifurcation trees for the different harmonic amplitudes, and the corresponding quantity levels of harmonic amplitudes are different. To avoid abundant illustrations, other harmonic amplitudes will not be presented. Herein, harmonic amplitude A_{20} is presented in Fig.6.7(x), which is similar to the harmonic amplitude A_2 but the corresponding quality level is $A_{20} \sim 3 \times 10^{-12}$ for $\Omega \in (1.0, 2.0)$. For low frequency $\Omega \in (0.0, 1.0)$, the quantity level of harmonic amplitude A_{20} decay exponentially with excitation frequency. Further the harmonic amplitudes $A_{k/4}$ ($k = 81, 82, 83$), similar to $A_{k/4}$ ($k = 1, 2, 3$), are presented in Figs.6.7(xi)—(xiii), and the quantity levels of $A_{k/4}$ ($k = 81, 82, 83$) are about 2.5×10^{-12}, 3×10^{-7} and 1.5×10^{-12}, which are much lower than the quantity levels of $A_{k/4}$ ($k = 1, 2, 3$). In Fig.6.7(xiv), harmonic amplitude A_{21} decay also exponentially with excitation frequency Ω. That is, A_{21} changes from 10^0 to 10^{-14} for $\Omega \in (0.0, 1.6)$. In Figs.6.7(xv) and (xvi), harmonic amplitudes of A_{60} and A_{61} are presented. The overall maximum quantity levels for A_{60} and A_{61} are $A_{60} < 0.01$ and $A_{61} < 0.024$. However for $\Omega \in (0.6, 2.0)$, $A_{60} \sim 10^{-14}$ and $A_{61} \sim 10^{-14}$. For $\Omega \in (0.0, 0.6)$, the quantity level of harmonic amplitude A_{60} and A_{61} decay exponentially with excitation frequency Ω. From the above discussion, the periodic motion, for $\Omega > 1.0$, one can use about 80 harmonic terms to approximate period-1, period-2 and period-4 mo-

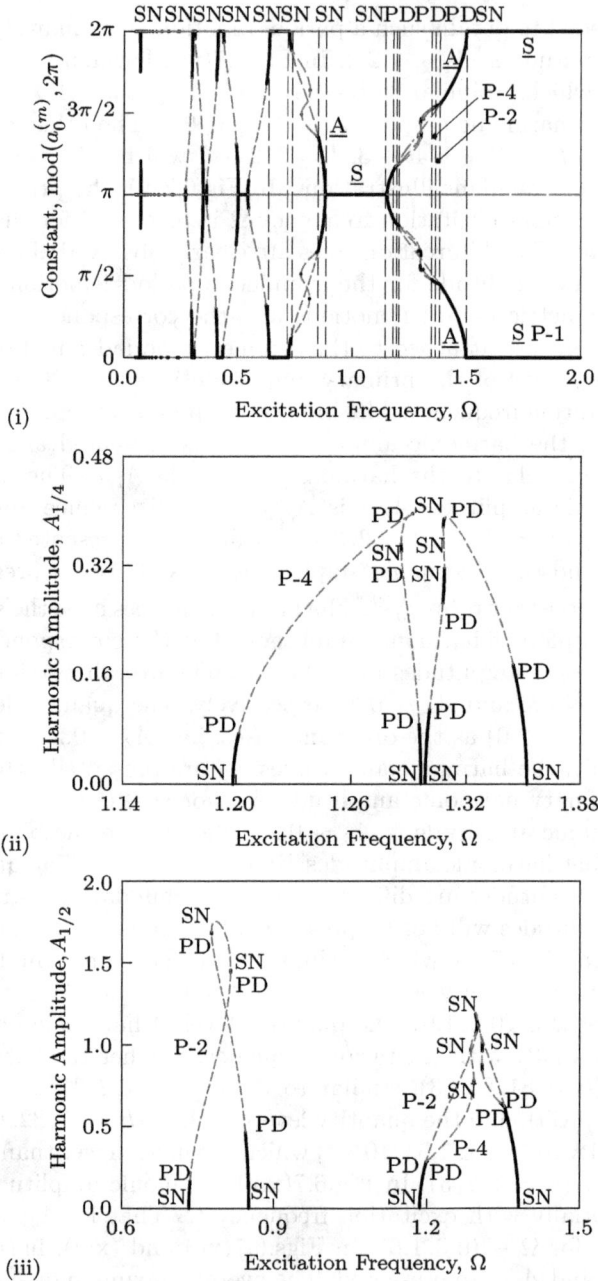

Fig. 6.7 Frequency-amplitude characteristics for bifurcation trees of period-1 to period-4 motions: (i) $a_0^{(m)}$ ($m = 1, 2, 4$). (ii)—(xvi) $A_{k/m}$ ($m = 4, k = 1, 2, 3, 4; 6, 8, 12, 16; 80, 81, \cdots, 84; 240, 244$); Parameters: ($\alpha = 1.5, \delta = 0.75, Q_0 = 5.0$).

(iv)

(v)

(vi)

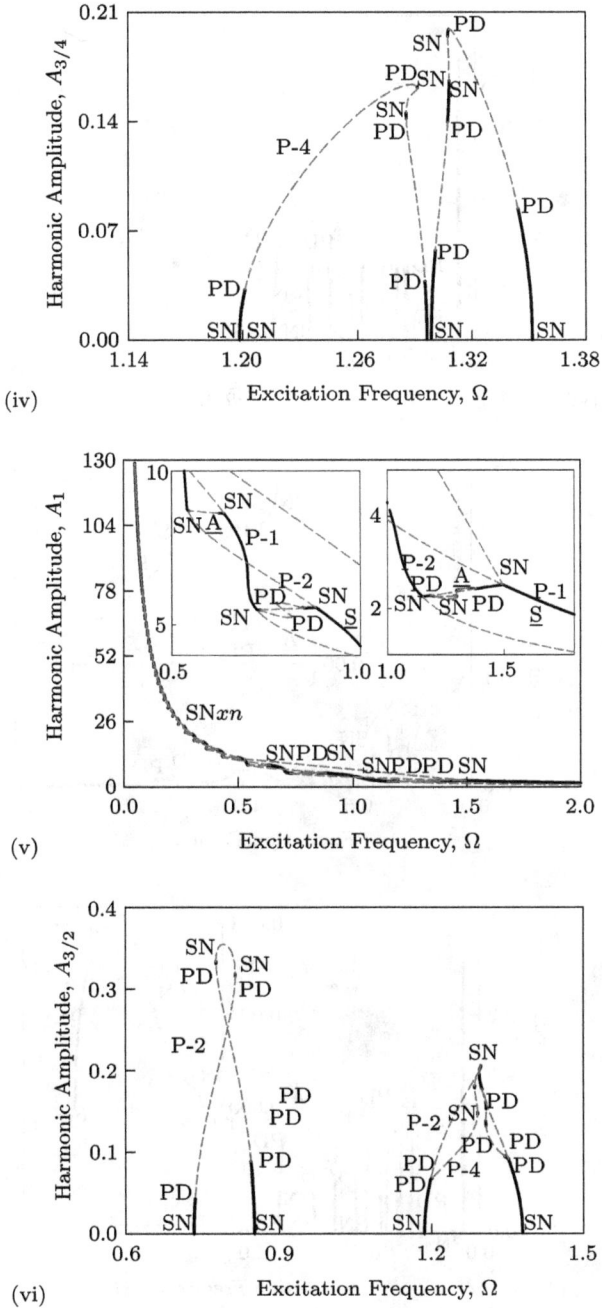

Fig. 6.7 Continued.

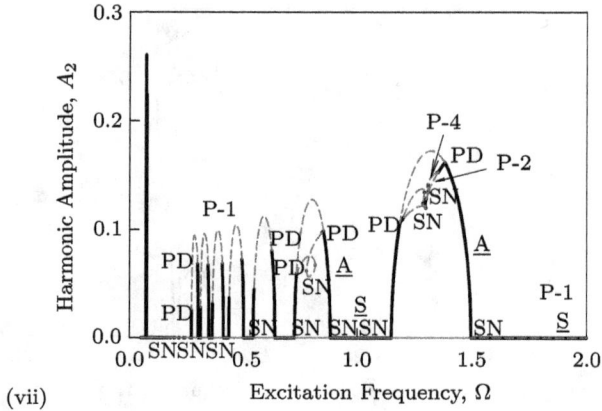

(vii)

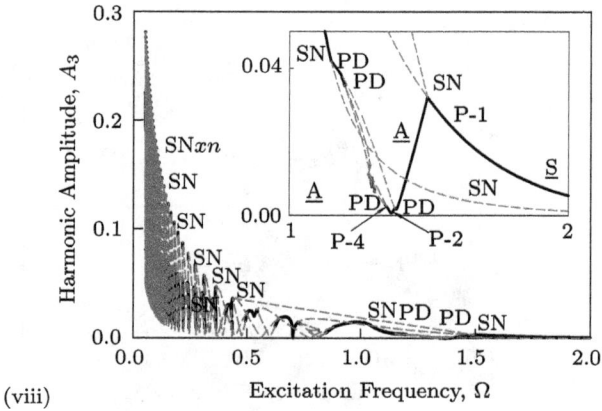

(viii)

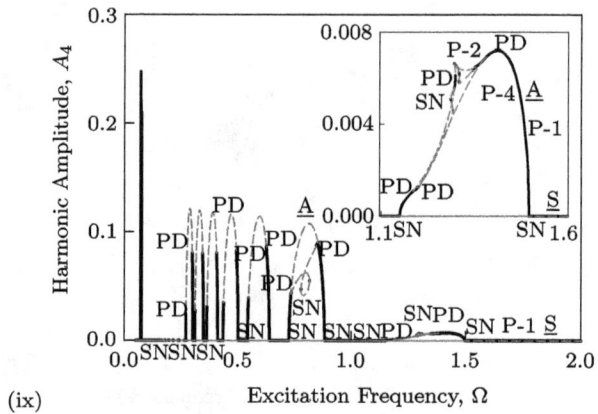

(ix)

Fig. 6.7 Continued.

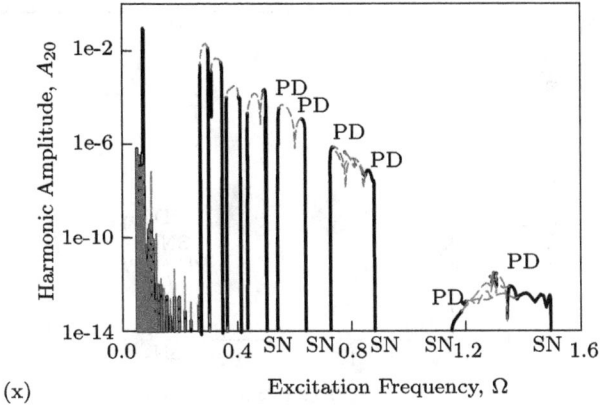

(x)

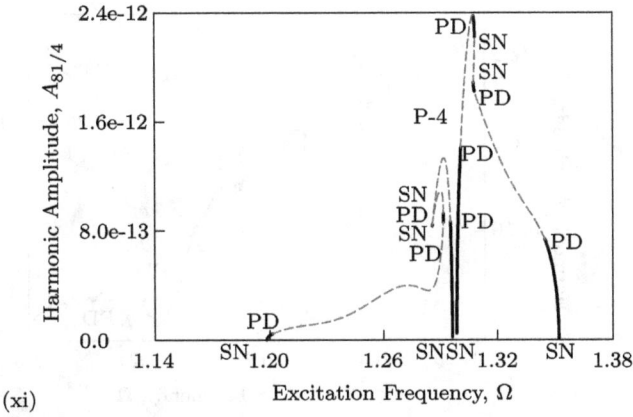

(xi)

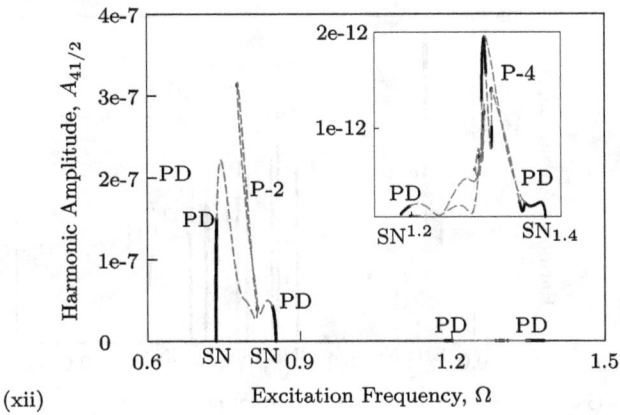

(xii)

Fig. 6.7 Continued.

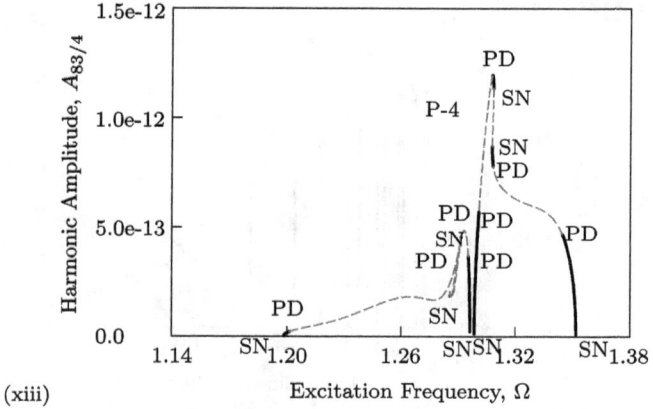

(xiii)

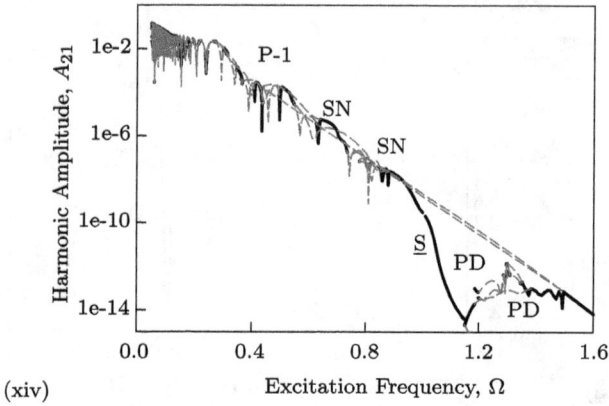

(xiv)

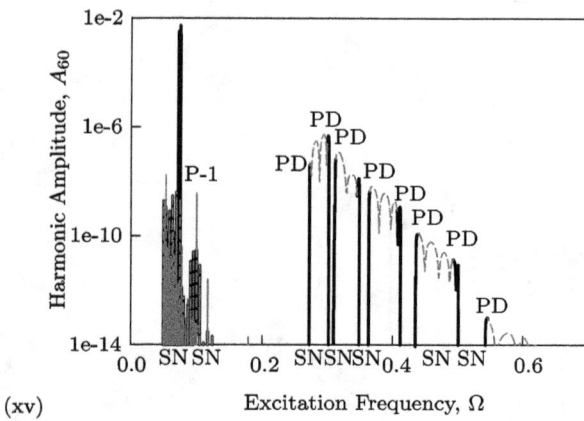

(xv)

Fig. 6.7 Continued.

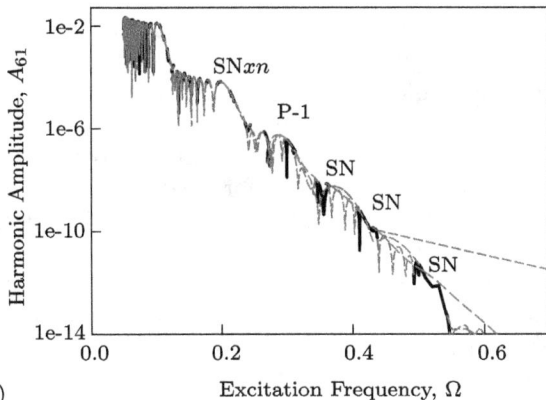

(xvi)

Fig. 6.7 Continued.

tions. For $\Omega < 1.0$ but not close to zero, one can use 250 harmonic terms to approximate period-1, period-2, and period-4 motions. For $\Omega \approx 0.0$, infinite harmonic terms should be adopted to approximate the periodic motions. For harmonic phases, $\varphi_{k/2^l}^R = \mathrm{mod}\,(\varphi_{k/2^l}^B + (k(1+2r)/2^l+1)\pi,\, 2\pi) \in [0, 2\pi)$ with $t_0 = rT$ $(r \in \{0, 1, \cdots, 2^l - 1\})$.

6.3.2 Period-3 to period-6 motions

The bifurcation trees of period-3 motion to chaos are presented in Fig.6.8 through period-3 to period-6 motions. The given parameters are also $\alpha = 1.5$, $\delta = 0.75$, $Q_0 = 5.0$. The constant terms $a_0^{(m)}$ $(m = 3, 6)$ are presented in Fig.6.8(i) for the solution center at $\mathrm{mod}(a_0^{(m)}, l\pi) = 0 (l = 0, 1, 2, \cdots)$. The bifurcation trees are clearly observed. For the solution center on the left side of $\mathrm{mod}(a_0^{(m)}, l\pi) = 0$, $\mathrm{mod}(l\pi - a_0^{(m)}, 2\pi) = \mathrm{mod}(a_0^{(m)R} - l\pi, 2\pi), (l = 0, 1, 2, \ldots)$. For the symmetric period-m motion, $\mathrm{mod}(a_0^{(m)}, l\pi) = 0$, labeled by "S". However, for the asymmetric period-m motion, $\mathrm{mod}(a_0^{(m)}, l\pi) \neq 0$, labeled by "A". For the symmetric period-3 motions, the saddle-node bifurcations occur at $\Omega \approx 0.763920, 0.801850, 0.744290, 0.759830$. For such saddle-node bifurcations, the asymmetric periodic motions appear, and the symmetric period-3 motions change from stable to unstable solutions or from unstable to stable solutions. When the asymmetric period-3 motion experiences a period-doubling bifurcations, the period-6 motions will appear and the asymmetric period-3 motion is from the stable to unstable solution. The frequencies of $\Omega \approx 0.745990, 0.756100$ are not only for the period-doubling bifurcations of the asymmetric period-3 motions but also for the saddle-node bifurcations of the period-6 motion appearance. When the period-6 motion possesses a period-doubling bifurcation, the period-12 motion appears and the period-6 motion is from the stable to unstable solution. The frequencies of $\Omega \approx 0.746450, 0.755716$ are for the period-doubling bifurcations of period-6 motions and for the saddle-node bifurcations of the period-12 motions appearance. In Fig.6.8(ii), the harmonic amplitudes $A_{1/6}$ are presented. For period-3 motions, $A_{1/6} = 0$. The bifurcation points are clearly observed, and the quantity levels of the harmonic amplitudes for period-6 motion are $A_{1/6} \sim 0.07$. In Fig.6.8(iii), the harmonic amplitudes $A_{1/3}$ for period-6 and period-3 motions are presented. For the branch of the right side ($\Omega \in (0.763920, 0.801850)$), only the symmetric period-3 motions are presented because the range of stable asymmetric period-3 and period-6 motions is very small. The quantity of the harmonic amplitudes $A_{1/3}$ are $A_{1/3} \in (0.9, 3.0)$. In Fig.6.8(iv), the harmonic amplitude $A_{1/2}$ is presented for period-6 motion only, which is similar to the harmonic amplitude $A_{1/6}$. The quantity level of such a harmonic amplitude is $A_{1/2} \sim 0.09$. In Fig.6.8(v), the harmonic amplitudes $A_{2/3}$ for period-3 and period-6 motions are presented, which is similar to $a_0^{(m)}$ ($m = 3, 6$). For the symmetric period-3 motions, $A_{2/3} = 0$. For asymmetric period-3 and period-6 motions, the harmonic amplitudes of $A_{2/3}$ possess the quantity level of $A_{2/3} \sim 0.4$. In Fig.6.8(vi), the harmonic amplitude $A_{5/6}$ is presented for the period-6 motion only, which is similar to the harmonic amplitude $A_{1/6}$. The quantity level of such a harmonic amplitude is $A_{5/6} \sim 0.08$. The other harmonic amplitudes $A_{k/6}$ ($\mathrm{mod}(k, 6) \neq 0$, $k = 6, 7, \cdots$) will not be presented herein for less abundant illustrations. In Fig.6.8(vii), the primary harmonic amplitudes A_1 varing with excitation frequency Ω are presented for the period-3 to period-6 motion. The bifurcation trees are clearly observed. The entire skeleton of frequency-amplitudes for the symmetric period-3 motion is presented, and the asymmetric period-3 motion and the period-6 motion are attached to the symmetric period-3 motion. The quantity level of the primary amplitude is in the range of $A_1 \in (5.40, 6.09)$, which varies with excitation frequency. The bifurcation points are listed in Tables 6.6 and 6.7. In Fig. 6.8(viii), the harmonic amplitude A_2 is presented. The bifurcation trees are similar to the harmonic amplitude $A_{2/3}$. The quantity level of A_2 is $A_2 \sim 0.04$. In Figs.6.8(ix) and (xi), the harmonic amplitudes A_3 and A_5 are presented, which are similar to the primary harmonic amplitude A_1. The bifurcation trees for such harmonic amplitudes have the similar structures for the different harmonic amplitudes, but the corresponding quantity levels of harmonic amplitudes are different. The overall maximum quantities for such two harmonic amplitudes are $A_3 \in (0.02, 0.10)$, and $A_5 \in (0.01, 0.06)$. In Fig.6.8(x), the harmonic amplitude A_4 is presented, similar to the harmonic amplitude A_2. The quality level of such a harmonic amplitude is $A_4 \sim 0.016$. To look into the higher order harmonic effects, the harmonic term A_{20} is presented in Fig.6.8(xii), which is similar to the harmonic amplitude of A_2 but the corresponding quality level is $A_{20} \sim 3 \times 10^{-7}$. Further the harmonic amplitudes of $A_{k/6}$ ($k = 121, 122, \cdots, 126$) similar to $A_{k/6}$ ($k = 1, 2, \cdots, 6$) are presented in Figs.6.8(xiii)—(xviii), and the quantity

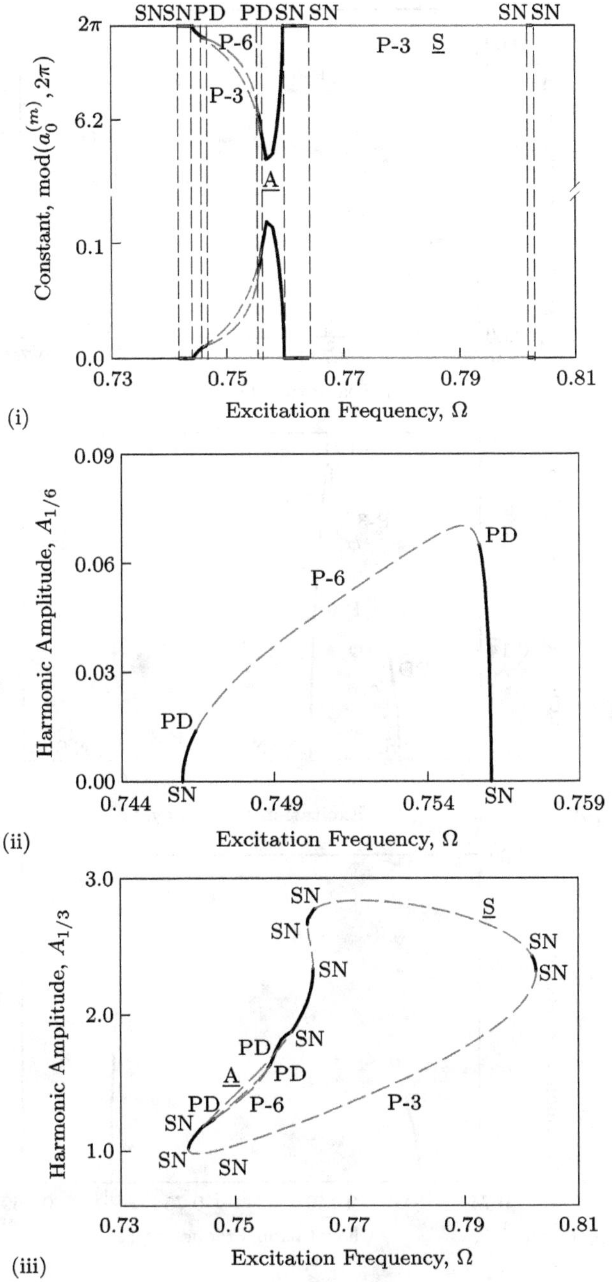

Fig. 6.8 Frequency-amplitude characteristics for bifurcation trees of period-3 to period-6 motions: (i) $a_0^{(m)}$ ($m = 3, 6$). (ii)—(xviii) $A_{k/m}$ ($m = 6$, $k = 1, 2, \cdots, 6$; $12, 18, \cdots, 30; 120, 121, \cdots, 126$); Parameters: ($\alpha = 1.5, \delta = 0.75, Q_0 = 5.0$).

(iv)

(v)

(vi)

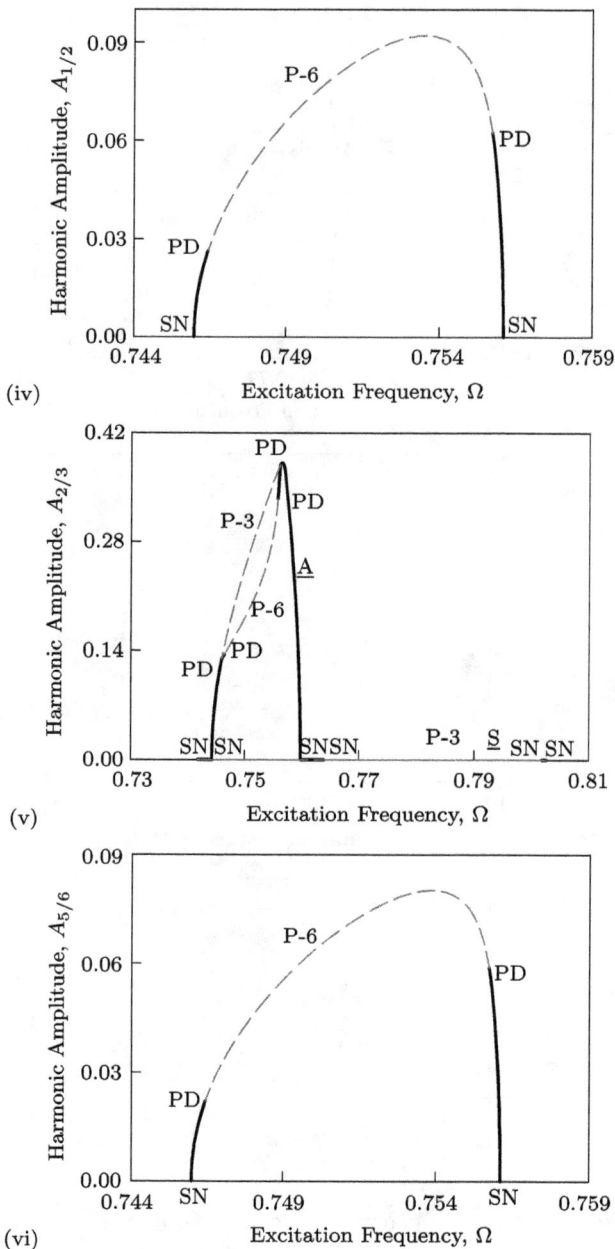

Fig. 6.8 Continued.

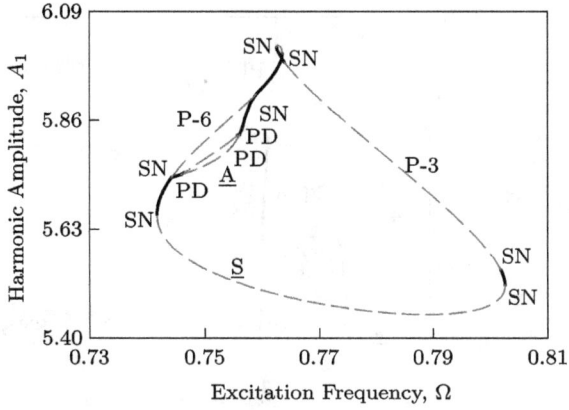

(vii)

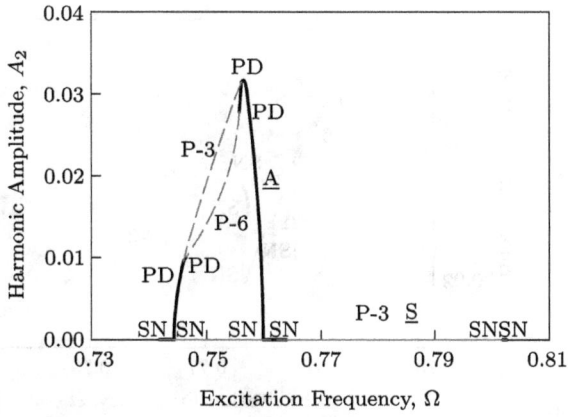

(viii)

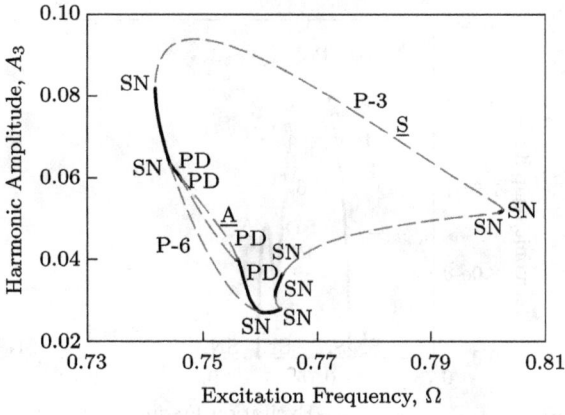

(ix)

Fig. 6.8 Continued.

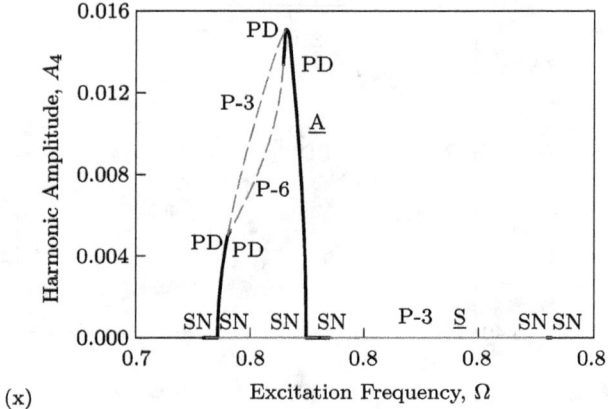

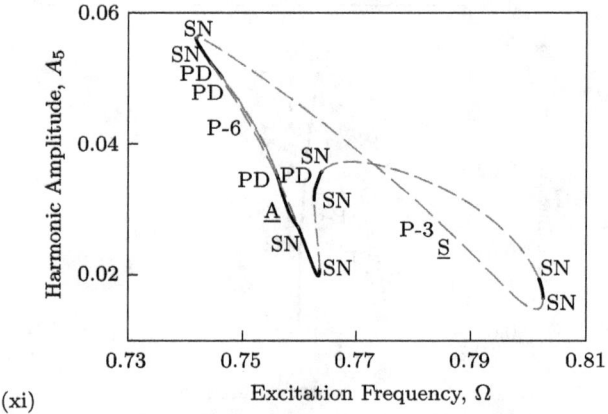

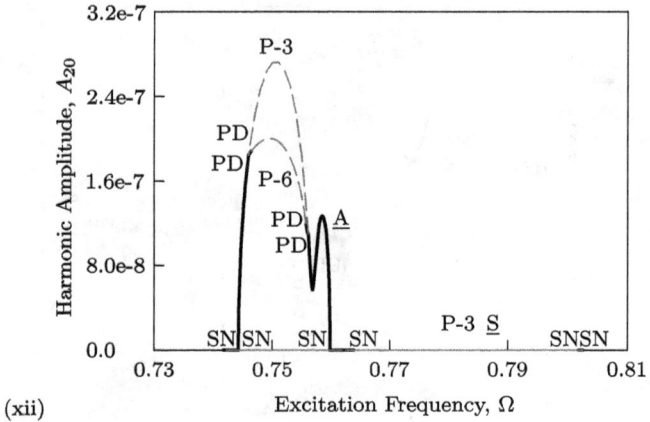

Fig. 6.8 Continued.

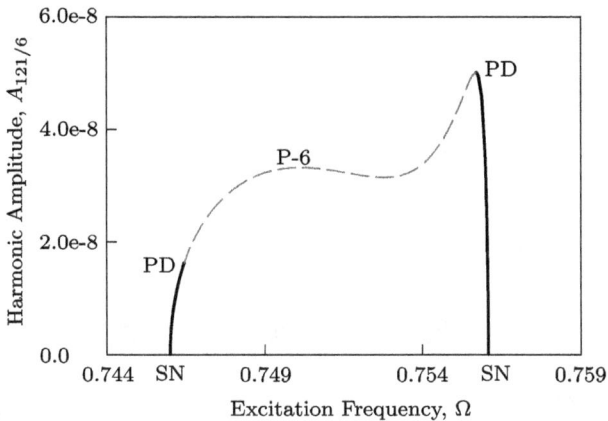

(xiii)

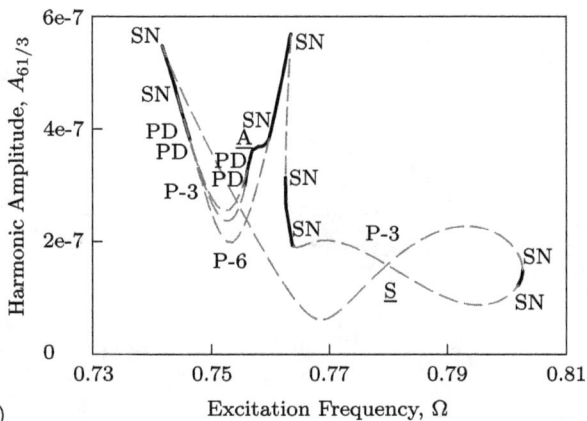

(xiv)

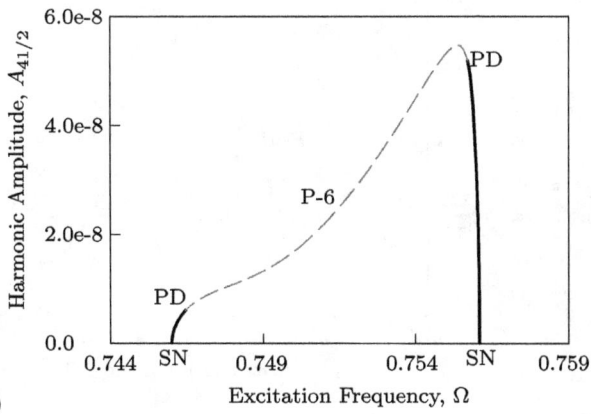

(xv)

Fig. 6.8 Continued.

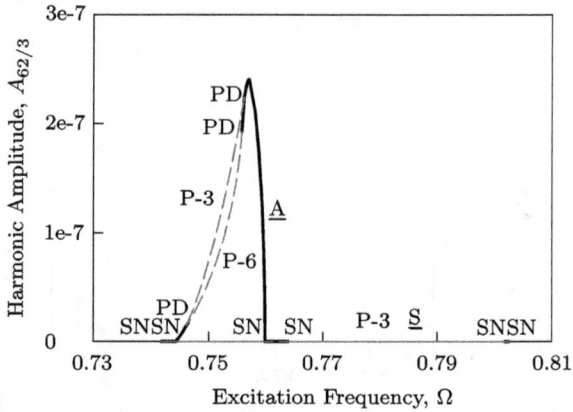

(xvi)

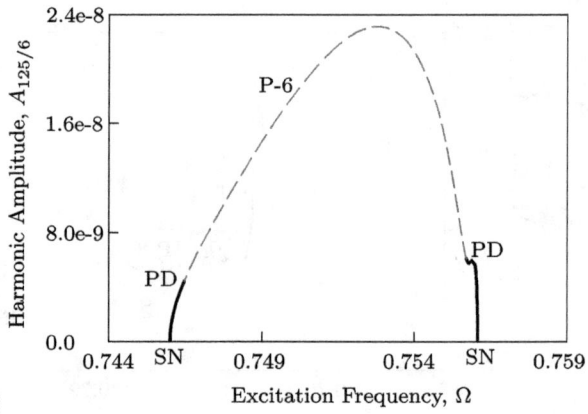

(xvii)

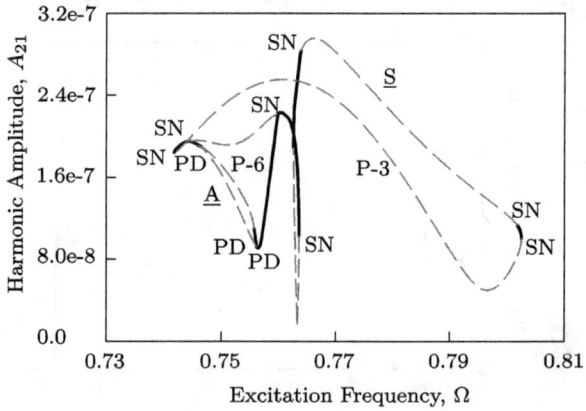

(xviii)

Fig. 6.8 Continued.

levels of $A_{k/6}$ $(k = 121, 122, \cdots, 126)$ are about 10^{-7}, which is much lower than the quantity levels of $A_{k/6}$ $(k = 1, 2, \cdots, 6)$. The harmonic phases have a following relation of $\varphi_{k/2^l m}^R = \mod(\varphi_{k/2^l m}^B + (k(m+2r)/2^l m + 1)\pi, 2\pi) \in [0, 2\pi)$ with $t_0 = rT$ for $r \in \{0, 1, \cdots, 2^l m - 1\}$ and $m = 3$.

6.3.3 Symmetric to asymmetric period-5 motions

The bifurcation trees of period-5 motion to chaos will be presented through the symmetric and asymmetric period-5 motions, as presented in Fig.6.9. The given parameters still are $\alpha = 1.5$, $\delta = 0.75$, $Q_0 = 5.0$. The constant term $a_0^{(m)}$ $(m = 5)$ is presented in Fig.6.9(i) for the solution center at $\mod(a_0^{(m)}, l\pi) = 0 (l = 0, 1, 2, \cdots)$. For the solution center on the left side of $\mod(a_0^{(m)}, l\pi) = 0$, $\mod(l\pi - a_0^{(m)L}, 2\pi) = \mod(a_0^{(m)R} - l\pi, 2\pi)$, $(l = 0, 1, 2, \cdots)$. For the symmetric period-m motion, $\mod(a_0^{(m)}, l\pi) = 0 (l = 0, 1, 2, \cdots)$, labeled by "S". However, for asymmetric period-m motion, $\mod(a_0^{(m)}, l\pi) \neq 0$, labeled by "A". For the symmetric period-5 motion to an asymmetric period-5 motion, the saddle-node bifurcation occur at $\Omega \approx 1.230530$ and 1.264160. For such saddle-node bifurcation, the asymmetric period-5 motions appear, and the symmetric period-5 motions change from stable to unstable solutions or from unstable to stable solutions. When the asymmetric period-5 motion experiences a period-doubling bifurcations, the period-10 motions will appear and the asymmetric period-5 motion is from the stable to unstable solution. The frequencies of $\Omega \approx 1.231350$, 1.262742 are not only for the period-doubling bifurcations of the asymmetric period-5 motions but also for the saddle-node bifurcations of the period-10 motions. For the branch of the right side ($\Omega \in (1.231350, 1.262742)$), only the period-5 motions are presented because the ranges of the stable asymmetric period-10 motion are very tiny. In Fig.6.9(ii), the harmonic amplitude $A_{1/5}$ for the period-5 motion is presented. The bifurcation points are clearly observed, and the quantity level of the harmonic amplitude for period-5 motion is $A_{1/5} \in (1.0, 1.7)$. In Fig.6.9(iii), the harmonic amplitude $A_{2/5}$ for period-6 motions is presented, which is similar to the constant $a_0^{(5)}$. The quantity level of the harmonic amplitude is $A_{2/5} \sim 0.15$. In Fig.6.9(iv), the harmonic amplitude $A_{3/5}$ is presented for period-5 motion, which is similar to the harmonic amplitude $A_{1/5}$. The quantity level of such harmonic amplitude is $A_{3/5} \in (0.4, 0.8)$. In Fig.6.9(v), the harmonic amplitude $A_{4/5}$ similar to $A_{2/5}$ period-5 motion is presented. For symmetric period-5 motion, $A_{4/5} = 0$. For asymmetric period-5 motion, the harmonic amplitude $A_{4/5}$ possesses the quantity level of $A_{4/5} \sim 0.06$. The other harmonic amplitudes $A_{k/5}$ $(\mod(k, 5) \neq 0, k = 5, 6, \cdots)$ will not be presented herein for less abundant illustrations. In Fig.6.9(vi), the primary harmonic amplitude A_1 versus excitation frequency Ω is presented for the period-5 motion. The bifurcation trees are clearly observed. The skeleton of frequency-amplitude for the symmetric period-5 motion is presented. The quantity level of the primary amplitude is $A_1 \in (2.19, 2.31)$, which varies with the excitation frequency. The bifurcation points are presented in Tables 6.8 and 6.9. In Fig.6.9(vii), the harmonic amplitude A_2 is presented. The bifurcation trees are similar to the harmonic amplitude $A_{2/5}$. The quantity level of A_2 is $A_2 \sim 0.012$. In Figs.6.9(viii), (x), (xii) and (xiv), the harmonic amplitudes $A_{k/5}$ ($k = 15, 25, 35, 45$) are presented which are similar to primary harmonic amplitude A_1. The bifurcation trees for such harmonic amplitudes have the similar structures for the different harmonic amplitudes but the corresponding quantity levels of harmonic amplitudes are different. The overall maximum quantities for such two harmonic amplitudes are $A_3 \sim 2 \times 10^{-2}$, $A_5 \sim 10^{-3}$, $A_7 \sim 10^{-4}$, and $A_7 \sim 5 \times 10^{-6}$, respectively. In Figs.6.9(ix), (xi) and (xiii), the harmonic amplitude $A_{k/5}$ ($k = 20, 30, 40$) similar to primary harmonic amplitude A_2 are presented. The quality levels of such harmonic amplitudes are $A_4 \sim 6 \times 10^{-4}$, $A_6 \sim 6 \times 10^{-5}$, and $A_8 \sim 2.5 \times 10^{-6}$. Similarly, other harmonic amplitudes can be similarly presented.

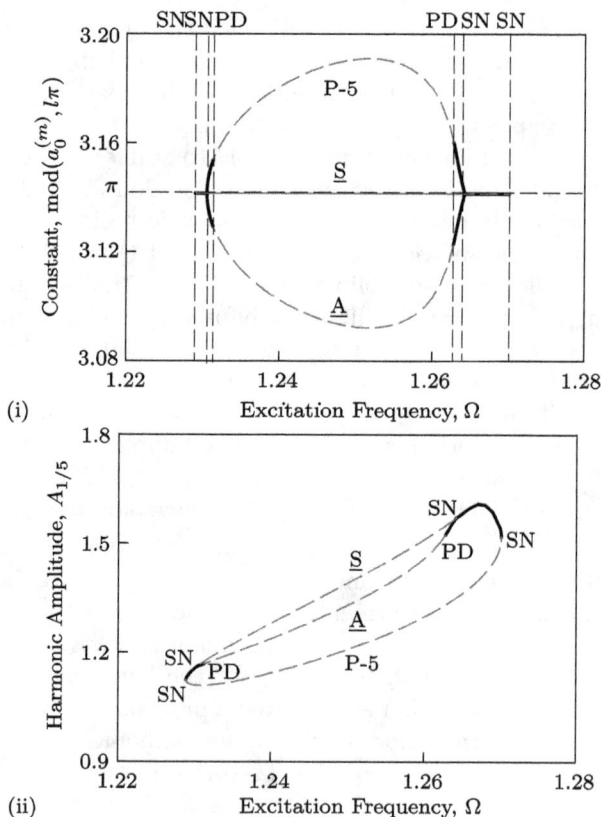

Fig. 6.9 Frequency-amplitude characteristics for bifurcation trees of period-5 motions: (i) $a_0^{(m)}$ ($m = 5$). (ii)—(xiv) $A_{k/m}$ ($m = 5, k = 1, 2, \cdots, 5; 10, 15, \cdots, 45$); Parameters: ($\alpha = 1.5, \delta = 0.75, Q_0 = 5.0$).

(iii)

(iv)

(v)

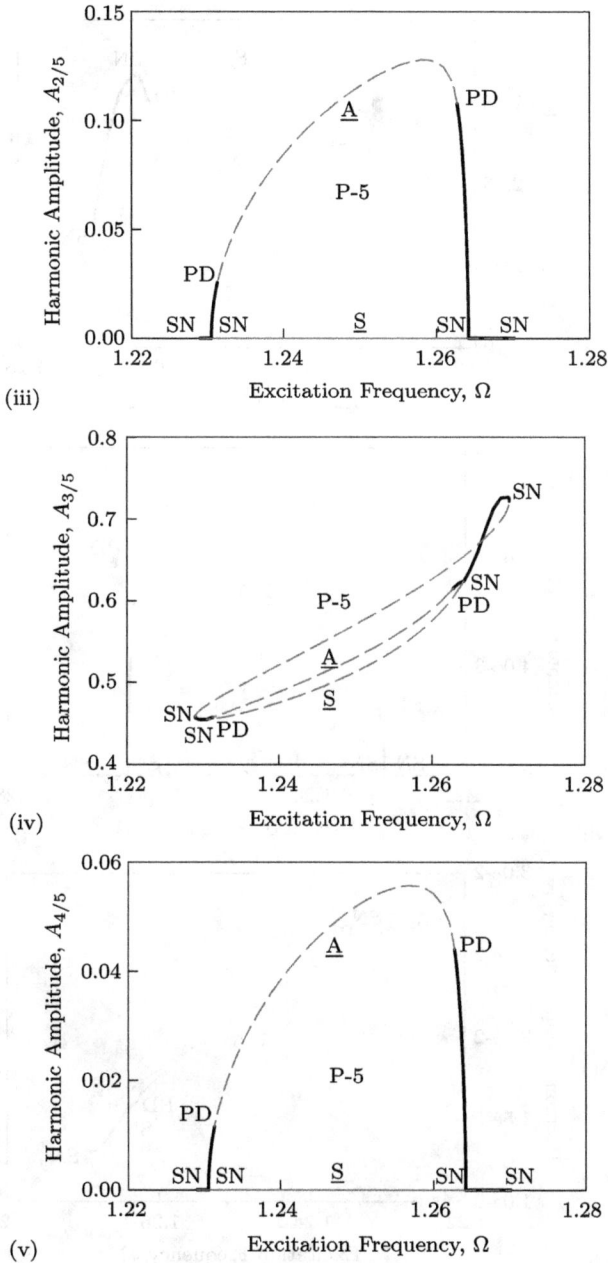

Fig. 6.9 Continued.

(vi)

(vii)

(viii)

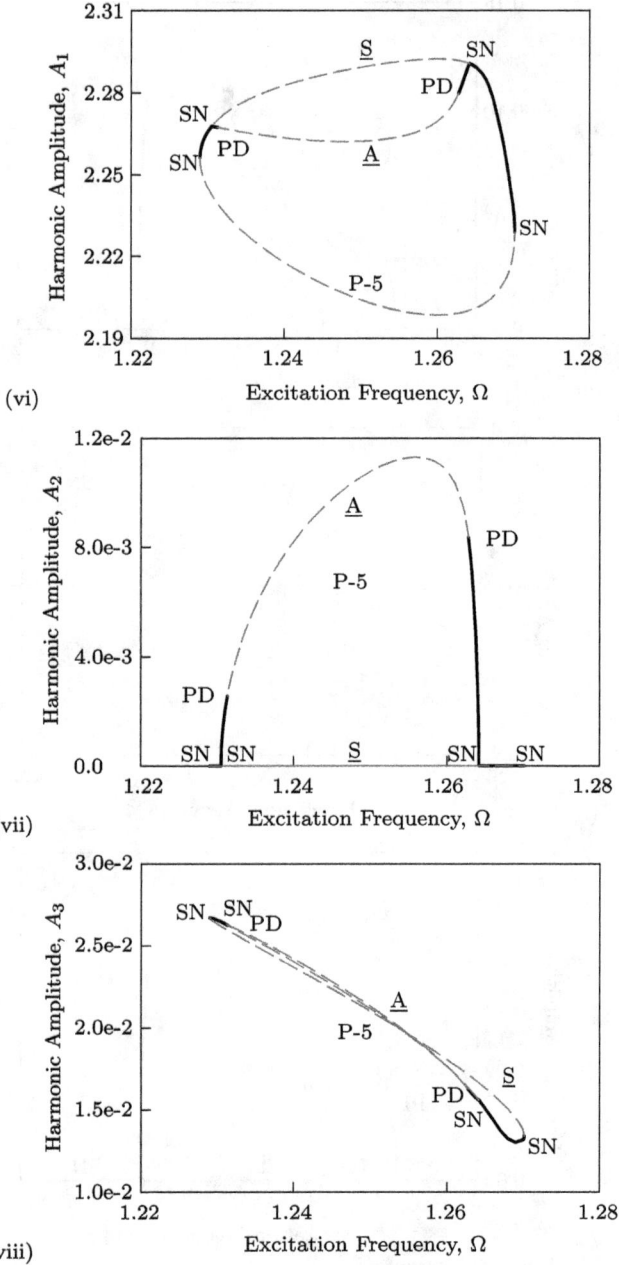

Fig. 6.9 Continued.

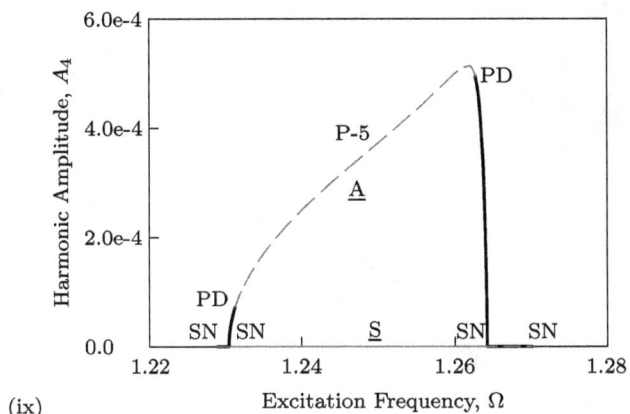

(ix)

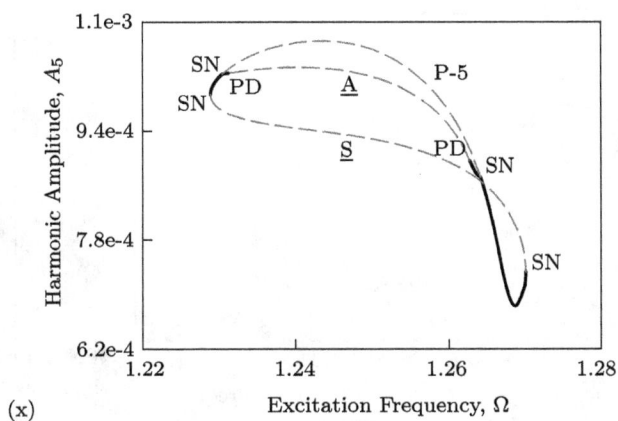

(x)

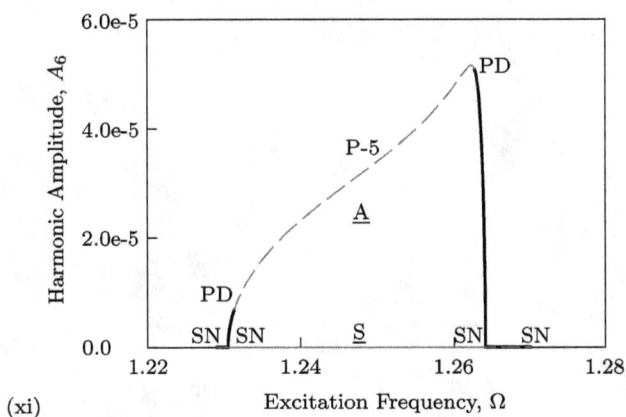

(xi)

Fig. 6.9 Continued.

(xii)

(xiii)

(xiv)

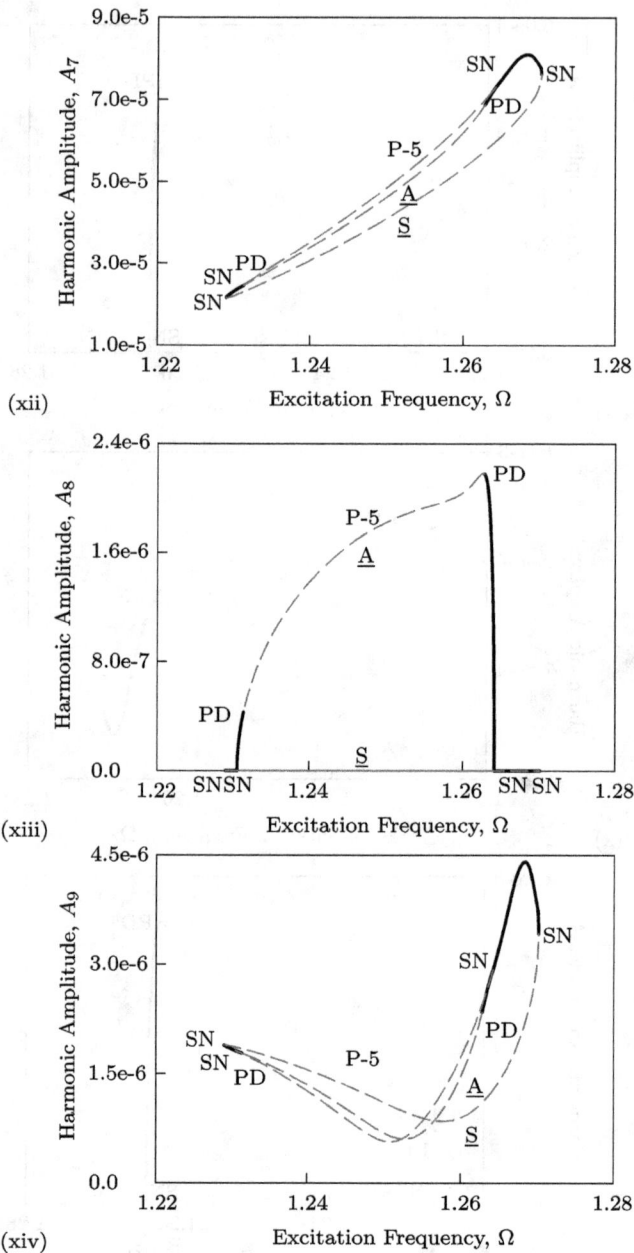

Fig. 6.9 Continued.

6.4 Bifurcation trees varying with excitation amplitude

If a period-m motion is determined by $x_k = x_{k+mN}$ with $y_k = y_{k+mN}$, then such a period-m motion has a center point and such a periodic motion can be called the non-travelable period-m motion. However, if a period-m motion is determined by $\mathrm{mod}(x_k, 2\pi) = \mathrm{mod}(x_{k+mN}, 2\pi)$ with $y_k = y_{k+mN}$ but $x_k \neq x_{k+mN}$, such a periodic motion does not have a center. Thus, such a period-m motion can be called the travelable period-m motion. For a better description of periodic motions in pendulum, the non-travelable and travelable periodic motions are defined as follows.

Definition 6.1 For a period-m motion of dynamical system in Eq.(5.1) for N-nodes per period where $N = T/h$ with time step h, if

$$x_k = x_{k+mN} \quad \text{and} \quad y_k = y_{k+mN}, \tag{6.32}$$

then such a period-m motion is called the non-travelable period-m motion in the dynamical system.

Definition 6.2 For a period-m motion of dynamical system in Eq.(5.1) for N-nodes per period where $N = T/h$ with time step h, if

$$\mathrm{mod}(x_k, 2\pi) = \mathrm{mod}(x_{k+mN}, 2\pi) \quad \text{with} \quad x_k \neq x_{k+mN} \quad \text{and} \quad y_k = y_{k+mN}, \tag{6.33}$$

then such a period-m motion is called the travelable period-m motion in the dynamical system.

Excitation amplitude effects on bifurcation trees for non-travelable period-1 motions to chaos and non-travelable period-3 motions to chaos will be discussed first. The bifurcation trees from travelable period-1 motions to chaos and travelable period-2 motions to chaos in pendulum are also presented. The parameters of the periodically forced pendulum are $\alpha = 1.5, \delta = 0.75$, and $\Omega = 1.0$.

6.4.1 Non-travelable period-1 motions to chaos

As in Guo and Luo (2016), the bifurcation trees varying with excitation amplitude for non-travelable period-1 motions to chaos will be presented through the analytical predictions of period-1 to period-4 motions. Displacement and velocity of the periodic nodes $\mathrm{mod}(x_{\mathrm{mod}(k,N)}, 2\pi)$ and $y_{\mathrm{mod}(k,N)}$ for $\mathrm{mod}(k, N) = 0$ are predicted. A global view of bifurcation trees of period-1 to period-4 motions in the periodically excited pendulum is illustrated in Figs.6.10(a) and (b). For a better illustration of bifurcations trees, three main zoomed windows of $Q_0 \in (2.0, 5.0), (5.0, 9.0)$, and $(9.5, 12.0)$, are shown in Figs.6.10 (c,d), (e,f) and (g,h), respectively. The global view of bifurcation trees of period-1 to period-4 motions lies in the range of $Q_0 \in (0, \infty)$. There are two branches of symmetric period-1 motions for $Q_0 \in (0, \infty)$. The two symmetric period-1 motions alternatively switch between stable and unstable motions, and the two branches form a double helix shape together, as observed obviously in Fig.6.10(b). All the unstable symmetric period-1 motions are enclosed by saddle-node bifurcations. Some of the unstable symmetric periodic motions are relative to coexisting asymmetric motions, while the others are related to jumping phenomena. The ranges for unstable symmetric period-1 motions and the corresponding saddle-node bifurcation points for asymmetric period-1 motions and jumping phenomena are tabulated in Table 6.10. The unstable asymmetric period-1, period-2 and period-4 motions and the corresponding bifurcation points are tabulated in Tables 6.11—6.14.

Table 6.10 Excitation amplitudes for unstable symmetric period-1 motions with jumping ($\alpha = 2.0, \delta = 0.75, \Omega = 1.0$)

Branch 1		
Q_0	SN (L)	SN (R)
—	N/A	N/A
Branch 2		
Q_0	SN (L)	SN (R)
(4.7742,4.8417)	4.7742	4.8417

Note: L and R denote left and right. SN is for saddle-node bifurcation.

Table 6.11 Excitation amplitudes for unstable symmetric to asymmetric period-1 motions ($\alpha = 2.0, \delta = 0.75, \Omega = 1.0$)

Branch 1 (Jumping)		
Q_0	SN (L)	SN (R)
(2.2710, 8.0800)	2.2710	8.0800
(9.9950, 15.5920)	9.9950	15.5920
(17.9680, 20)	17.9680	
Branch 2 (Jumping)		
Q_0	SN (L)	SN (R)
(0, 4.6628)	—	4.6628
(6.0000, 11.7910)	6.0000	11.7910
(13.9870, 19.4330)	13.9870	19.4330

Note: L and R denote left and right. SN is for saddle-node bifurcations.

Table 6.12 Excitation amplitudes for unstable asymmetric period-1 to period-2 motions ($\alpha = 2.0, \delta = 0.75, \Omega = 1.0$)

Q_0	PD (L)	PD (R)
(2.4392, 4.6146)	2.4392	4.6146
(6.3290, 7.9330)	6.3290	7.9330
(10.4080, 11.5470)	10.4080	11.5470
(14.4980, 15.2370)	14.4980	15.2370
(18.6600, 18.8900)	18.6600	18.8900

Note: L and R denote left and right. PD is for period-doubling bifurcation.

Table 6.13 Excitation amplitudes for unstable period-2 motions ($\alpha = 2.0, \delta = 0.75, \Omega = 1.0$)

With Period-4 Motions		
Q_0	PD (L)	PD (R)
(2.474226, 4.298323)	2.474226	4.298323
(2.705312, 4.603100)	2.705312	4.603100
(6.405792, 7.376442)	6.405792	7.376442
(6.847842, 7.893200)	6.847842	7.893200
(10.526300, 10.983100)	10.526300	10.983100
(10.939600, 11.468000)	10.939600	11.468000
With Jumping Phenomena		
Q_0	SN (L)	SN (R)
(2.705311, 4.298326)	2.705311	4.298326
(6.847542, 7.376994)	6.847542	7.376994
(10.932196, 10.993312)	10.932196	10.993312

Note: L and R denote left and right. SN and PD are for saddle-node and period-doubling bifurcations.

Table 6.14 Excitation amplitudes for unstable period-4 motions ($\alpha = 2.0, \delta = 0.75, \Omega = 1.0$)

With Period-8 Motions		
Q_0	PD (L)	PD (R)
(6.833940, 7.376155)	6.833940	7.376155
(6.422600, 7.362210)	6.422600	7.362210
(6.847993, 7.392316)	6.847993	7.392316
(6.853290, 7.884400)	6.853290	7.884400
(10.887250, 10.977490)	10.887250	10.977490
(10.555200, 10.936050)	10.555200	10.936050
(10.943370, 11.038580)	10.943370	11.038580
(10.972310, 11.448800)	10.972310	11.448800
With Jumping Phenomena		
Q_0	SN (L)	SN (R)
(6.833938, 7.362212)	6.833938	7.362212
(6.853289, 7.392316)	6.853289	7.392316
(10.886778, 10.936429)	10.886778	10.936429
(10.972133, 11.038790)	10.972133	11.038790

Note: L and R denote left and right. SN and PD are for saddle-node and period-doubling bifurcations.

The symmetric period-1 motion is stable for the range of $Q_0 \in (0.0, 2.271)$. The corresponding saddle-node bifurcation occurs at $Q_0 \approx 2.271$ and the symmetric period-1 motion becomes unstable. The unstable motion is connected to a pair of stable asymmetric period-1 motions. The two asymmetric period-1 motions exist for the range of $Q_0 \in (2.2710, 4.6628)$. The asymmetric period-1 motions start from the saddle-node bifurcation at $Q_0 \approx 2.271$ from one branch of the symmetric period-1 motion and end at the saddle-node bifurcation at $Q_0 \approx 4.6628$ from the other branch of the symmetric period-1 motion, as shown in Figs.6.10(c) and (d). The asymmetric period-1 motions possess a pair of period-doubling bifurcations at $Q_0 \approx 2.4392$ and $Q_0 \approx 4.6146$. The induced period-2 motions exist for $Q_0 \in (2.4392, 4.6146)$. The period-2 motions possess two pairs of period-doubling bifurcations at $Q_0 \in (2.4744226, 4.298323)$ and $\{2.705312, 4.603100\}$for period-4 motions. The period-4, period-8, and other periodic motions are not illustrated due to the very tiny stable ranges. Continuously, such a bifurcation tree of period-1 motions will reach to chaos. Two jumping phenomena also exist for the period-2 motions, and the corresponding unstable motions are enclosed by the two saddle-node bifurcations at $Q_0 \approx 2.705311$ and 4.298326. After the saddle-node bifurcation at $Q_0 \approx 4.6628$, the symmetric period-1 motion is stable for $Q_0 \in (4.6628, 4.8417)$. The jumping phenomena occurs at $Q_0 \approx 4.7742$ and 4.8417. The unstable period-2 motion enclosed by the two saddle-node bifurcations exists for $Q_0 \in (4.7742, 4.8417)$.

After the jumping phenomenon, the symmetric period-1 motion is stable for $Q_0 \in (4.7742, 6.0000)$. At $Q_0 \approx 6.0000$, the saddle-node bifurcation starts a new pair of asymmetric period-1 motions, which ends at the saddle-node bifurcation from a different branch at $Q_0 \approx 8.0800$, as shown in Figs.6.10(e) and (f). The asymmetric period-1 motions for $Q_0 \in (6.0000, 8.0800)$ possess a pair of period-doubling bifurcations at $Q_0 \approx 6.329$ and 7.933 that enclose both the period-2 motions and unstable asymmetric period-1 motions. The period-2 motion has two jumping phenomena with the saddle-node bifurcations at $Q_0 \approx 6.847542$ and 7.376994. Before and after the jumping phenomena, the period-2 motions have two pairs of period-doubling bifurcations at $Q_0 \approx 6.405792, 7.376442$ and $Q_0 \approx 6.847842, 7.893200$ for period-4 motions. The period-4 motions in the range of $Q_0 \in (6.405792, 7.376442)$ have two jumping phenomena at the saddle-node bifurcations of $Q_0 \approx 6.833938$ and 7.362212. Two pairs of further period-doubling bifurcations are located at $Q_0 \approx 6.833940, 7.376155$ and $Q_0 \approx 6.422600, 7.362210$, introducing period-8 motions, and further period-doubling bifurcations to period-16 motions, and continuation to chaos. On the other hand, the period-4 motions for $Q_0 \in (6.3847842, 7.893200)$ have two jumping phenomena enclosed by the SN bifurcations at $Q_0 \approx 6.853289$ and 7.392316. Again, two pairs of period-doubling bifurcations of the period-4 motion exist at $Q_0 \approx 6.847993, 7.392316$ and $Q_0 \approx 6.85329, 7.88440$, which lead to period-8 motions and further period-doubling bifurcations to period-16 motions, and continuation to chaos.

The period-8, period-16 to motions are not shown due to the very tiny stable range.

As the excitation amplitude increases, another pair of asymmetric period-1 motions exists for $Q_0 \in (9.995, 11.791)$, as shown in Figs.6.10(g) and (h). The asymmetric period-1 motions start from the saddle-node bifurcations on one branch at $Q_0 \approx 9.995$ and end until the saddle-node bifurcations on the other branch at $Q_0 \approx 11.791$. The asymmetric period-1 motions have a pair of period-doubling bifurcations at $Q_0 \approx 10.408$ and 11.547, which yield both the stable period-2 motions and unstable asymmetric period-1 motions. The period-2 motions possess two jumping phenomena at $Q_0 \approx 10.932196$ and 10.993312 with the saddle-node bifurcations. Before and after the jumping phenomena, there are two pairs of period-doubling bifurcations at $Q_0 \in \{10.5263, 10.9831\}$ and $Q_0 \in \{10.9396, 11.4680\}$ where produce period-4 motions. The period-4 motions in the range of $Q_0 \in (10.5263, 10.9831)$ have two jumping phenomena at $Q_0 \approx 10.886778$ and 10.936429 by the saddle-node bifurcations. Two pairs of period-doubling bifurcations are located at $Q_0 \in \{10.88725, 10.97749\}$ and $Q_0 \in \{10.55520, 10.93605\}$, inducing period-8 motions and further period-doubling bifurcations to period-16 motions, and continuation to chaos. On the other hand, the period-4 motions for $Q_0 \in (10.9396, 11.4680)$ have a jumping phenomenon enclosed by the saddle-node bifurcations at $Q_0 \approx 10.97231$, and 11.038790. Again, two pairs of period-doubling bifurcations exist for this period-4 motions at $Q_0 \approx 10.94337, 11.03858$ and $Q_0 \approx 10.97231, 11.44880$, leading to period-8 motions and further period-doubling bifurcations to period-16 motions, and continuation to chaos. Once again, the period-8, period-16 motions to chaos are also not shown due to the very tiny stable range.

The bifurcations trees become simple with further increasing Q_0. For $Q_0 \in (13.987, 15.592)$, a pair of asymmetric period-1 motions exists. Similar to previous asymmetric period-1 motions, the asymmetric period-1 motions start with the saddle-node bifurcation at $Q_0 \approx 13.987$ from one branch of the symmetric period-1 motions, and end with another saddle-node bifurcation at $Q_0 \approx 15.592$ from a different branch. Such asymmetric period-1 motions have a pair of period-doubling bifurcations at $Q_0 \approx 14.498$ and 15.237. The period-2 motions are stable between the two period-doubling bifurcations. Similarly, For $Q_0 \in (17.968, 19.433)$, a pair of asymmetric period-1 motions exist. The asymmetric period-1 motions start with the saddle-node bifurcation of $Q_0 \approx 17.968$ from one branch of the symmetric period-1 motions, and end with another saddle-node bifurcation of $Q_0 \approx 19.433$ from a different branch. The asymmetric period-1 motions have a pair of period-doubling bifurcations at $Q_0 \approx 18.66$ and 18.89. The period-2 motions enclosed by the two period-doubling bifurcations are stable.

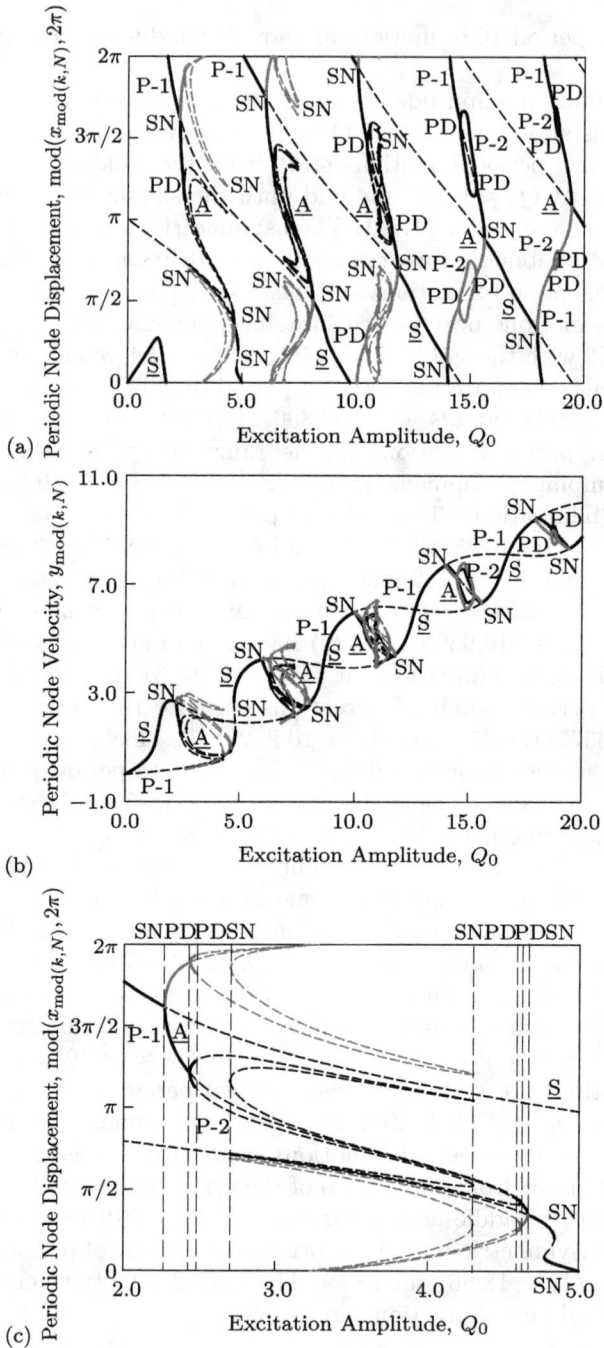

Fig. 6.10 Bifurcation trees of period-1 to period-4 motions varying with excitation amplitude Q_0 through node displacement $x_{\mathrm{mod}\,(k,N)}$ and node velocity $y_{\mathrm{mod}\,(k,N)}$: (a,b) Global view ($Q_0 \in (0, 20]$), (c,d) Zoom-1 view ($Q_0 \in (2.0, 5.0)$), (e,f) Zoom-2 view ($Q_0 \in (5.5, 8.5)$): (g, h) Zoom-3 view ($Q_0 \in (9.5, 12.0)$). ($\alpha = 2.0, \delta = 0.75, \Omega = 1.0$). $\mathrm{mod}(k, N) = 0$.

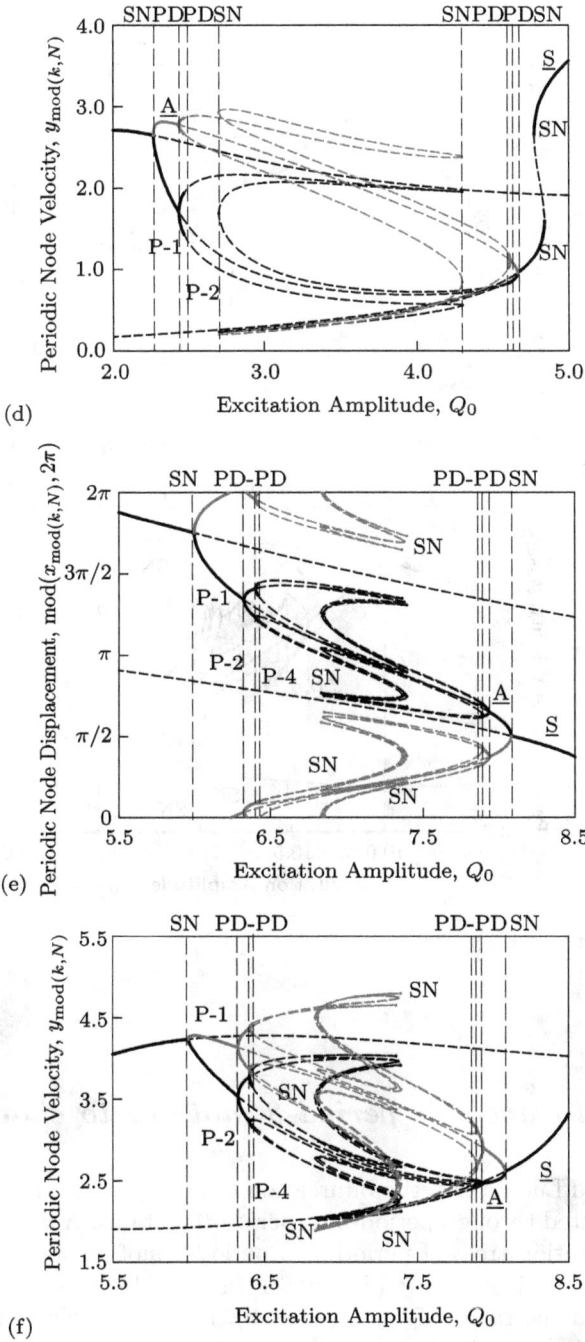

Fig. 6.10 Continued.

Fig. 6.10 Continued.

6.4.2 Non-travelable period-3 motions to chaos

As in Guo and Luo (2016), the bifurcation trees of period-3 motions to chaos will be presented through period-3 to period-6 motions. A global view of analytical bifurcation trees of period-3 to period-6 motions in the periodically excited pendulum is illustrated in Fig.6.11(a) and (b). There are three bifurcation trees in the range of $Q_0 \in (3.5, 8.0)$. For a clear view of bifurcations trees, the zoomed windows of $Q_0 \in (3.55, 4.09)$, $(6.2, 7.1)$ and $(7.01, 7.90)$ are shown in Figs.6.11(c,d), (e,f) and (g,h), respectively. The ranges for unstable symmetric period-3 motions and the corresponding saddle-node bifurcations for asymmetric period-3 motion and jumping phenomena are tabulated in

Table 6.15. The ranges for unstable asymmetric period-3 and period-6 motions and the corresponding period-doubling and saddle-node bifurcations are listed in Table 6.16. In Figs.6.11(c) and (d), the symmetric period-3 motion is in the range of $Q_0 \in (3.5922, 4.0839)$, with two saddle-node bifurcations at $Q_0 \approx 3.5922, 4.0839$ for symmetric period-3 motion appearance. The symmetric period-3 motion has two saddle-node bifurcations at $Q_0 \approx 3.7970, 4.0620$ for onset of asymmetric period-3 motions. Thus, the asymmetric period-3 motion is in the range of $Q_0 \in (3.7970, 4.0620)$ with two period-doubling bifurcations at $Q_0 \approx 3.8616, 4.0437$ for period-6 motion. The period-6 motion exists in the range of $Q_0 \in (3.8616, 4.0437)$ with the two saddle-node bifurcations at $Q_0 \approx 3.8745, 4.0382$ for period-12 motion. If the period-doubling bifurcations continuosly occur, the period-24 motions to chaos may exist. The ranges for stable period-6 motions are very short. Thus, period-12 and higher order periodic motions on this bifurcation will not be presented. In Figs.6.11(e) and (f), a symmetric period-3 motion exists in range of $Q_0 \in (6.3110, 7.01410)$ with saddle-node bifurcations at $Q_0 \approx 6.3110, 7.01410$ for its own appearance. The symmetric period-3 motion has two saddle-node bifurcations at $Q_0 \approx 6.3590, 7.0071$ for asymmetric period-3 motions. In Figs.6.11(g) and (h), the symmetric period-3 motion exists in range of $Q_0 \in (7.0193, 7.8626)$ with two saddle-node bifurcations at $Q_0 \approx 7.0193, 7.8626$ for it own appearance. The symmetric period-3 motion has two saddle-node bifurcations at $Q_0 \approx 7.6693, 7.6779$ for two jumping phenomena, there are three solutions in the corresponding range. In addition, the symmetric period-3 motion possesses two period-doubling bifurcations at $Q_0 \approx 7.0270, 7.6562$ for onset of asymmetric period-3 motions. Thus, the asymmetric period-3 motions exist in the range of $Q_0 \in (7.0270, 7.6562)$. The asymmetric period-3 motions have two period-doubling bifurcations at $Q_0 \approx 7.0309, 7.6470$ for period-6 motion. Thus, the period-6 motions exist in range of $Q_0 \in (7.0309, 7.6470)$.

Table 6.15 Excitation amplitudes for unstable symmetric period-3 motions ($\alpha = 2.0, \delta = 0.75, \Omega = 1.0$)

With asymmetric Period-3 Motions		
Q_0	SN (L)	SN (R)
(3.7970, 4.0620)	3.7970	4.0620
(6.3590, 7.0071)	6.3590	7.0071
(7.0270, 7.6562)	7.0270	7.6562
With Jumping Phenomena		
Q_0	SN (L)	SN (R)
(3.5922, 4.0839)	3.5922	4.0839
(6.3110, 7.0140)	6.3110	7.0140
(7.6693, 7.6779)	7.6693	7.6779
(7.0193, 7.8626)	7.0193	7.8626

Note: L and R denote left and right. SN is for saddle-node bifurcation.

Table 6.16 Excitation amplitudes for unstable asymmetric period-3 and period-6 motions ($\alpha = 2.0, \delta = 0.75, \Omega = 1.0$)

Asymmetric Period-3 Motions		
Q_0	PD (L)	PD (R)
(3.8616, 4.0437)	3.8616	4.0437
(7.0309, 7.6470)	7.0309	7.6470
Period-6 Motions		
Q_0	PD (L)	PD (R)
(3.8745, 4.0382)	3.8745	4.0382

Note: L and R denote left and right. PD is for period-doubling bifurcation.

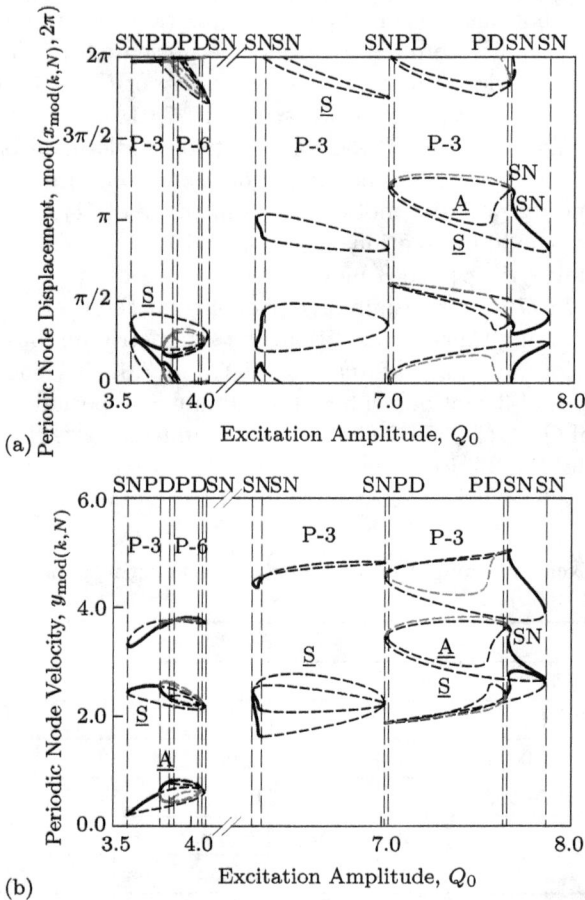

(a)

(b)

Fig. 6.11 Bifurcation trees of period-3 motions to chaos through displacement $\mathrm{mod}(x_{\mathrm{mod}(k,N)}, 2\pi)$ and velocity $y_{\mathrm{mod}(k,N)}$: (a,b) Global view for $Q_0 \in (3.5, 8.0)$, (c,d) zoom-1 view for $Q_0 \in (3.55, 4.09)$, (e.f) zomm-2 view for $Q_0 \in (6.2, 7.1)$, (g,h) zoom-3 view for $Q_0 \in (7.01, 7.90)$. ($\alpha = 2.0, \delta = 0.75, \Omega = 1.0$). $\mathrm{mod}(k, N) = 0$.

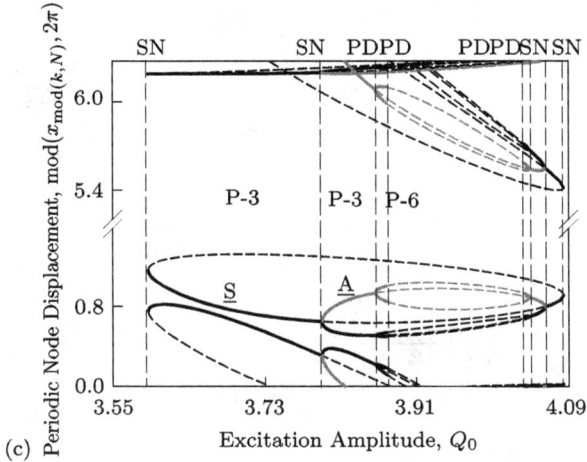

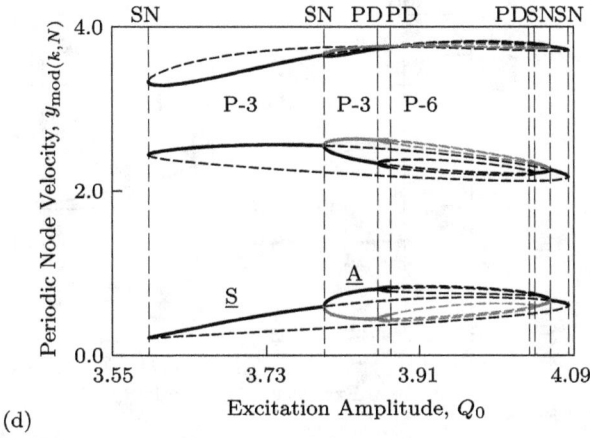

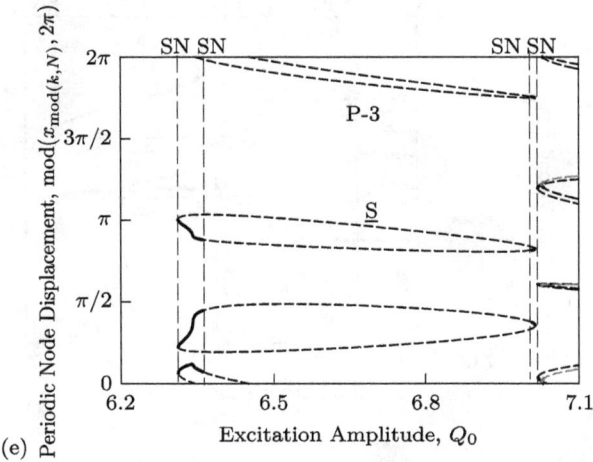

Fig. 6.11 Continued.

(f)

(g)

(h)

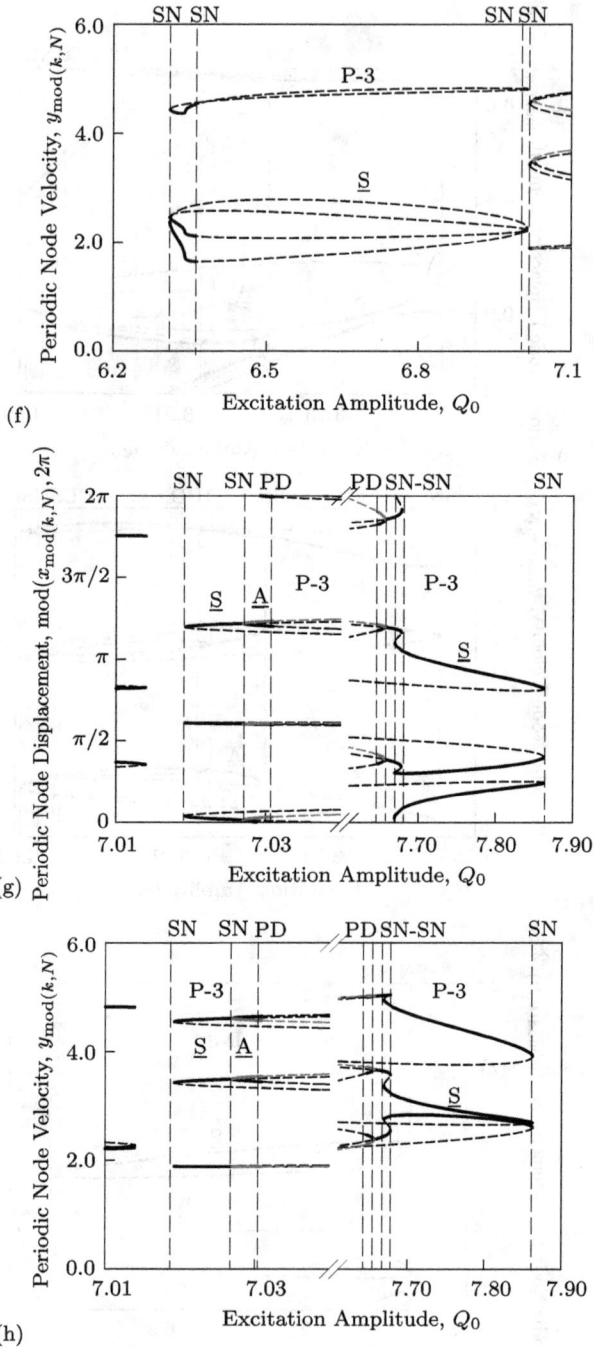

Fig. 6.11 Continued.

6.4.3 Travelable period-1 motions to chaos

As before, the bifurcation trees of periodic motions to chaos are based on the non-travelable periodic motions. The independent bifurcation trees of travelable period-1 motions to chaos will be presented through period-1 to period-2 motions. A global view of analytical bifurcation trees of travelable period-1 to period-2 motions in the periodically excited pendulum is illustrated in Figs.6.12 (a) and (b). The bifurcation tree does not start from symmetric period-1 motions but begins from the travelable asymmetric period-1 motion. In addition, such a bifurcation tree of period-1 motion to chaos is independent, which will not go from zero to infinite excitation amplitudes. For a clear view of bifurcations trees, the zoomed windows of $Q_0 \in (4.380, 4.400), (3.18, 3.27), (2.65, 2.77)$, and $(2.312, 2.320)$ are shown in Figs.6.12(c,d)—(i,j), respectively. The ranges for unstable asymmetric period-1 motions and the corresponding saddle-node bifurcations for asymmetric period-2 motion and jumping phenomena are tabulated in Table 6.17. The ranges for unstable asymmetric period-2 motion and the corresponding period-doubling and saddle-node bifurcations are listed in Table 6.18. The travelable asymmetric period-1 motion is in the range of $Q_0 \in (2.3127, 4.3950)$, with two saddle-node bifurcations at $Q_0 \approx 2.3127, 4.3950$ for the asymmetric period-1 motion appearance. The stable period-1 motions are in the range of $Q_0 \in (2.3127, 2.3163), (2.7590, 3.1900)$, and $(4.3894, 4.3950)$. The asymmetric period-1 motion has two pairs of period-doubling bifurcations at $Q_0 \in \{2.3163, 2.7590\}$ and $Q_0 \in \{3.1900, 4.3894\}$ for onset of asymmetric period-2 motions. Thus, the travelable asymmetric period-2 motion is in two ranges of $Q_0 \in (2.3163, 2.7590)$ and $(3.1900, 4.3894)$. During the two ranges of excitation amplitudes, the travelable period-2 motions possess three pairs of period-doubling bifurcations at $Q_0 \in \{2.3183, 2.6883\}, \{2.6645, 2.7270\}$ and $\{3.2418, 4.3864\}$ for onset of period-4 motions and two pairs of saddle-node bifurcations for jumping phenomena at

Table 6.17 Excitation amplitudes for unstable asymmetric period-1 motions ($\alpha = 2.0, \delta = 0.75, \Omega = 1.0$)

with period-2 motion		
Q_0	PD (L)	PD (R)
(2.3163,2.7590)	2.3163	2.7590
(3.1900, 4.3894)	3.1900	4.3894
with jumping phenomena		
Q_0	SN (L)	SN (R)
(2.3127, 4.3950)	2.3127	4.3950

Note: L and R denote left and right. SN and PD are for saddle-node and period-doubling bifurcations.

$Q_0 \in \{2.6592, 2.6886\}$ and $\{3.2394, 3.2605\}$. For the jumping phenomena, the three solutions are observed during the range of $Q_0 \in (2.6592, 2.6886)$ and $(3.2394, 3.2605)$. The two stable and one unstable solutions of travelable period-2 motions can be observed. Since the ranges of stable travelable period-1 and period-2 motions are very tiny, the zoomed views in Figs.6.12(c,d)–(i,j) are given for better illustrations.

Table 6.18 Excitation amplitudes for unstable period-2 motions ($\alpha = 2.0, \delta = 0.75, \Omega = 1.0$)

With period-4 motion		
Q_0	PD (L)	PD (R)
(2.3183,2.6883)	2.3183	2.6883
(2.6645, 2.7270)	2.6645	2.7270
(3.2418, 4.3864)	3.2418	4.3864
With jumping phenomena		
Q_0	SN (L)	SN (R)
(2.6592, 2.6886)	2.6592	2.6886
(3.2394, 3.2605)	3.2394	3.2605

Note: L and R denote left and right. SN and PD are for saddle-node and period-doubling bifurcations.

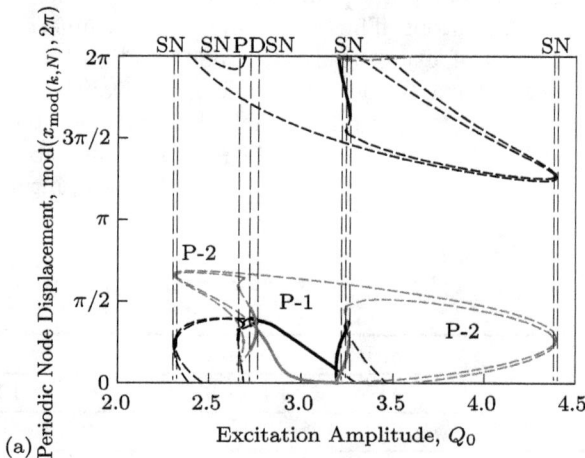

(a)

Fig. 6.12 Independent bifurcation trees of travalable period-1 motions to chaos through displacement $\mathrm{mod}(x_{\mathrm{mod}(k,N)}, 2\pi)$ and velocity $y_{\mathrm{mod}(k,N)}$: (a,b) Global view ($Q_0 \in (2.0, 4.5)$), (c,d) zoom-1 view ($Q_0 \in (4.380, 4.440)$), (e,f) Zoom-2 view ($Q_0 \in (3.18, 3.27)$), (g,h) Zoom-3 view ($Q_0 \in (2.65, 2.77)$), (i,j) Zoom-4 view ($Q_0 \in (2.312, 2.322)$), ($\alpha = 2.0, \delta = 0.75, \Omega = 1.0$). $\mathrm{mod}(k, N) = 0$.

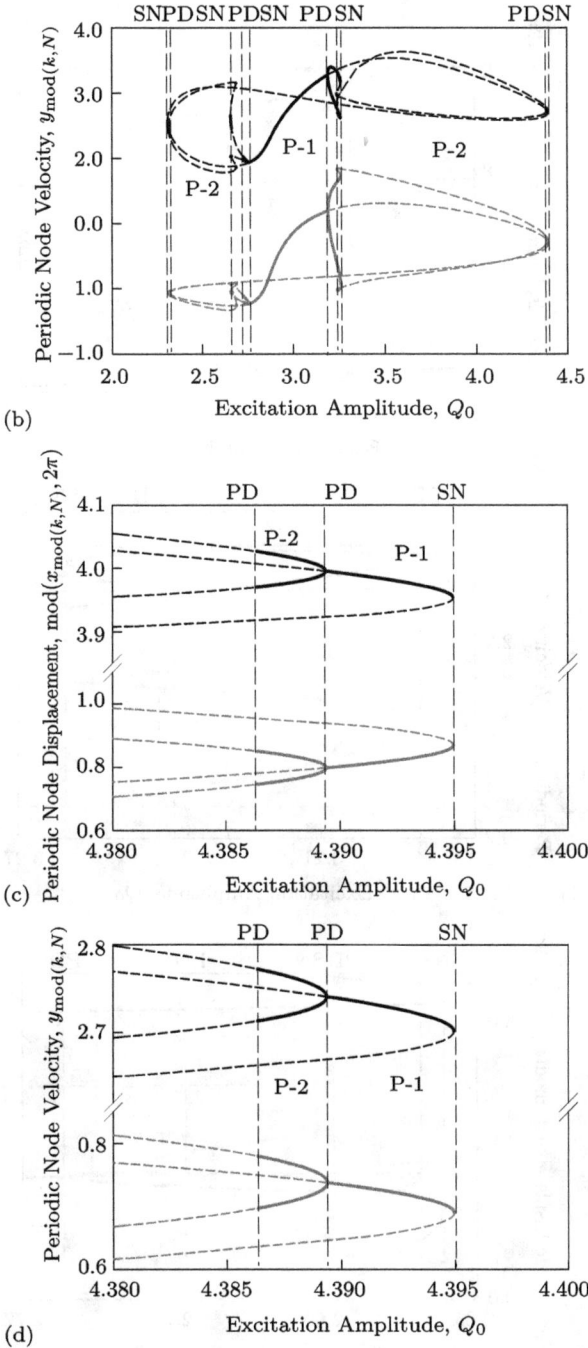

(b)

(c)

(d)

Fig. 6.12 Continued.

(e)

(f)

(g)

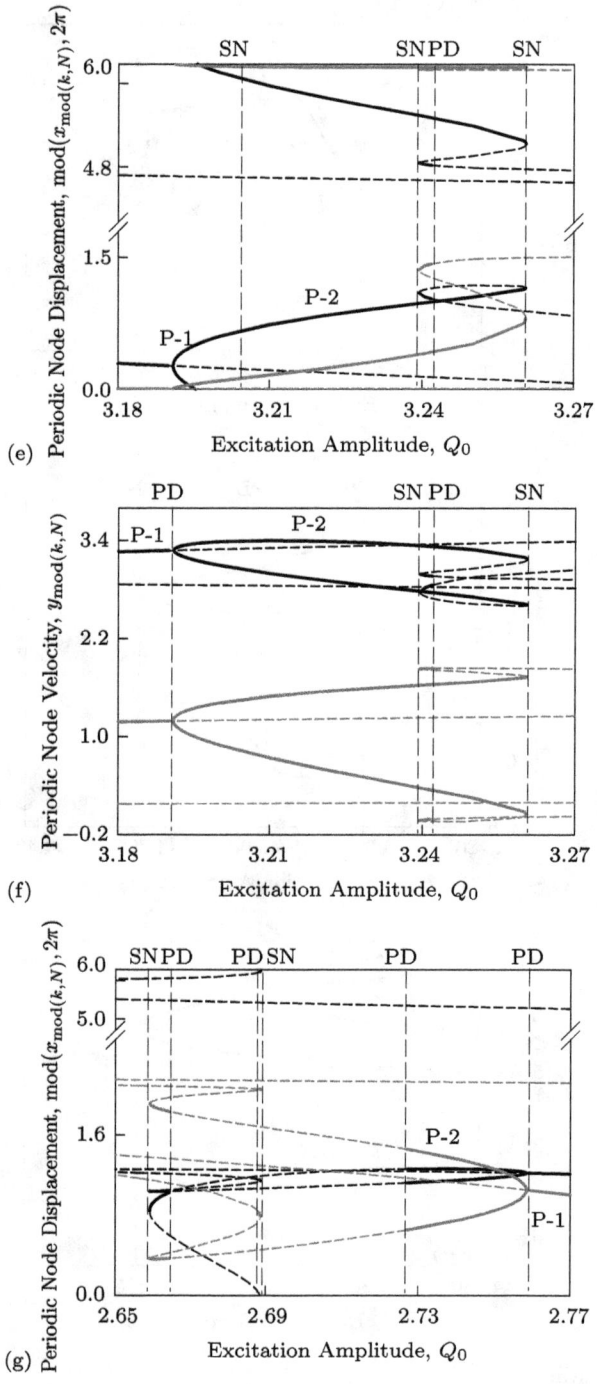

Fig. 6.12 Continued.

(h)

(i)

(j)

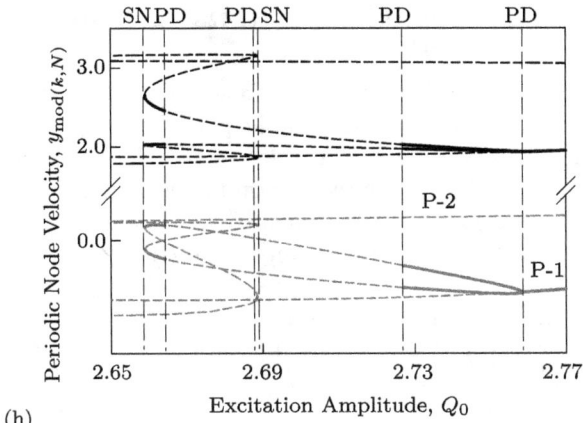

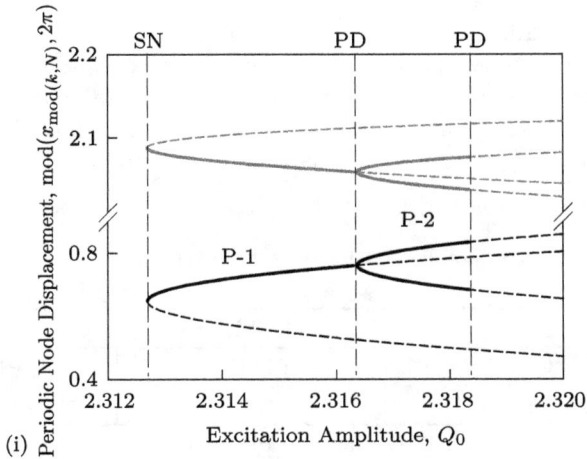

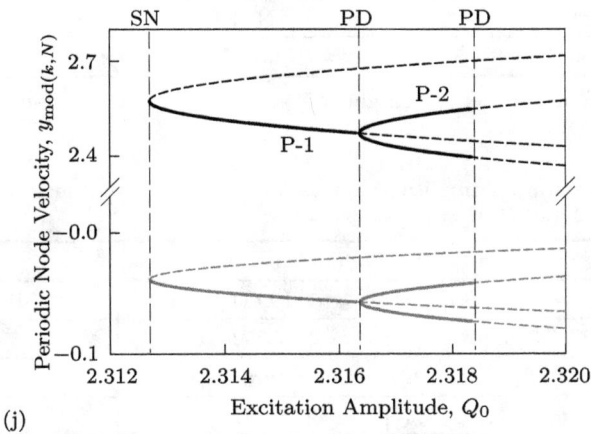

Fig. 6.12 Continued.

6.4.4 Travelable period-2 motions to chaos

The independent bifurcation trees of travelable period-2 motions to chaos will be presented through the travelable period-2 to period-8 motions. A global view of analytical bifurcation trees of travelable period-2 to period-8 motions in the periodically excited pendulum is illustrated in Figs.6.13(a) and (b). The bifurcation tree is independent and starts from the travelable asymmetric period-2 motion without any period-1 motions. For a clear view of bifurcations trees, the zoomed windows of $Q_0 \in (7.32, 7.40)$ and $(7.22, 7.26)$ are shown in Figs.6.13(c,d) and (e,f), respectively.

The ranges for unstable asymmetric period-2 motions and the corresponding saddle-node bifurcations for appearance of asymmetric period-2 motions are tabulated in Table 6.19. The ranges for unstable asymmetric period-4 and period-8 motions and the corresponding period-doubling bifurcations are listed in Table 6.20. The travelable asymmetric period-2 motion is in $Q_0 \in (7.2289, 7.3899)$ with two saddle-node bifurcations at $Q_0 \approx 7.2289$ and 7.3899 for appearance of asymmetric period-2 motions. The stable travelable period-2 motions are in the ranges of $Q_0 \in (7.2289, 7.2427)$ and $(7.3541, 7.3899)$. The asymmetric period-2 motion has a pair of period-doubling bifurcations

Table 6.19 Excitation amplitudes for unstable asymmetric period-2 motions ($\alpha = 2.0, \delta = 0.75, \Omega = 1.0$)

With Period-4 Motions		
Q_0	PD (L)	PD (R)
(7.2427, 7.3541)	7.2427	7.3541
With Jumping Phenomena		
Q_0	SN (L)	SN (R)
(7.2289, 7.3899)	2.3127	7.3899

Note: L and R denote left and right. SN and PD are for saddle-node and period-doubling bifurcations.

Table 6.20 Excitation amplitudes for unstable period-4 and unstable period-8 motions ($\alpha = 2.0, \delta = 0.75, \Omega = 1.0$)

Period-4 motions		
Q_0	PD (L)	PD (R)
(7.2524, 7.3364)	7.2524	7.3364
Period-8 motions		
Q_0	SN (L)	SN (R)
(7.2552, 7.3322)	7.2552	7.3322

Note: L and R denote left and right. SN and PD are for saddle-node and period-doubling bifurcations.

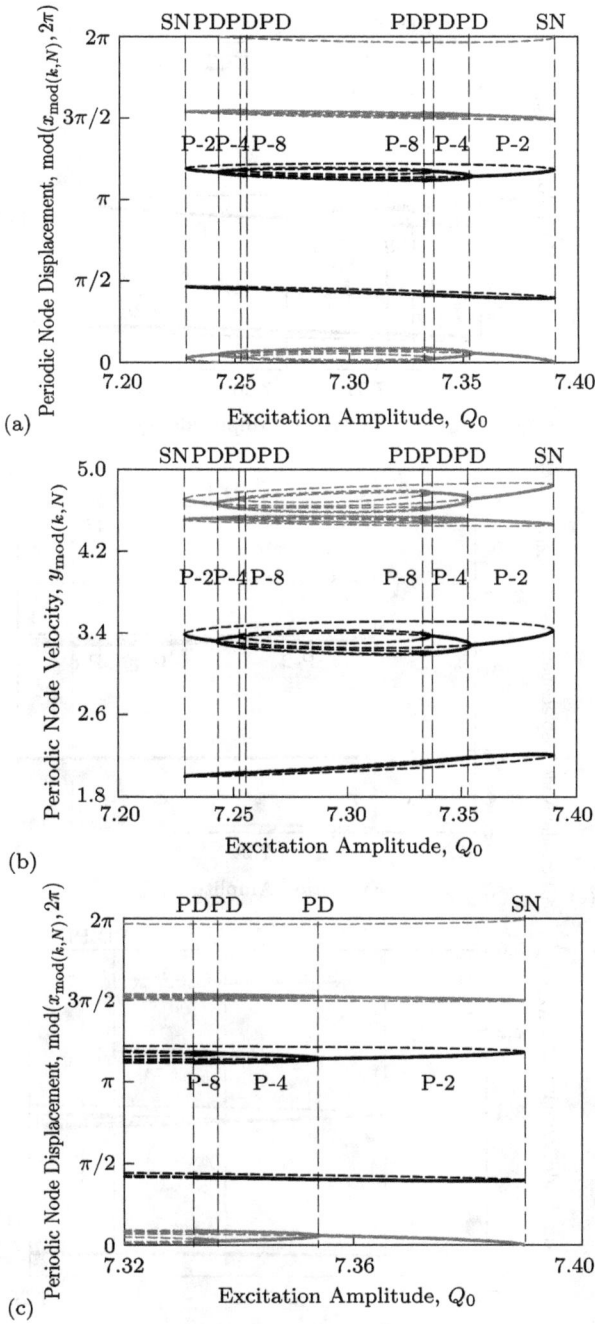

Fig. 6.13 Bifurcation trees of travelable period-2 motions to chaos by mod($x_{\mathrm{mod}(k,N)}$, 2π) and velocity $y_{\mathrm{mod}(k,N)}$: (a,b) Global view $Q_0 \in (7.20, 7.40)$, (c,d) Zoom-1 view $Q_0 \in (7.32, 7.40)$, (e,f) zoom-2 view $Q_0 \in (7.22, 7.26)$, ($\alpha = 2.0, \delta = 0.75, \Omega = 1.0$). mod($k, N$) = 0.

(d)

(e)

(f)

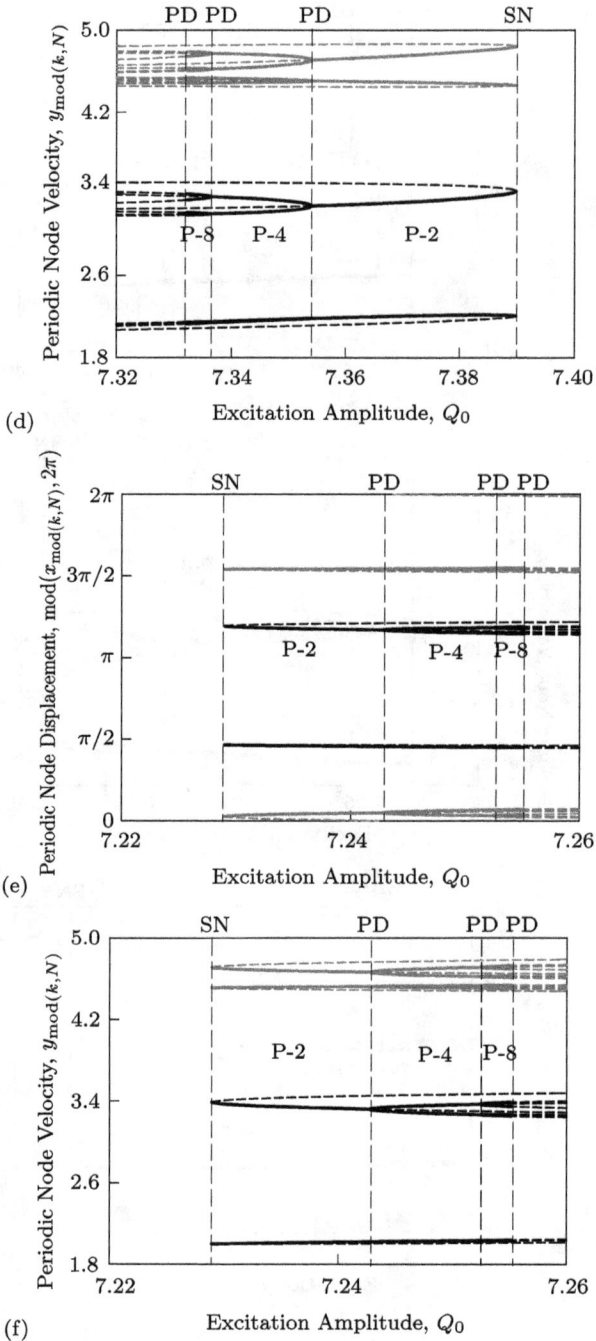

Fig. 6.13 Continued.

at $Q_0 \in \{7.2427, 7.3541\}$ for onset of asymmetric period-4 motions. The travelable period-4 motions exist in the range of $Q_0 \in (7.2427, 7.3541)$. The unstable period-4 motion lies in the range of $Q_0 \in (7.2524, 7.3364)$. The stable period-4 motions are in the two segments of $Q_0 \in (7.2427, 7.2524)$ and $(7.3364, 7.3541)$. In the range of $Q_0 \in (7.2524, 7.3364)$, a travelable period-8 motion exists. Such a period-8 motion has two period-doubling bifurcations at $Q_0 \in \{7.2552, 7.3322\}$ for appearance of a travelable period-16 motion.

6.5 Numerical simulations

In this section, numerical and analytical results of periodic motions in periodically forced pendulum will be presented with different excitation frequencies or amplitudes. For numerical simulations, initial conditions of periodic motions will be obtained from the analytical predictions. The numerical and analytical results are presented by solid curves and circular hollow symbols, respectively. The initial node and periodic nodes are depicted by green circular symbols. The travelable and non-travelable periodic motions will be presented herein and the corresponding harmonic amplitudes and phases will be also presented.

6.5.1 Non-travelable periodic motions

There are two types of non-travelable period-m motions with centers determined by the corresponding constants $\mathrm{mod}(a_0^{(m)}, 2\pi)$ near $\mathrm{mod}(a_0^{(m)}, 2\pi) = \pi$ and $\mathrm{mod}(a_0^{(m)}, 2\pi) = 0$ or 2π in the finite Fourier series. For non-travelable period-m motions, $x_k^{(m)} = x_{k+mN}^{(m)}$. However, for travelable period-m motions, $x_k^{(m)} \neq x_{k+mN}^{(m)}$ but $\mathrm{mod}(x_k^{(m)}, 2\pi) = \mathrm{mod}(x_{k+mN}^{(m)}, 2\pi)$.

6.5.1.1 Period-1 to period-4 motions

(A) *Periodic motion with center near* $\mathrm{mod}(a_0^{(m)}, 2\pi) = \pi$
Consider parameters of $\alpha = 1.5, \delta = 0.75, Q_0 = 5.0$ for non-travelable periodic motions with center near $\mathrm{mod}(a_0^{(m)}, 2\pi) = \pi$. The symmetric period-1 motion is considered first, and then asymmetric period-1, period-2 and period-4 motions will be presented herein.

For $\Omega = 1.0$, a symmetric period-1 motion is presented in Fig.6.14. The initial condition of this motion is $(x_0, y_0) \approx (0.8046, 1.4237)$ from the analytical prediction. Such a symmetric periodic motion is centered at $(\pi, 0)$. The time-history of displacement for the symmetric period-1 motion is presented in

Fig.6.14(a). The trajectory of the symmetric period-1 motion in phase space is illustrated in Fig.6.14(b). The symmetric period-1 motion is symmetric to itself about $(\pi, 0)$ in phase space. The harmonic amplitudes of the symmetric period-1 motion are presented in Fig.6.14(c). The corresponding constant is $a_0 = \pi$. If $\mod(x_0, 2\pi) \approx 0.8046$ and the same velocity of $y_0 \approx 1.4237$, there are infinite solutions of period-1 motions with $\mod(a_0, 2\pi) = \pi$. For the symmetric perod-1 motion, $A_{(2l-1)} \neq 0$ but $A_{2l} = 0(l = 1, 2, \cdots)$. The main harmonic amplitudes are $A_1 \approx 2.5854$ and $A_3 \approx 0.0631$. The other harmonic amplitudes are $A_{(2l-1)} \in (10^{-12}, 10^{-2})$ for $l = 3, 4, \cdots, 10$. In Fig.6.14(d), harmonic phases lie in $\varphi_{2l-1} \in [0, 2\pi)$ for $l = 1, 2, \cdots, 10$.

With the saddle-node bifurcation, the non-travelable asymmetric pierod-1 motion exists with center near $\mod(a_0, 2\pi) = \pi$. Thus, two non-travelable asymmetric period-1 motions are presented in Fig.6.15 for $\Omega = 1.16$. The initial conditions of the asymmetric period-1 motions are $(x_0, y_0) \approx (0.7574,$

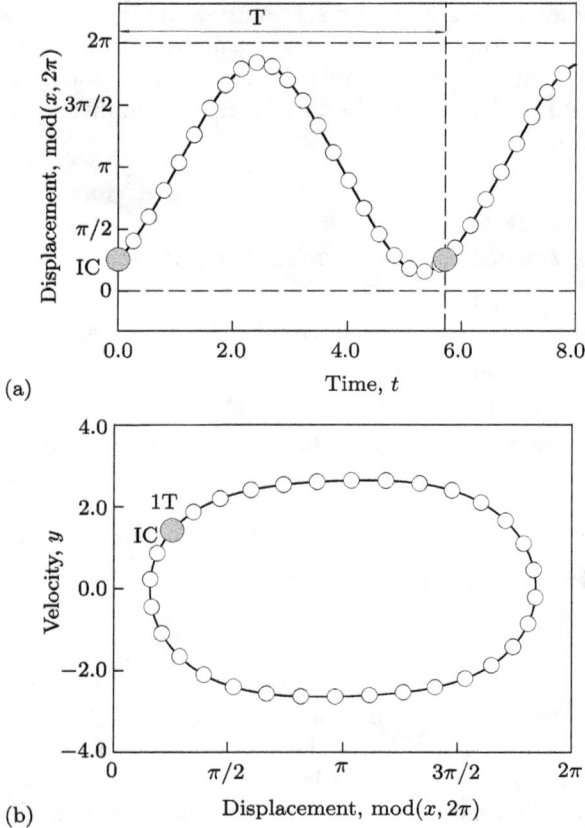

(a)

(b)

Fig. 6.14 Symmetric period-1 motions ($\Omega = 1.0$): (a) displacement, (b) trajectory, (c) harmonic amplitude, (d) harmonic phase. Initial condition: $(x_0, y_0) \approx (0.8046, 1.4237)$. Parameters ($\alpha = 1.5, \delta = 0.75, Q_0 = 5.0$).

(c)

(d)

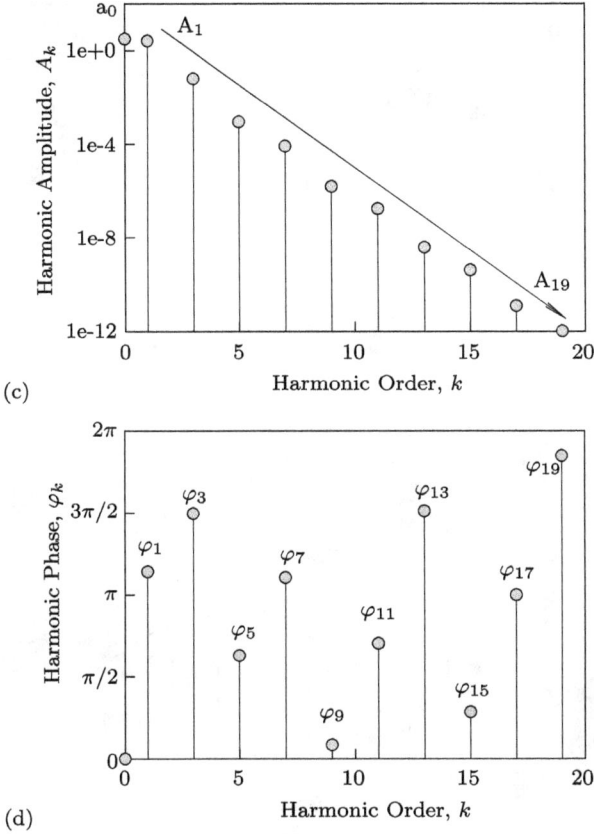

Fig. 6.14 Continued.

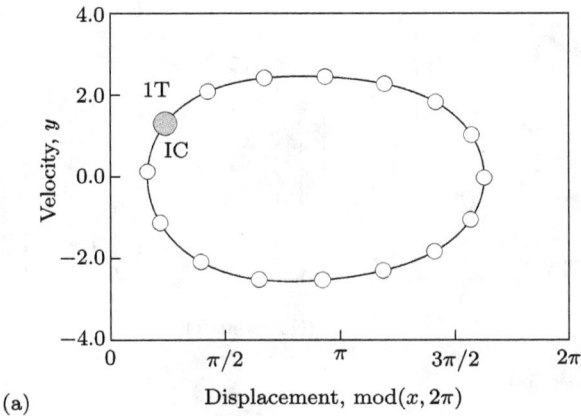

(a)

Fig. 6.15 Asymmetric period-1 motions ($\Omega = 1.16$): (a) trajectory of black branch, initial condition of $(x_0, y_0) \approx (0.7574, 1.3007)$; (b) trajectory of red branch, initial condition of $(x_0, y_0) \approx (1.3585, 1.0565)$. Parameters ($\alpha = 1.5, \delta = 0.75, Q_0 = 5.0$).

(b)

(c)

(d)

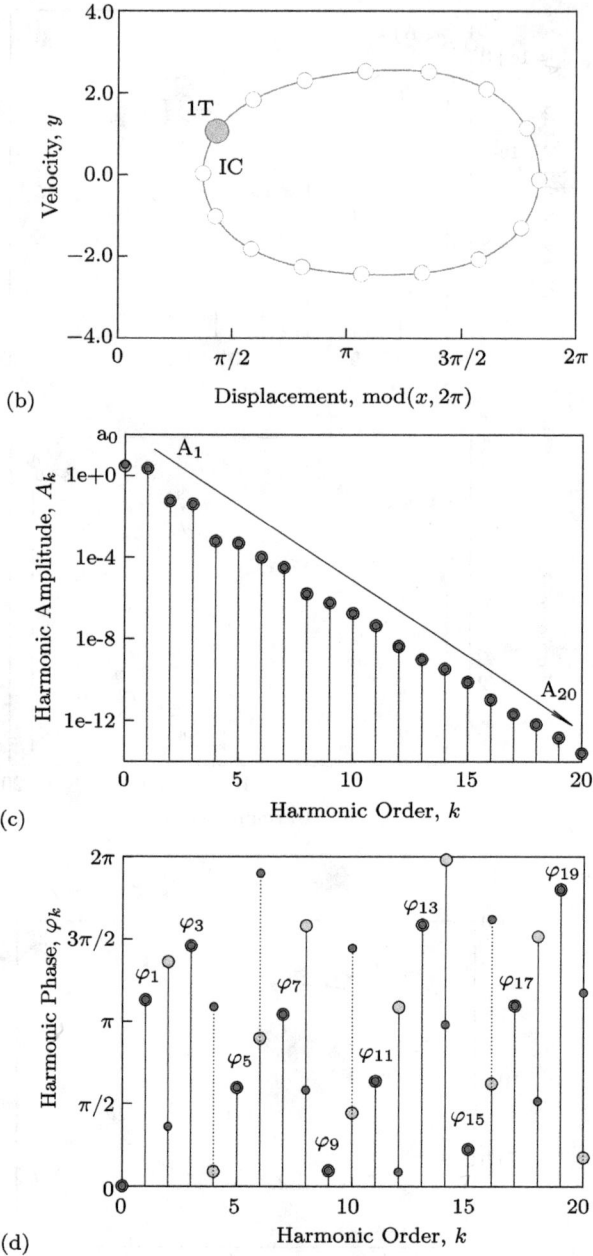

Fig. 6.15 Continued.

1.3007) (black) and $(x_0, y_0) \approx (1.3585, 1.0565)$ (red) from the analytical prediction. The trajectories of the two asymmetric period-1 motion in phase space are illustrated in Figs.6.15(a) and (b) on the left and right sides of $\mathrm{mod}(x, 2\pi) = \pi$, respectively. The harmonic amplitudes of the two non-travelable asymmetric period-1 motions are presented in Fig.6.15(c). The two constants satisfy $\pi - \mathrm{mod}\,(a_0^B, 2\pi) \approx 0.2761 \approx \mathrm{mod}\,(a_0^R, 2\pi) - \pi$, which is for asymmetry of $\mathrm{mod}(x, 2\pi) = \pi$. The harmonic amplitudes for two asymmetric period-1 motions are of the same. For the two asymmetric perod-1 motions, $A_{(2l-1)} \neq 0$ and $A_{2l} \neq 0 (l = 1, 2, \cdots)$. The main harmonic amplitudes are $A_1 \approx 2.2632, A_2 \approx 0.0586$, and $A_3 \approx 0.0410$. The other harmonic amplitudes are $A_k \in (10^{-14}, 10^{-2})$ for $k = 4, 5, \cdots, 20$. In Fig.6.15(d), harmonic phases for two asymmetric period-1 motions are presented and distributed in the range of $\varphi_k \in [0, 2\pi)$ with $\varphi_k^R = \mathrm{mod}(\varphi_k^B + (k+1)\pi, 2\pi) \in [0, 2\pi)$ for $k = 1, 2, \cdots, 20$.

After the period-doubling bifurcation of the non-travelable asymmetric period-1 motion, the non-travelable period-2 motions will appear. Consider excitation frequency of $\Omega = 1.19$ for a pair of two non-travelable asymmetric period-2 motions. The two trajectories, harmonic amplitude and phase of the asymmetric period-2 motions are presented in Fig.6.16. The corresponding initial conditions are $(x_0, y_0) \approx (0.6108, 1.3309)$ for the black branch and $(x_0, y_0) \approx (1.5784, 0.9840)$ for the red branch from the analytical prediction. The trajectories of the two asymmetric period-2 motion in phase space are illustrated in Figs.6.16(a) and (b) on the left and right sides of $\mathrm{mod}(x, 2\pi) = \pi$, respectively. Harmonic amplitudes of the two asymmetric period-2 motions are presented in Fig.6.16(c). The two constants satisfy $\pi - \mathrm{mod}(a_0^{(2)B}, 2\pi) \approx 0.5506 \approx \mathrm{mod}(a_0^{(2)R}, 2\pi) - \pi$, which is for asym-

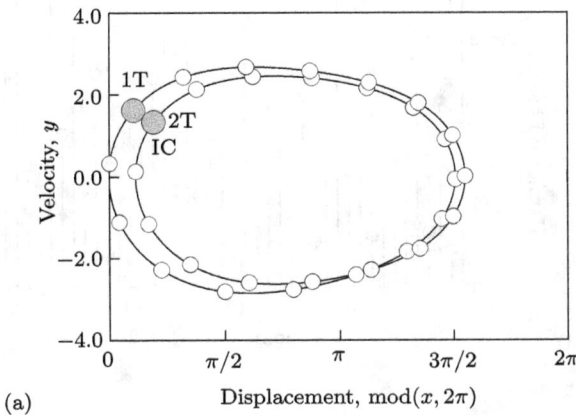

(a)

Fig. 6.16 Asymmetric period-2 motions ($\Omega = 1.19$): (a) phase portrait of black branch, initial condition: $(x_0, y_0) \approx (0.6108, 1.3309)$; (b) phase portrait of red branch, initial condition: $(x_0, y_0) \approx (1.5784, 0.9840)$. Parameters ($\alpha = 1.5, \delta = 0.75, Q_0 = 5.0$).

(b)

(c)

(d)

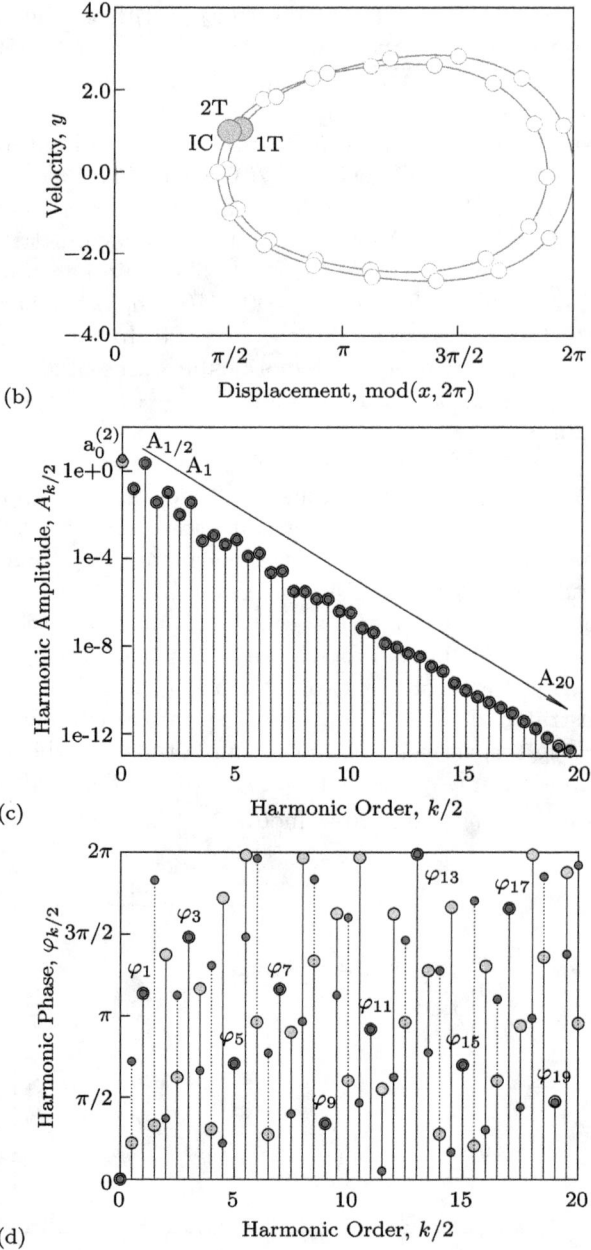

Fig. 6.16 Continued.

metry of $\text{mod}(x, 2\pi) = \pi$. The harmonic amplitudes for two asymmetric period-2 motions are of the same. For the two asymmetric perod-2 motions, the main harmonic amplitudes are $A_{1/2} \approx 0.1600$, $A_1 \approx 2.2761$, $A_{3/2} \approx 0.0383$, $A_2 \approx 0.1076$, $A_{5/2} \approx 0.0102$, and $A_3 \approx 0.0373$. The other harmonic amplitudes are $A_{k/2} \in (10^{-14}, 10^{-2})$ for $k = 7, 8, \cdots, 40$. The harmonic phases distributed in $\varphi_{k/2} \in [0, 2\pi)$ ($k = 1, 2, \cdots, 40$) for the two period-2 motions are presented in Fig.6.16(d), and there is a relation of $\varphi_{k/2}^R = \text{mod}(\varphi_{k/2}^B + (k(1+2r)/2+1)\pi, 2\pi)$, $r = 1$ with $t_0 = rT$ ($r \in \{0, 1\}$).

After the period-doubling bifurcation of the non-travelable asymmetric period-2 motion, the non-travelable period-4 motions will exist. Consider the excitation frequency of $\Omega = 1.20$ for a pair of two non-travelable asymmetric period-4 motions. The two trajectories, harmonic amplitudes and phases of the asymmetric period-4 motions are presented in Fig.6.17. The initial conditions are $(x_0, y_0) \approx (0.7730, 1.1834)$ for the black branch and $(x_0, y_0) \approx (1.8227, 1.0649)$ for the red branch. The trajectories of the two asymmetric period-4 motion in phase space are illustrated in Figs.6.17(a) and (b) on the left and right sides of $\text{mod}(x, 2\pi) = \pi$, respectively. The harmonic amplitudes of the two non-travelable asymmetric period-4 motions are placed in Fig.6.17(c). The two constants satisfy $\pi - \text{mod}(a_0^{(4)B}, 2\pi) \approx 0.5864 \approx \text{mod}(a_0^{(4)R}, 2\pi) - \pi$, which is for asymmetry of $\text{mod}(x, 2\pi) = \pi$. The harmonic amplitudes for two asymmetric period-4 motions are of the same. For the two asymmetric perod-4 motions, the main harmonic amplitudes are $A_{1/4} \approx 0.0551$, $A_{1/2} \approx 0.2861$, $A_{3/4} \approx 0.0239$, $A_1 \approx 2.2647$, $A_{5/4} \approx 0.0137$, $A_{3/2} \approx 0.0677$, $A_{7/4} \approx 6.0897e-3$, $A_2 \approx 0.1108$, $A_{9/4} \approx 4.6969e-3$, $A_{5/2} \approx 0.0173$, $A_{11/4} \approx 8.6895e-4$, and $A_3 \approx 0.0352$. The other harmonic amplitudes are $A_{k/4} \in (10^{-14}, 10^{-2})$ for $k = 12, 13, \cdots, 80$. In Fig.6.17(d), harmonic phases are presented with $\varphi_{k/2}^R = \text{mod}(\varphi_{k/2}^B + (k(2r+1)/4+1)\pi, 2\pi) \in [0, 2\pi)$, $k = 1, 2, \cdots, 80$ and $r = 0$ with $t_0 = rT$ ($r \in \{0, 1\}$).

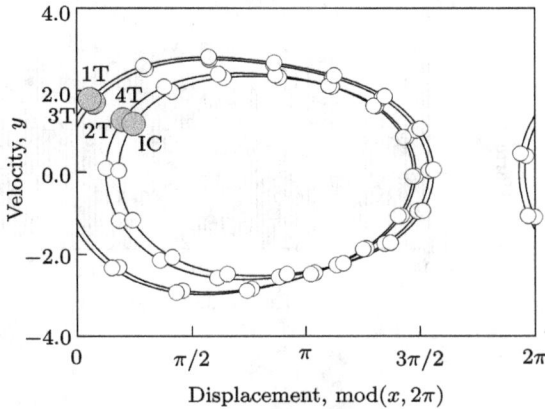

(a)

Displacement, $\text{mod}(x, 2\pi)$

Fig. 6.17 Asymmetric period-4 motions ($\Omega = 1.20$): (a) phase portrait of black branch, initial condition: $(x_0, y_0) \approx (0.7730, 1.1834)$; (b) phase portrait of red branch, initial condition: $(x_0, y_0) \approx (1.8227, 1.0649)$. Parameters ($\alpha = 1.5$, $\delta = 0.75$, $Q_0 = 5.0$).

(b)

(c)

(d)

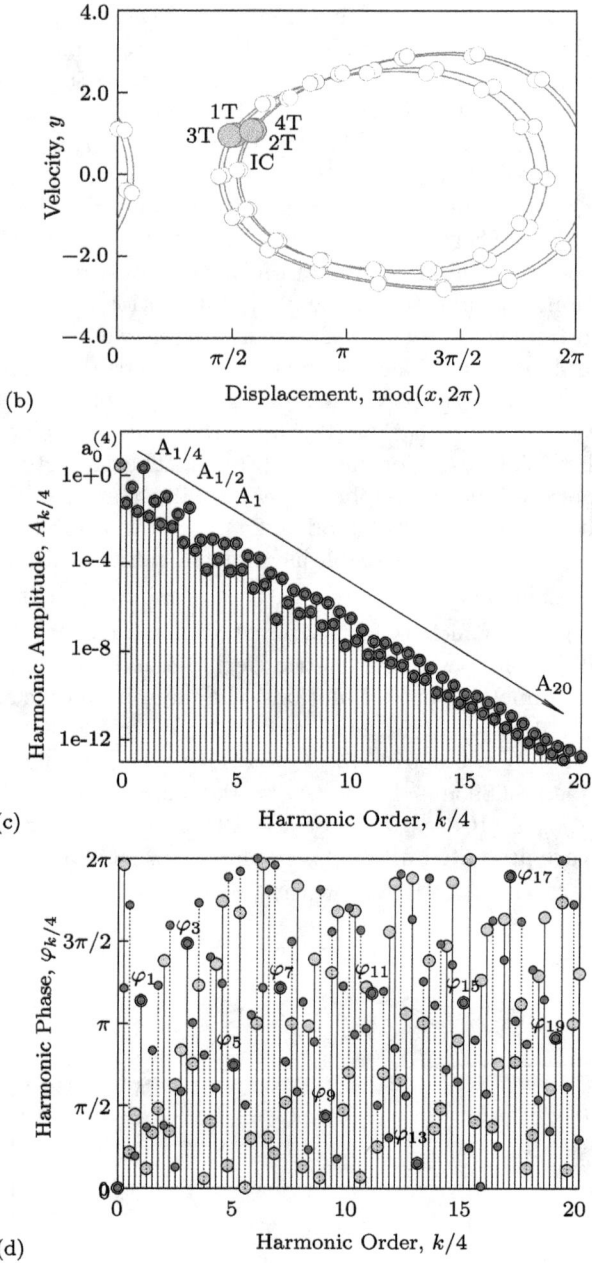

Fig. 6.17 Continued.

(B) *Periodic motions with center near* $\mathrm{mod}(a_0^{(m)}, 2\pi) = 2\pi$

Consider parameters of $\alpha = 1.5$, $\delta = 0.75$, $Q_0 = 5.0$ for non-travelable periodic motions with center near $\mathrm{mod}(a_0^{(m)}, 2\pi) = 2\pi$. The symmetric period-1 motion is considered first, and then asymmetric period-1, period-2 and period-4 motions will be presented herein.

For $\Omega = 1.5$, a symmetric period-1 motion is presented in Fig.6.18 for one of the two branches in the bifurcation trees. The initial condition of the symmetric period-1 motion is $(x_0, y_0) \approx (4.2051, 1.9880)$ from the analytical prediction. Such a symmetric periodic motion is centered at $(2\pi, 0)$. The time history of displacement for the symmetric period-1 motion is presented in Fig.6.18(a). The trajectory of the symmetric period-1 motion in phase space is illustrated in Fig.6.18(b). The symmetric period-1 motion is symmetric to itself about $(2\pi, 0)$ in phase space. The harmonic amplitude of the non-travelable symmetric period-1 motion is presented in Fig.6.18(c). The

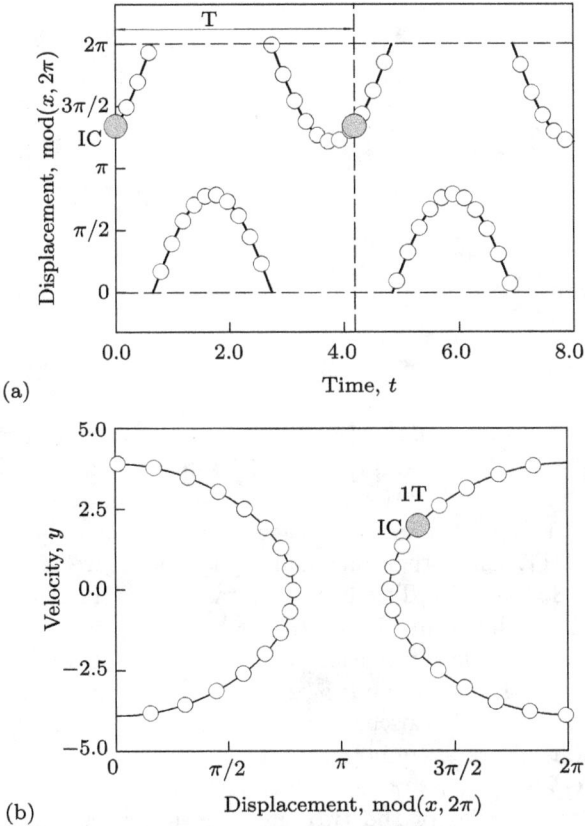

(a)

(b)

Fig. 6.18 Symmetric period-1 motions ($\Omega = 1.50$): (a) displacement, (c) trajectory, (b) harmonic amplitude, (d) harmonic phases. Initial condition: $(x_0, y_0) \approx (4.2051, 1.9880)$. Parameters ($\alpha = 1.5, \delta = 0.75, Q_0 = 5.0$).

(c)

(d)

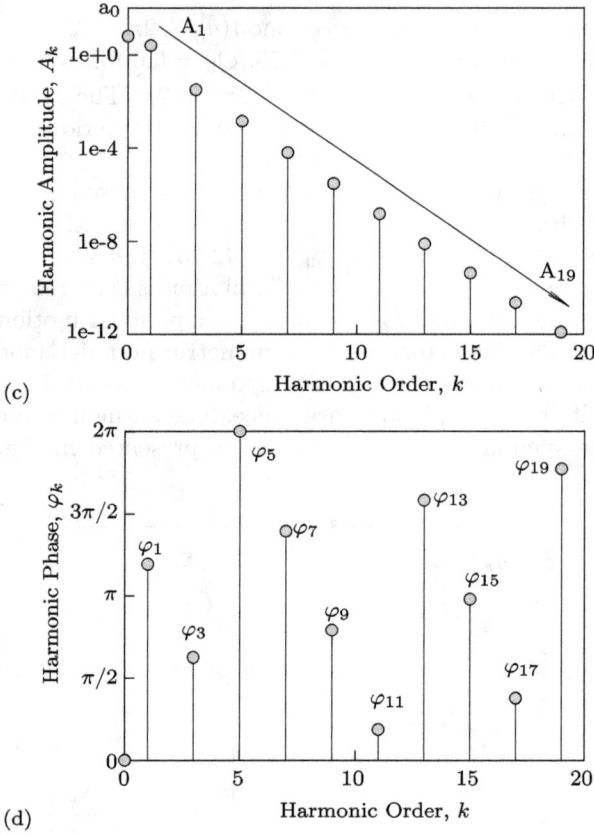

Fig. 6.18 Continued.

corresponding constant is $a_0 = 2\pi$. If $\mod(x_0, 2\pi) \approx 4.2051$ with the same velocity of $y_0 \approx 1.9880$, there are also infinite solutions of period-1 motions with $\mod(a_0, 2\pi) = 2\pi$. For the symmetric perod-1 motion, $A_{(2l-1)} \neq 0$ but $A_{2l} = 0$ $(l = 1, 2, \cdots)$. The main harmonic amplitudes are $A_1 \approx 2.5039$ and $A_3 \approx 0.0317$. The other harmonic amplitudes are $A_{(2l-1)} \in (10^{-12}, 10^{-2})$ for $l = 3, 4, \cdots, 10$. The harmonic phases are in $\varphi_{2l-1} \in [0, 2\pi)$ for $l = 1, 2, \cdots, 10$, as shown in Fig.6.18(d).

With the saddle-node bifurcation, the non-travelable asymmetric pierod-1 motion exists with center $\mod(a_0, 2\pi)$ near 2π. Thus, two non-travelable asymmetric period-1 motions are presented in Fig.6.19 for $\Omega = 1.40$. The initial conditions of the asymmetric period-1 motions are $(x_0, y_0) \approx (5.4339, 2.1523)$ (black) and $(x_0, y_0) \approx (2.9642, 1.3703)$ (red) from the analytical prediction. The trajectories of the two asymmetric period-1 motion in phase space are illustrated in Figs.6.19(a) and (b) on the left and right sides of $\mod(x, 2\pi) = 2\pi$, respectively. The harmonic amplitudes of the two asymmetric period-1 motions are presented in Fig.6.19(c). The two constants sat-

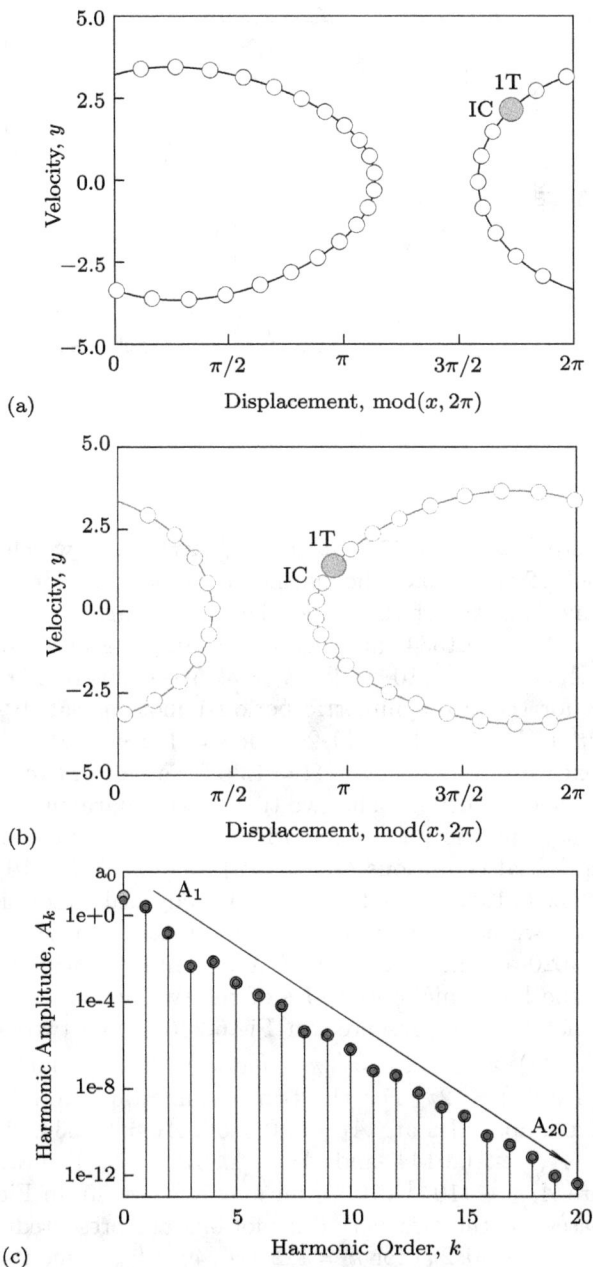

Fig. 6.19 Asymmetric period-1 motions ($\Omega = 1.40$): (a) trajectory of black branch (Right), IC: $(x_0, y_0) \approx (5.4339, 2.1523)$; (b) trajectory of red branch (left), IC: $(x_0, y_0) \approx (2.9642, 1.3703)$, (c) harmonic amplitudes, (d) harmonic phases. Parameters ($\alpha = 1.5, \delta = 0.75, Q_0 = 5.0$).

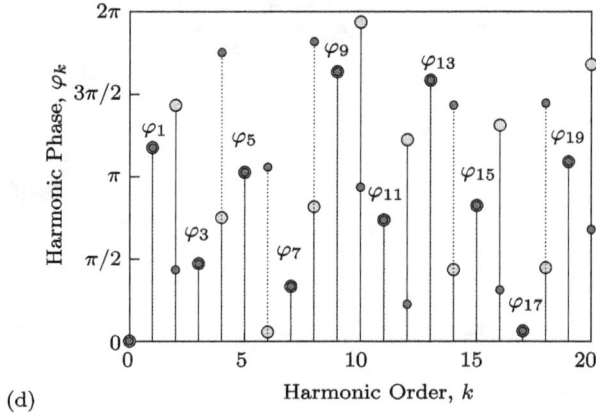

(d)

Fig. 6.19 Continued.

isfy $2\pi - \text{mod}(a_0^B, 2\pi) \approx 1.2726 \approx \text{mod}(a_0^R, 2\pi) - 2\pi$, which is for asymmetry of $\text{mod}(x, 2\pi) = 2\pi$. The harmonic amplitudes for two asymmetric period-1 motions are of the same. The main harmonic amplitudes are $A_1 \approx 2.4410$, $A_2 \approx 0.1534$, and $A_3 \approx 0.0410$. The other harmonic amplitudes are $A_k \in (10^{-14}, 10^{-2})$ for $k = 4, 5, \cdots, 20$. In Fig.6.19(d), harmonic phases for the two symmetric period-1 motions satisfy a relation of $\varphi_k^R = \text{mod}(\varphi_k^B + (k+1)\pi, 2\pi) \in [0, 2\pi)$ for $k = 1, 2, \cdots, 20$.

Consider excitation frequency of $\Omega = 1.36$ for a pair of two non-travelable asymmetric period-2 motions. The two trajectories, harmonic amplitude and phase of the asymmetric period-2 motions are presented in Fig.6.20. The corresponding initial conditions are $(x_0, y_0) \approx (5.9797, 1.7619)$ (black) and $(x_0, y_0) \approx (2.5288, 1.0577)$ (red) from the analytical prediction. The trajectories of the two asymmetric period-2 motions in phase space are illustrated in Figs.6.20(a) and (b) on the left and right sides of $\text{mod}(x, 2\pi) = 2\pi$, respectively. The harmonic amplitudes of the two non-travelable asymmetric period-2 motions are presented in Fig.6.20(c). The two constants satisfy $\text{mod}(a_0^{(2)B}, 2\pi) \approx 1.2726 \approx 2\pi - \text{mod}(a_0^{(2)R}, 2\pi)$, which is for asymmetry of $\text{mod}(x, 2\pi) = 2\pi$. For the two asymmetric perod-2 motions, the main harmonic amplitudes are $A_{1/2} \approx 0.5206$, $A_1 \approx 2.4334$, $A_{3/2} \approx 0.0766$, $A_2 \approx 0.1558$, $A_{5/2} \approx 0.0144$, and $A_3 \approx 1.3072e - 3$. The other harmonic amplitudes are $A_{k/2} \in (10^{-14}, 10^{-2})$ for $k = 7, 8, \cdots, 40$. In Fig.6.20(d), the harmonic phases for the two period-2 motions are presented and they are distributed in $\varphi_{k/2} \in [0, 2\pi)$ for $k = 1, 2, \cdots, 40$. $\varphi_{k/2}^R = \text{mod}(\varphi_{k/2}^B + (k(1 + 2r)/2 + 1)\pi, 2\pi), r = 1$ with $t_0 = rT$ ($r \in \{0, 1\}$).

On the bifurcation trees of non-travelable period-1 motions to chaos, after the period-doubling bifurcation of the non-travelable asymmetric period-2 motion, the non-travelable period-4 motions will exist. Consider the excitation frequency of $\Omega = 1.35$ for a pair of two non-travelable asymmetric

(a)

(b)

(c)

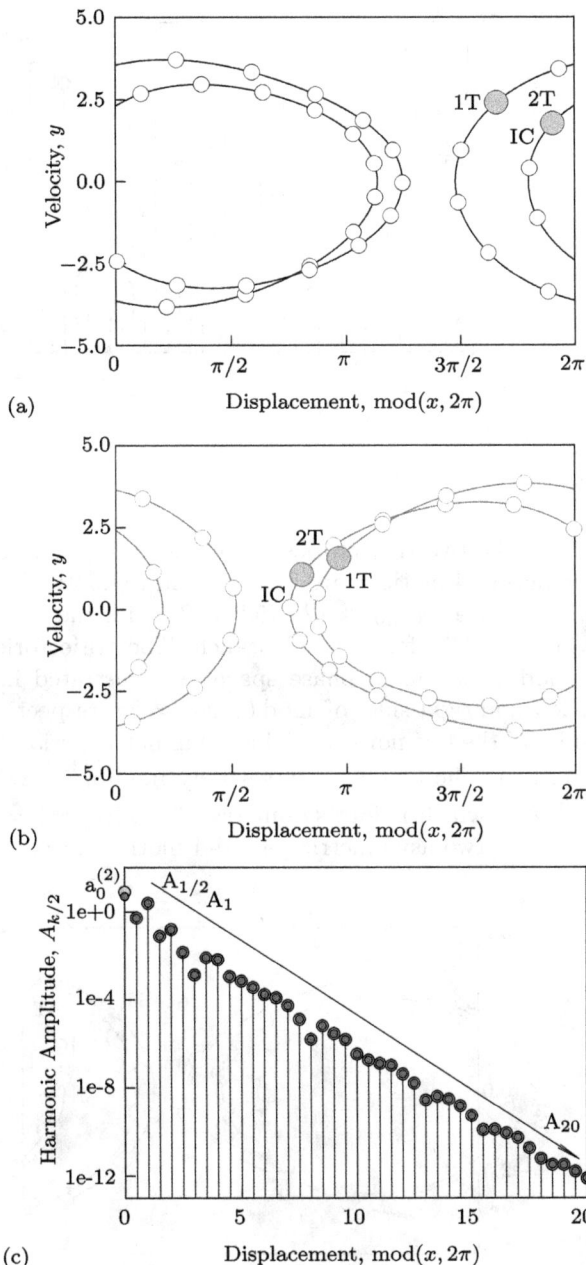

Fig. 6.20 Asymmetric period-2 motions ($\Omega = 1.36$): (a) trajectory of black branch (right), IC: $(x_0, y_0) \approx (5.9797, 1.7619)$; (b) trajectory of red branch (left), IC: $(x_0, y_0) \approx (2.5288, 1.0577)$; (c) harmonic amplitudes, (d) harmonic phases. Parameters ($\alpha = 1.5, \delta = 0.75, Q_0 = 5.0$).

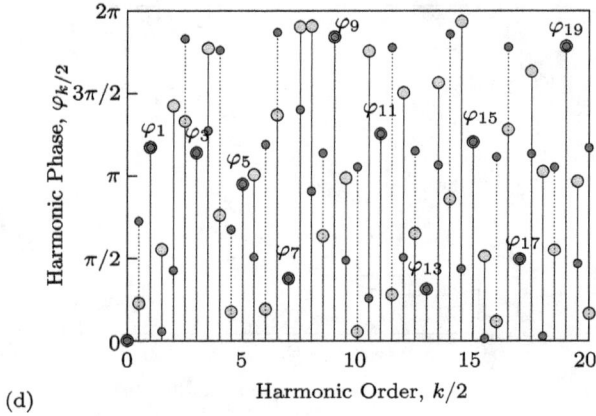

(d)

Fig. 6.20 Continued.

period-4 motions. The two trajectories, harmonic amplitudes and phases of the asymmetric period-4 motions are presented in Fig.6.21. The corresponding initial conditions are $(x_0, y_0) \approx (2.4152, 0.9947)$ for the black branch and $(x_0, y_0) \approx (2.4152, 0.9947)$ for the red branch. The trajectories of the two asymmetric period-4 motion in phase space are illustrated in Figs.6.21(a) and (b) on the left and right sides of mod $(x, 2\pi) = 2\pi$, respectively. The harmonic amplitudes of the two non-travelable asymmetric period-2 motions are placed in Fig.6.21(c). The two constants satisfy mod$(a_0^{(4)B}, 2\pi) \approx 1.4732 \approx 2\pi - \mathrm{mod}(a_0^{(4)R}, 2\pi)$, which is for asymmetry of mod$(x, 2\pi) = 2\pi$. The harmonic amplitudes for two asymmetric period-4 motions are of the same. For

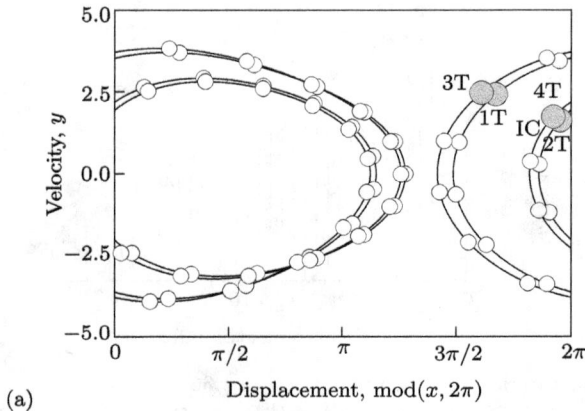

(a)

Fig. 6.21 Asymmetric period-4 motions ($\Omega = 1.35$): (a) trajectory of black branch (right), initial condition: $(x_0, y_0) \approx (6.0373, 1.7324)$; (b) trajectory of red branch (left), initial condition: $(x_0, y_0) \approx (2.4152, 0.9947)$, (c) harmonic amplitudes, (d) harmonic phases. Parameters ($\alpha = 1.5, \delta = 0.75, Q_0 = 5.0$).

(b)

(c)

(d)

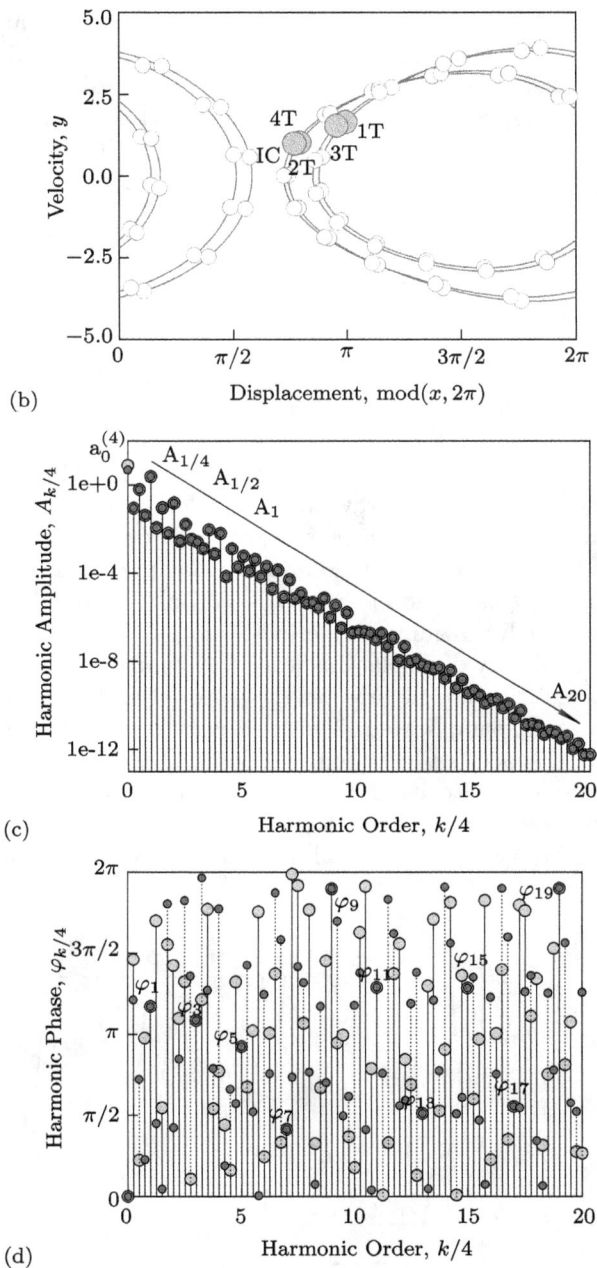

Fig. 6.21 Continued.

the two asymmetric perod-4 motions, the main harmonic amplitudes are $A_{1/4} \approx 0.0888$, $A_{1/2} \approx 0.6255$, $A_{3/4} \approx 0.0424$, $A_1 \approx 2.4353$, $A_{5/4} \approx 0.0117$, $A_{3/2} \approx 0.0924$, $A_{7/4} \approx 6.6020\mathrm{e}-3$, $A_2 \approx 0.1530$, $A_{9/4} \approx 2.8235\mathrm{e}-3$, $A_{5/2} \approx 0.0166$, $A_{11/4} \approx 3.4517\mathrm{e}-3$, and $A_3 \approx 2.5941\mathrm{e}-3$. The other harmonic amplitudes are $A_{k/4} \in (10^{-14}, 10^{-2})$ for $k = 12, 13, \cdots, 80$. In Fig.6.21(d), the corresponding harmonic phases are presented, which are distributed in the range $[0, 2\pi)$ for $\varphi_{k/4}^R = \mathrm{mod}\,(\varphi_{k/4}^B + (k(1+2r)/4+1)\pi, 2\pi)$, $k = 1, 2, \cdots, 80$ and $r = 1$ with $t_0 = rT$ ($r \in \{0, 1, 2, 3\}$).

6.5.1.2 Period-3 to period-6 motions

In this section, symmetric and asymmetric period-3 motions and period-6 motions will be presented for illustration of complexity in different bifurcation trees. Such a bifurcation tree is closed and independent. The periodic motion on the bifurcation tree is non-travelable. The center for periodic motions of $\mathrm{mod}\,(x, 2\pi)$ in such bifurcation tree is near $(2\pi, 0)$ or $(\pi, 0)$.

For $\Omega = 0.762$, a symmetric period-3 motion is presented in Fig.6.22. The initial condition of this motion is $(x_0, y_0) \approx (3.7268, 2.9670)$ from the analytical prediction. Such a symmetric period-3 motion is centered at $(2\pi, 0)$ or $(0, 0)$. The time history of displacement for the symmetric period-3 motion is presented in Fig.6.22(a). Compared to the period-1 motion, the period-3 motion becomes much complicated. The trajectory of the symmetric period-3 motion in phase space is illustrated in Fig.6.22(b). The symmetric period-3 motion is symmetric to itself about $(2\pi, 0)$ in phase space. Three cycles around the center of $(2\pi, 0)$ or $(0, 0)$ are formed. The harmonic amplitudes of the non-travelable symmetric period-3 motion are presented in Fig.6.22(c).

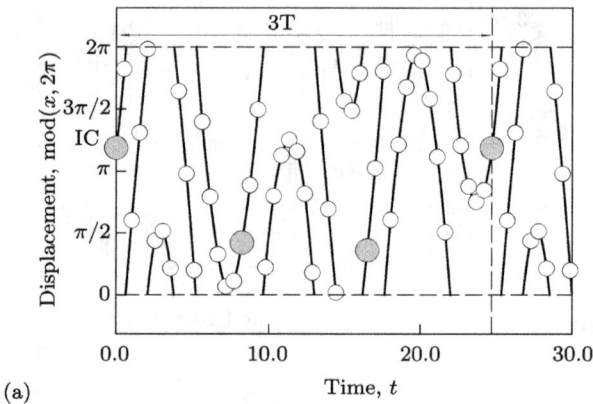

(a)

Fig. 6.22 Symmetric period-3 motions ($\Omega = 0.762$): (a) Displacement, (b) trajectory, (c) harmonic amplitudes, (d) harmonic phases. Parameters ($\alpha = 1.5, \delta = 0.75, Q_0 = 5.0$). Initial condition: $(x_0, y_0) \approx (3.7268, 2.9670)$.

(b)

(c)

(d)

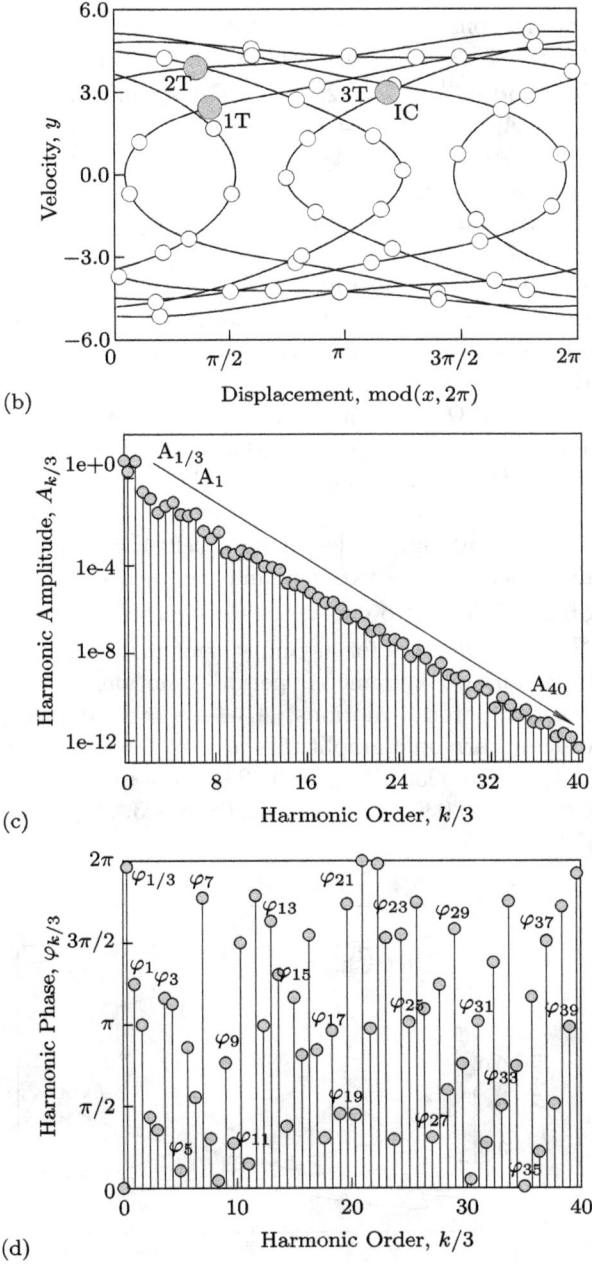

Fig. 6.22 Continued.

The corresponding constant is $a_0^{(3)} = 2\pi$. If $\mathrm{mod}(x_0, 2\pi) \approx 3.7268$ with the same velocity of $y_0 \approx 2.9670$, there are infinite solutions of period-3 motions with $\mathrm{mod}(a_0^{(3)}, 2\pi) = 2\pi$ or 0. For symmetric perod-3 motions, $A_{(2l-1)/3} \neq 0$ but $A_{(2l)/3} = 0$ $(l = 1, 2, \cdots)$. The main harmonic amplitudes are $A_{1/3} \approx 2.0410$, $A_1 \approx 5.9571$, $A_{5/3} \approx 0.2446$, $A_{7/3} \approx 0.1183$, $A_3 \approx 0.0271$, $A_{11/3} \approx 0.0545$, $A_5 \approx 0.0222$, $A_{17/3} \approx 0.0193$, $A_{19/3} \approx 0.0239$, and $A_7 \approx 3.7577\mathrm{e}-3$. The other harmonic amplitudes are $A_{(2l-1)/3} \in (10^{-12}, 10^{-2})$ for $l = 22, 23, \cdots, 60$. The harmonic phases are distributed in $\varphi_{(2l-1)/3} \in [0, 2\pi)$ for $l = 1, 2, \cdots, 60$, as shown in Fig.6.22(d).

With the saddle-node bifurcation of symmetric period-3 motion, the non-travelable asymmetric pierod-3 motion exists with center near $\mathrm{mod}(a_0^{(3)}, 2\pi) = 2\pi$. Thus, a pair of two non-travelable asymmetric period-3 motions is presented in Fig.6.23 for $\Omega = 0.758$. The initial conditions of the asymmetric period-3 motions are $(x_0, y_0) \approx (1.3009, 3.4547)$ (black) and $(x_0, y_0) \approx (1.1077, 3.9509)$ (red) from the analytical prediction. The trajectories of the two asymmetric period-3 motion in phase space are illustrated in Figs.6.23(a) and (b) on the left and right sides of $\mathrm{mod}(x, 2\pi) = 2\pi$, respectively. The harmonic amplitudes of the two asymmetric period-3 motions are presented in Fig.6.23(c). The two constants satisfy $\mathrm{mod}(a_0^{(3)B}, 2\pi) \approx 0.1137 \approx 2\pi - \mathrm{mod}(a_0^{(3)R}, 2\pi)$, which is for asymmetry of $\mathrm{mod}(x, 2\pi) = 2\pi$. The harmonic amplitudes for two asymmetric period-3 motions are of the same. For the two asymmetric perod-3 motions, $A_{(2l-1)/3} \neq 0$ and $A_{(2l)/3} \neq 0$ $(l = 1, 2, \cdots)$. The main harmonic amplitudes are $A_{1/3} \approx 1.8145$, $A_{2/3} \approx 0.2944$, $A_1 \approx 5.8967$, $A_{4/3} \approx 0.0936$, $A_{5/3} \approx 0.2235$, $A_2 \approx 0.0243$, $A_{7/3} \approx 0.1138$, $A_{8/3} \approx 0.0204$, $A_3 \approx 0.0309$, $A_{10/3} \approx 7.2685\mathrm{e}-3$, $A_{11/3} \approx 0.0574$, $A_4 \approx$

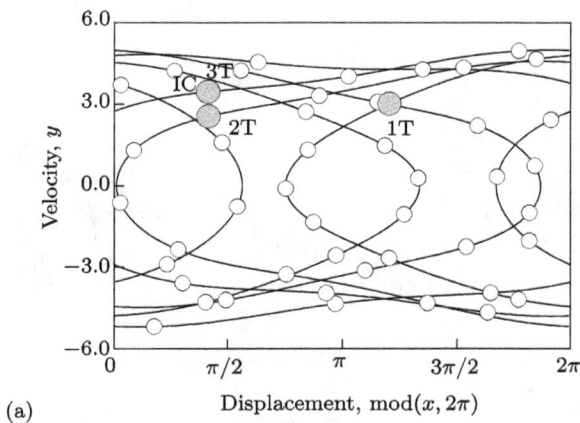

(a)

Fig. 6.23 Asymmetric period-3 motions ($\Omega = 0.758$): (a) trajectory of black branch (right), initial condition: $(x_0, y_0) \approx (1.3009, 3.4547)$; (b) trajectory of red branch (left), initial condition: $(x_0, y_0) \approx (1.1077, 3.9509)$, (c) harmonic amplitudes, (d) harmonic phases. Parameters ($\alpha = 1.5, \delta = 0.75, Q_0 = 5.0$).

(b)

(c)

(d)

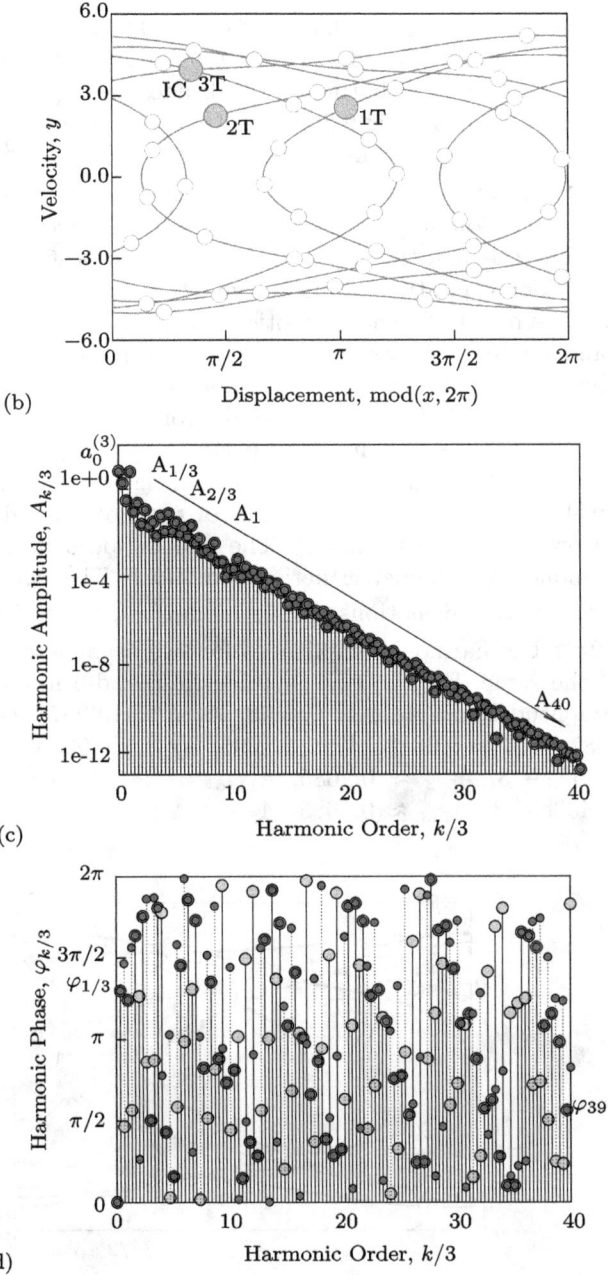

Fig. 6.23 Continued.

0.0112, $A_{13/3} \approx 0.0752$, $A_{14/3} \approx 0.0116$, $A_{15} \approx 0.0298$, $A_{16/3} \approx 8.4148e - 3$, $A_{17/3} \approx 0.0160$, $A_6 \approx 6.3730e - 3$, $A_{19/3} \approx 0.0215$, $A_{20/3} \approx 3.3613e - 3$, and $A_7 \approx 5.2751e - 3$. The other harmonic amplitudes are $A_{k/3} \in (10^{-14}, 10^{-2})$ for $k = 21, 22, \cdots, 120$. In Fig.6.23(d), harmonic phases are presented for $\varphi_{k/3} \in [0, 2\pi)$ with $\varphi_{k/(2^l m)}^R = \mod(\varphi_{k/(2^l m)}^B + k((m + 2r)/(2^l m) + 1)\pi, 2\pi)$ for $l = 0$, $m = 3$, $r = 0$ and $t_0 = rT$ for $r \in \{0, 1, \cdots, 2^l m - 1\}$.

After the period-doubling bifurcation of the non-travelable asymmetric period-3 motion, the non-travelable asymmetric period-6 motions will be obtained. Consider excitation frequency of $\Omega = 0.756$ for a pair of two non-travelable asymmetric period-6 motions. The two trajectories, harmonic amplitude and phase of the asymmetric period-6 motions are presented in Fig.6.24. The corresponding initial conditions are $(x_0, y_0) \approx (1.3241, 3.2092)$ (black) and $(x_0, y_0) \approx (1.2328, 3.7937)$ (red) from the analytical prediction. The trajectories of the two asymmetric period-6 motion in phase space are illustrated in Figs.6.24(a) and (b) on the left and right sides of $\mod(x, 2\pi) = 2\pi$, respectively. The harmonic amplitudes of the two asymmetric period-6 motions are presented in Fig.6.24(c). The two period-6 motions are close to the corresponding asymmetric period-3 motions. The two constants for the asymmetric period-6 motions satisfy $\mod(a_0^{(6)B}, 2\pi) \approx 0.0905 \approx 2\pi - \mod(a_0^{(6)R}, 2\pi)$. The harmonic amplitudes for two asymmetric period-6 motions are the same. For the two asymmetric perod-6 motions, the main harmonic amplitudes are $A_{1/6} \approx 0.0403$, $A_{1/3} \approx 1.6297$, $A_{1/2} \approx 0.0328$, $A_{2/3} \approx 0.3681$, $A_{5/6} \approx 0.0318$, $A_1 \approx 5.8296$, $A_{7/6} \approx 0.0215$, $A_{4/3} \approx 0.1164$, $A_{3/2} \approx 3.7209e - 3$, $A_{5/3} \approx 0.2085$, $A_{11/6} \approx 2.8707e - 3$, $A_2 \approx 0.0305$, $A_{13/6} \approx 1.8433e - 3$, $A_{7/3} \approx 0.1083$, $A_{5/2} \approx 2.9228e - 3$, $A_{8/3} \approx 0.0321$,

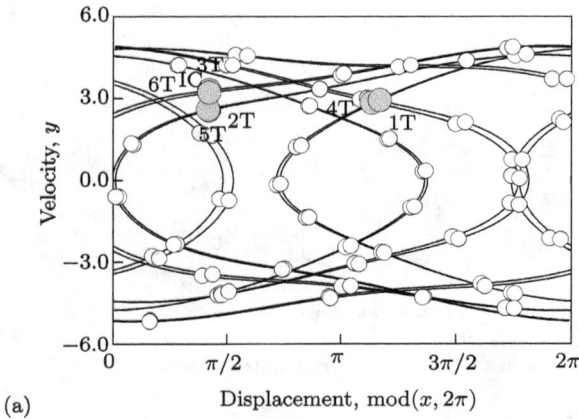

(a)

Fig. 6.24 Asymmetric period-6 motions ($\Omega = 0.756$): (a) trajectory of black branch (right), initial condition: $(x_0, y_0) \approx (1.3241, 3.2092)$; (b) trajectory of red branch (left), initial condition: $(x_0, y_0) \approx (1.2328, 3.7937)$, (c) harmonic amplitudes, (d) harmonic phases. Parameters ($\alpha = 1.5, \delta = 0.75, Q_0 = 5.0$).

(b)

(c)

(d)

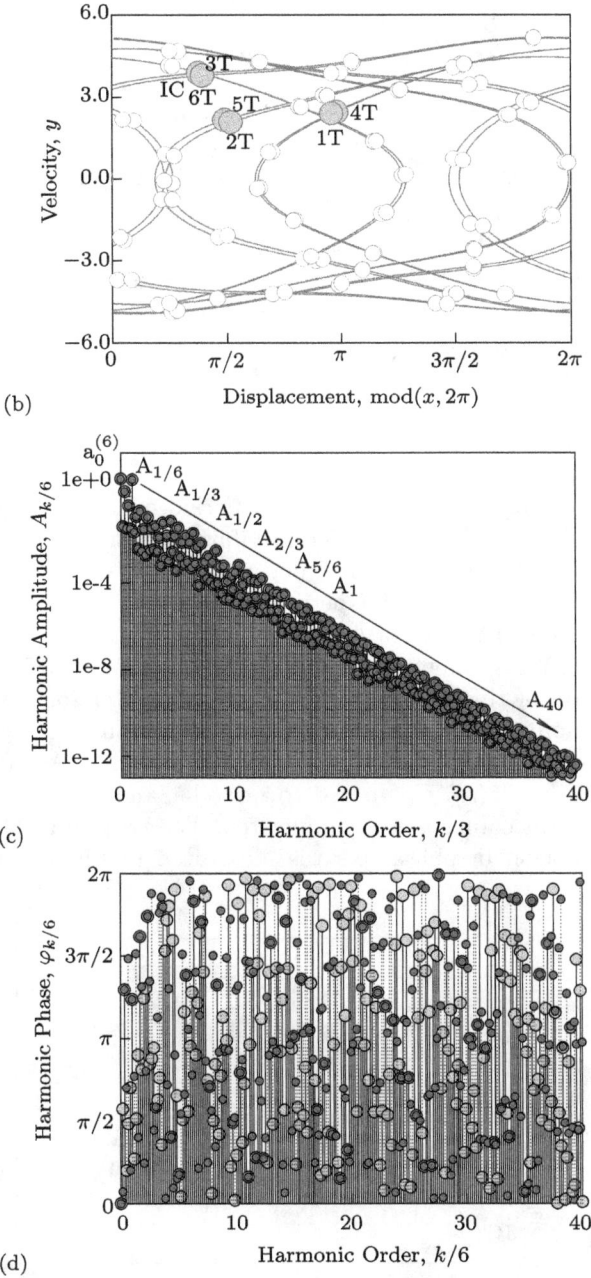

Fig. 6.24 Continued.

$A_{17/6} \approx 3.3253e - 3$, $A_3 \approx 0.0399$, $A_{19/6} \approx 2.8363e - 3$, $A_{10/3} \approx 0.0122$, $A_{7/2} \approx 6.4339e - 4$, $A_{11/3} \approx 0.0598$, $A_{23/6} \approx 1.2937e - 3$, $A_4 \approx 1.2937e - 3$, $A_{25/6} \approx 1.0601e - 3$, $A_{13/3} \approx 0.0711$, $A_{9/2} \approx 2.5821e - 3$, $A_{14/3} \approx 0.0139$, $A_{29/6} \approx 3.6914e - 4$, $A_5 \approx 0.0353$, $A_{31/6} \approx 1.5019e - 3$, $A_{16/3} \approx 7.1537e - 3$, $A_{11/2} \approx 7.1515e - 4$, $A_{17/3} \approx 0.0142$, $A_{35/6} \approx 8.6713e - 4$, $A_6 \approx 7.0971e - 3$, $A_{37/6} \approx 4.8717e-4$, $A_{19/3} \approx 0.0193$, $A_{13/2} \approx 7.3324e-4$, $A_{20/3} \approx 4.3342e-3$, $A_{41/6} \approx 7.5179e - 5$, and $A_7 \approx 6.2794e - 3$. The other harmonic amplitudes are $A_{k/6} \in (10^{-13}, 10^{-2})$ for $k = 43, 44, \cdots, 240$. In Fig.6.24(d), harmonic phases are presented and distributed in $\varphi_{k/6} \in [0, 2\pi)$ for $k = 1, 2, \cdots, 240$. $\varphi^R_{k/(2^l m)} = \mod(\varphi^B_{k/(2^l m)} + k((m + 2r)/(2^l m) + 1)\pi, 2\pi)r = 0$, $m = 3, l = 1$ with $t_0 = rT$ ($r \in \{0, 1, \cdots, 2^l m - 1\}$).

6.5.1.3 Period-5 motions

In this section, symmetric and asymmetric period-5 motions will be presented for illustration of complexity in different bifurcation trees. Such a bifurcation tree is closed and independent. The periodic motion on the bifurcation tree is non-travelable. The center for periodic motions of $\mod(x, 2\pi)$ in such bifurcation tree is near $(2\pi, 0)$ or $(\pi, 0)$.

For $\Omega = 1.265$, a symmetric period-5 motion is presented in Fig.6.25. The initial condition of this motion is $(x_0, y_0) \approx (0.4253, 1.2591)$ from the analytical prediction. Such a symmetric period-5 motion is centered at $(\pi, 0)$. The time history of displacement for the symmetric period-5 motion is presented in Fig.6.25(a). Compared to period-1 and period-3 motions, the period-5 motion becomes much complicated. The trajectory of the symmetric period-5 motion in phase space is illustrated in Fig.6.25(b). The sym-

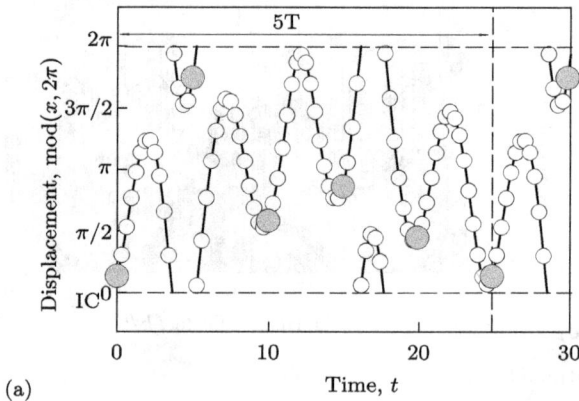

Fig. 6.25 Symmetric period-5 motions ($\Omega = 1.265$): (a) displacement, (b) trajectory, (c) harmonic amplitudes, (d) harmonic phases. Parameters ($\alpha = 1.5, \delta = 0.75, Q_0 = 5.0$). Initial condition: $(x_0, y_0) \approx (0.4253, 1.2591)$.

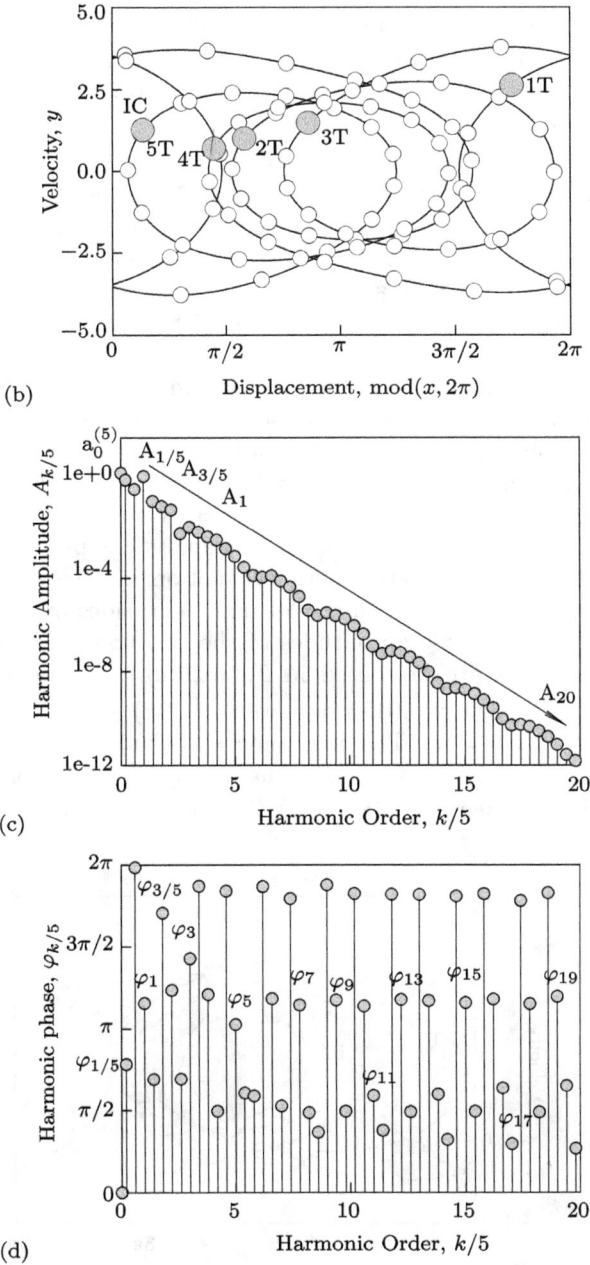

(b)

(c)

(d)

Fig. 6.25 Continued.

metric period-5 motion is symmetric to itself about $(\pi, 0)$ in phase space. Three cycles around the center of $(\pi, 0)$ are formed. The harmonic amplitude of the symmetric period-5 motion is presented in Fig.6.25(c). The corresponding constant is $a_0^{(5)} = \pi$. If $\text{mod}(x_0, 2\pi) \approx 0.4253$ and the same velocity of $y_0 \approx 1.2591$, there are infinite solutions of period-5 motions with $\text{mod}(a_0^{(5)}, 2\pi) = \pi$. For symmetric perod-5 motions, $A_{(2l-1)/5} \neq 0$ but $A_{(2l)/5} = 0$ $(l = 1, 2, \cdots)$. The main harmonic amplitudes are $A_{1/5} \approx 1.5831$, $A_{3/5} \approx 0.6387$, $A_1 \approx 2.2893$, $A_{7/5} \approx 0.1963$, $A_{9/5} \approx 0.1186$, $A_{11/5} \approx 0.0839$, $A_{13/5} \approx 7.8925\text{e} - 3$, and $A_3 \approx 0.0151$. The other harmonic amplitudes are $A_{(2l-1)/5} \in (10^{-12}, 10^{-2})$ for $l = 16, 17, \cdots, 100$. The harmonic phases are distributed in $\varphi_{(2l-1)/5} \in [0, 2\pi)$ for $l = 1, 2, \cdots, 100$, as shown in Fig.6.25(d).

With the saddle-node bifurcation of symmetric period-5 motion, the non-travelable asymmetric pierod-5 motion exists with center near mod $(a_0, 2\pi) = \pi$. Thus, a pair of two non-travelable asymmetric period-5 motions is presented in Fig.6.26 for $\Omega = 1.263$. The initial conditions of the asymmetric period-5 motions are $(x_0, y_0) \approx (1.7478, 0.9721)$ (black) and $(x_0, y_0) \approx (1.4191, 0.7131)$ (red) from the analytical prediction. The trajectories of the two asymmetric period-5 motion in phase space are illustrated in Figs.6.26(a) and (b) on the left and right sides of $\text{mod}(x, 2\pi) = \pi$, respectively. The harmonic amplitudes of the two non-travelable asymmetric period-5 motions are presented in Fig.6.26(c). The two constants satisfy $\pi - \text{mod}(a_0^{(5)B}, 2\pi) \approx 0.0155 \approx \text{mod}(a_0^{(5)R}, 2\pi) - \pi$, which is for asymmetry of $\text{mod}(x, 2\pi) = \pi$. The harmonic amplitudes for two asymmetric period-5 motions are of the same. For the two asymmetric perod-5 motions, $A_{(2l-1)/5} \neq 0$ and $A_{(2l)/5} \neq 0$ $(l = 1, 2, \cdots)$. The main harmonic ampli-

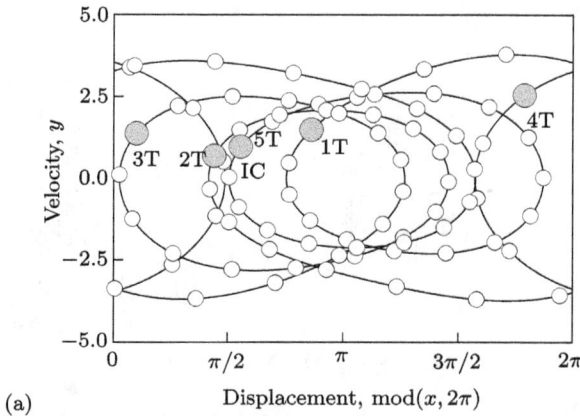

(a)

Fig. 6.26 Asymmetric period-5 motions $(\Omega = 1.263)$: (a) trajectory of black branch (right), IC: $(x_0, y_0) \approx (1.7478, 0.9721)$; (b) trajectory of red branch (left), IC: $(x_0, y_0) \approx (1.4191, 0.7131)$, (c) harmonic amplitudes, (d) harmonic phases. Parameters $(\alpha = 1.5, \delta = 0.75, Q_0 = 5.0)$.

(b)

(c)

(d)

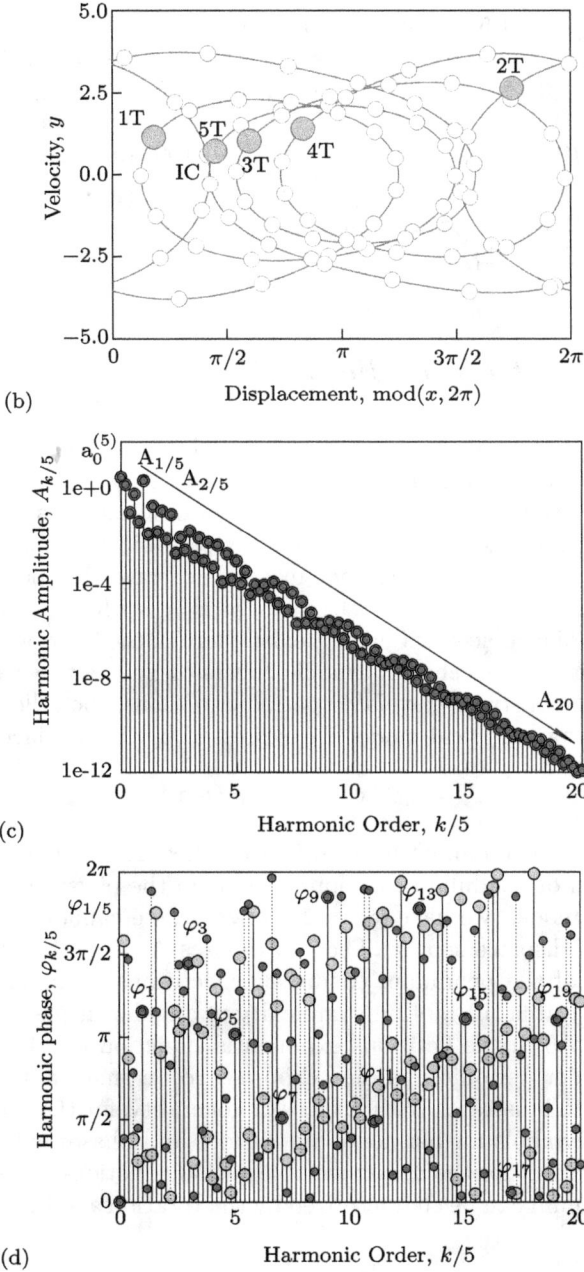

Fig. 6.26 Continued.

tudes are $A_{1/5} \approx 1.5329$, $A_{2/5} \approx 0.1013$, $A_{3/5} \approx 0.6173$, $A_{4/5} \approx 0.0410$, $A_1 \approx 2.2816$, $A_{6/5} \approx 0.0124$, $A_{7/5} \approx 0.1906$, $A_{8/5} \approx 0.0148$, $A_{9/5} \approx 0.1187$, $A_2 \approx 7.7201e - 3$, $A_{11/5} \approx 0.0838$, $A_{12/5} \approx 1.8895e - 3$, $A_{13/5} \approx 8.7981e - 3$, $A_{14/5} \approx 2.5159e - 3$, and $A_1 \approx 0.0161$. The other harmonic amplitudes are $A_{k/5} \in (10^{-14}, 10^{-2})$ for $k = 16, 17, \cdots, 100$. In Fig.6.23(d), harmonic phases are presented for $\varphi_{k/3} \in [0, 2\pi)$ with $\varphi_{k/(2^l m)}^R = \mathrm{mod}(\varphi_{k/(2^l m)}^B + k((m + 2r)/(2^l m) + 1))\pi, 2\pi)$ for $l = 0$, $m = 5$, $r = 2$ and $t_0 = rT$ for $r \in \{0, 1, \cdots, 2^l m - 1\}$.

6.5.2 Travelable periodic motions

In this section, travelable period-1 to period-2 motions will be presented for illustration of complexity in different bifurcation trees. Such a bifurcation tree is closed and independent. The periodic motions on the bifurcation tree are travelable. *No any centers* for periodic motions exist, and the symmetric periodic motions cannot exist. The travelable period-m motion has $\mathrm{mod}(x_0, 2\pi) = \mathrm{mod}(x_{mT}, 2\pi)$ but $x_0 \neq x_{mT}$ with $y_0 = y_{mT}$. Thus the Fourier series of displacement cannot exist. Herein, the Fourier series of velocity will be presented to show harmonic effects on such a travelable period-m motion. To demonstrate travelable periodic motions, the following formulae for coordinates in physical model are expressed by the displacement as

$$X = l_0 e^{at} \cos x, \quad Y = l_0 e^{at} \sin x \tag{6.34}$$

where l_0 is the pendulum length, e^{at} is the fictitious exponential function for illustration of pendulum rotation motions in the physical model. For the real physical model, we have $e^{at} = 1$. However, we cannot see the rotation complexity of displacement x. The coordinates X, Y in the physical model are locations of the pendulum. Using such fictitious exponential functions, we can easily watch the motion complexity of angular displacement, and the coefficient $a > 0$ in the fictitious exponential function is arbitrarily chosen. For higher-order periodic motions, choose a to be smaller. Otherwise, it is very difficult to illustrate. Without loss of generality, for the Fourier series of velocity, the symbols for harmonic amplitudes and phases will use the same as for displacement for the non-travelable periodic motions. The periodic motions in pendulum can be characterized by the rotation and libration numbers as follow:

$$(R_+ : R_- : L), \tag{6.35}$$

where R_+ is the number of positive rotations, R_- is the number of negative rotations, and L is the number of librations. The libration number consists of positive librations and negative librations. $L = L_+ + L_-$ and $L_+ = L_-$ for periodic motions. For non-travelable period-1 motions, we have $R_+ = R_-$.

Definition 6.3 If a periodic motion in pendulum possesses R_+ positive rotations, R_- negative rotations and L librations, then the periodic motion is called the $(R_+ : R_- : L)$-*periodic motion in pendulum*.

For following illustrations, consider the parameters

$$\alpha = 1.5, \quad \delta = 0.75, \quad \Omega = 1. \tag{6.36}$$

6.5.2.1 Travelable period-1 to period-2 motions

On the bifurcation trees of travelable pierod-1 motion to chaos, a pair of two non-travelable asymmetric period-1 motions is presented in Fig.6.27 for $Q_0 = 3.0$. The initial conditions of the travelable asymmetric period-1 motions are $(x_0, y_0) \approx (0.7380, 2.8316)$ (black) and $(x_0, y_0) \approx (0.0665, 0.8547)$ (red) from the analytical prediction. The trajectories of the two asymmetric period-1 motions in phase space are illustrated in Figs.6.27(a) and (b). The velocity harmonic amplitudes of the two travelable asymmetric period-1 motions are presented in Fig.6.27(c). The two velocity constants satisfy $a_0^B = -a_0^R = 1$, which is determined by angular velocity $\dot{x} = y$ and $y = \Omega = 1$. The velocity harmonic amplitudes for two travelable asymmetric period-1 motions are of the same. The main velocity harmonic amplitudes are $A_1 \approx 2.0164$, $A_2 \approx 0.6047$, $A_3 \approx 0.2348$, $A_4 \approx 0.0289$, and $A_5 \approx 0.0150$. The other velocity harmonic amplitudes are $A_k \in (10^{-14}, 10^{-2})$ for $k = 6, 7, \cdots, 30$. In Fig.6.27(d), velocity harmonic phases are presented for $\varphi_k \in [0, 2\pi)$ with $\varphi_k^R = \mathrm{mod}(\varphi_k^B + (k + 1)\pi, 2\pi)$ for $k = 1, 2, \cdots, 30$. To illustrate displacement complexity of the travelable period-1 motions, the trajectories of fictitious coordinates in (X, Y)-space are presented in Figs.6.27(e) and (f) for the black and red branches of the travelable period-1 motions. The parameters in fictitious exponential function are $l_0 = 1$, $a = 0.2$. We have $x_0 = 0.7380$ and $x_T = 7.0212$ for the black branch. The travelable period-1 motion with one cycle (2π) including one positive rotation plus two librations is on the *positive* direction of $x(t)$. This pendulum rotation is not simple positive and negative rotations. Such a positive travelable period-1 motion is of $(R_+ : R_- : L) = (1 : 0 : 2)$-kind. For the red branch, we have $x_0 = 0.0665$ and $x_T = -6.2167$ for the red branch. The travelable period-1 motion with one negative cycle (-2π) including one negative rotation plus two librations is on the *negative* direction of $x(t)$. The negative travelable period-1 motion is of $(R_+ : R_- : L) = (0 : 1 : 2)$-kind.

After the period-doubling bifurcation of the travelable asymmetric period-1 motion, the travelable asymmetric period-2 motions will be obtained. Consider excitation amplitude of $Q_0 = 3.2$ for a pair of two travelable asymmetric period-2 motions. The two trajectories, velocity harmonic amplitude and phase of the travelable asymmetric period-2 motions are presented in Fig.6.28. The corresponding initial conditions are $(x_0, y_0) \approx (6.1621, 3.3869)$ (black) and $(x_0, y_0) \approx (0.0977, 0.9019)$ (red) from the analytical prediction.

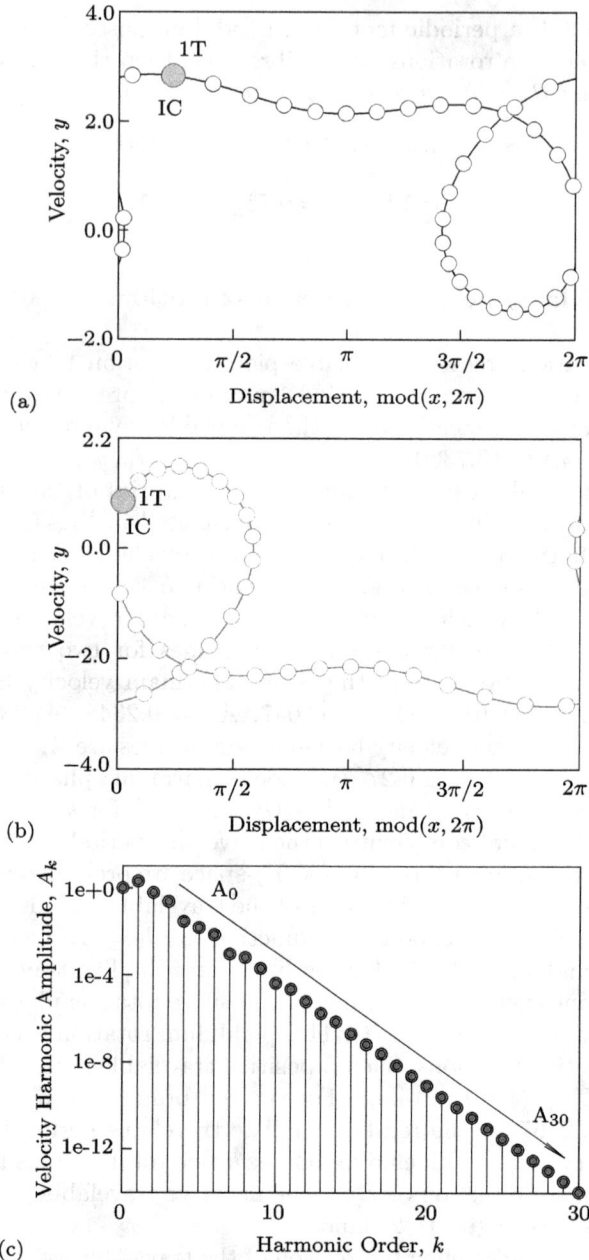

Fig. 6.27 Travelable asymmetric period-1 motions ($Q_0 = 3.0$): (a) trajectory of black branch (right), IC: $(x_0, y_0) \approx (0.7380, 2.8316)$; (b) trajectory of red branch (left), IC: $(x_0, y_0) \approx (0.0665, 0.8547)$, (c) harmonic amplitudes, (d) harmonic phases; (e) pendulum rotation pattern (black), (f) pendulum rotation pattern (red). Parameters ($\alpha = 1.5, \delta = 0.75, \Omega = 1$). $X = e^{0.2t} \cos x$, $Y = e^{0.2t} \sin x$.

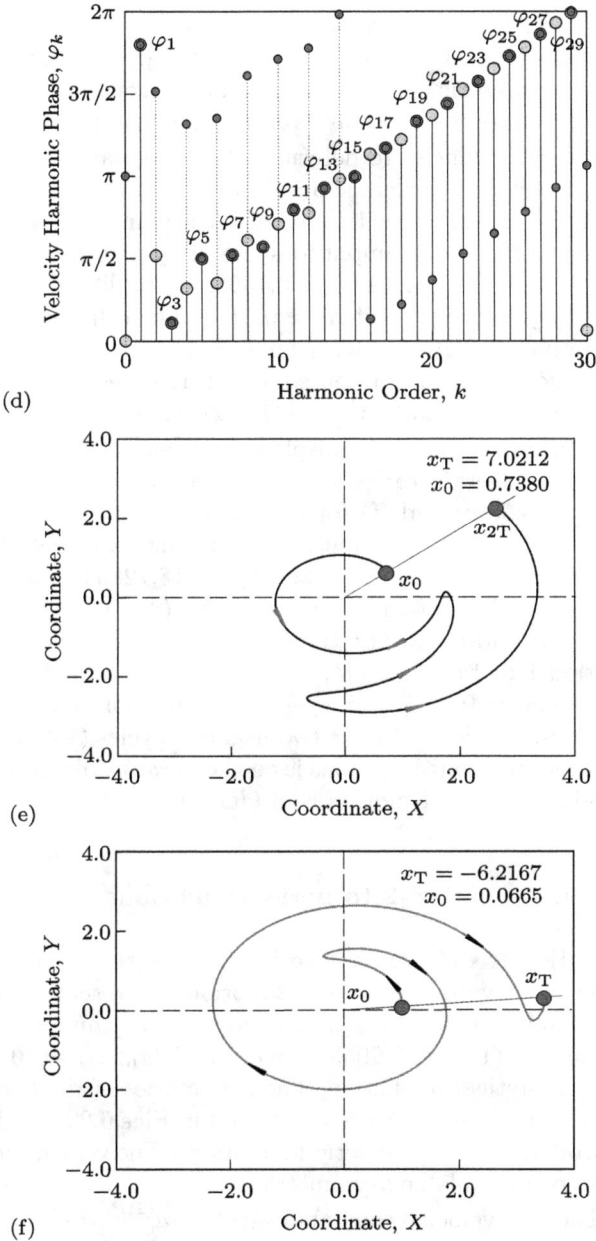

(d)

(e)

(f)

Fig. 6.27 Continued.

The trajectories of the two asymmetric period-2 motion in phase space are illustrated in Figs.6.28(a) and (b). The velocity harmonic amplitudes of the two travelable asymmetric period-2 motions are presented in Fig.6.28(c). The two velocity constants for the travelable asymmetric period-2 motions are $a_0^{(2)B} = -a_0^{(2)R} = 1$, which are determined by angular velocity $\dot{x} = y$ and $y = \Omega = 1$. The velocity harmonic amplitudes for two asymmetric period-2 motions are of the same. For the two travelable asymmetric period-2 motions, the main velocity harmonic amplitudes are $A_{1/2} \approx 0.1581$, $A_1 \approx 2.3728$, $A_{3/2} \approx 0.0674$, $A_2 \approx 0.4764$, $A_{5/2} \approx 0.1398$, $A_3 \approx 0.3130$, $A_{7/2} \approx 0.0367$, $A_4 \approx 0.0502$, $A_{9/2} \approx 0.0121$, and $A_5 \approx 0.0113$. The other velocity harmonic amplitudes are $A_{k/2} \in (10^{-13}, 10^{-2})$ for $k = 11, 12, \cdots, 60$. In Fig.6.28(d), velocity harmonic phases are presented and distributed in $\varphi_{k/2} \in [0, 2\pi)$ for $k = 1, 2, \cdots, 60$. $\varphi_{k/2}^R = \mathrm{mod}(\varphi_{k/2}^B + k(1 + 2r)/2 + 1)\pi, 2\pi)$ for $r = 0$ with $t_0 = rT$ ($r \in \{0, 1\}$). To illustrate displacement complexity of the travelable period-2 motions, the trajectories of fictitious coordinates in (X, Y)-space are presented in Figs.6.28(e) and (f) for the black and red branch of the travelable period-1 motions. The parameters in fictitious exponential function are $l_0 = 1$, $a = 0.1$. We have $x_0 = 6.1621$ and $x_{2T} = 18.7285$ for the black branch. The travelable period-2 motion with two cycles (4π) including two positive rotations plus four librations is on the *positive* direction of $x(t)$. The positive travelable period-1 motion is of $(R_+ : R_- : L) = (2 : 0 : 4)$-kind. For the red branch, we have $x_0 = 0.0977$ and $x_{2T} = -12.4687$ for the red branch. The travelable period-2 motion with the two negative cycles (-4π) including two negative rotations plus four librations is on the *negative* direction of $x(t)$. The negative travelable period-1 motion is of $(R_+ : R_- : L) = (0 : 2 : 4)$-kind.

6.5.2.2 Travelable period-2 to period-8 motions

On the bifurcation trees of travelable pierod-2 motion to chaos, a pair of two non-travelable asymmetric period-2 motions is presented in Fig.6.29 for $Q_0 = 7.37$. The initial conditions of the travelable asymmetric period-2 motions are $(x_0, y_0) \approx (1.2414, 2.2053)$ (black) and $(x_0, y_0) \approx (0.1376, 4.7541)$ (red) from the analytical prediction. The trajectories of the two asymmetric period-2 motion in phase space are illustrated in Figs.6.29(a) and (b), which are different from the travelable period-2 motions. The velocity harmonic amplitudes of the two travelable asymmetric period-2 motions are presented in Fig.6.29(c). The two velocity constants satisfy $-a_0^{(2)B} = a_0^{(2)R} = 0.5$, which is determined by the angular velocity $\dot{x} = y$ and $y = \Omega/2 = 0.5$. The velocity harmonic amplitudes for two travelable asymmetric period-2 motions are of the same. The main velocity harmonic amplitudes are $A_{1/2} \approx 0.4784$, $A_1 \approx 5.9068$, $A_{3/2} \approx 0.2423$, $A_2 \approx 0.3507$, $A_{5/2} \approx 0.0574$, $A_3 \approx 0.0720$, $A_{7/2} \approx 0.1281$, $A_4 \approx 0.0875$, $A_{9/2} \approx 0.2367$, $A_5 \approx 0.1149$, $A_{11/2} \approx 0.0469$, $A_6 \approx 0.0863$, $A_{13/2} \approx 0.0895$, $A_7 \approx 0.0307$, $A_{15/2} \approx 6.0042\mathrm{e} - 3$, $A_8 \approx 0.0175$, $A_{17/2} \approx 0.0128$, and $A_9 \approx 6.0044\mathrm{e} - 3$. The other veloc-

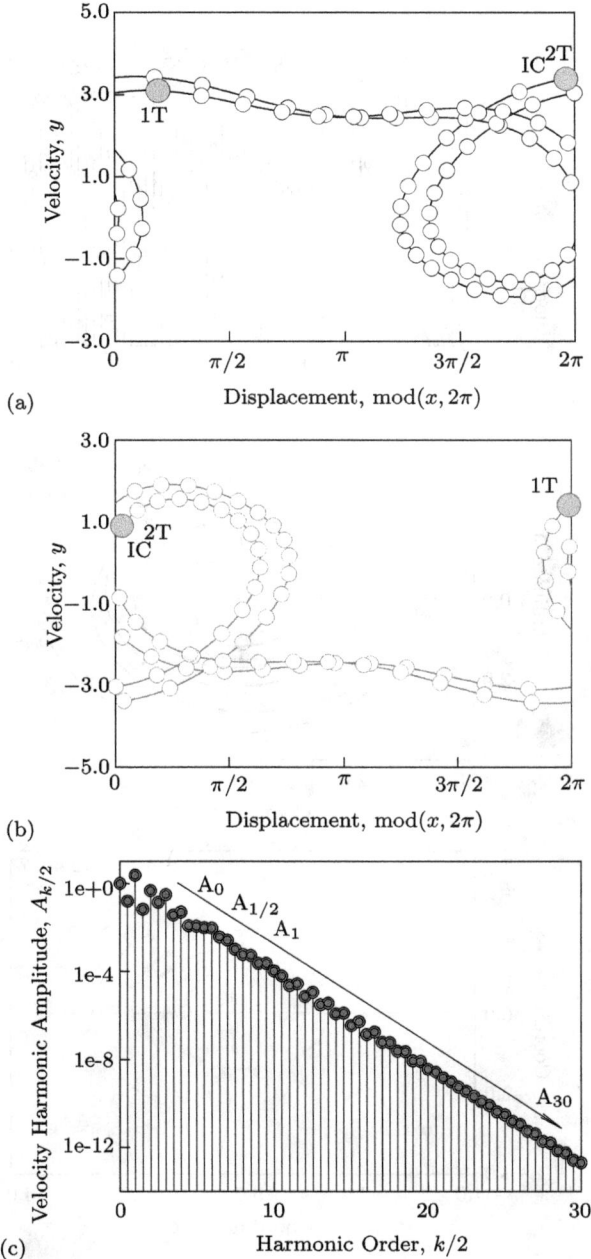

Fig. 6.28 Travelable asymmetric period-2 motions ($Q_0 = 3.2$): (a) trajectory of black branch (right), IC: $(x_0, y_0) \approx (6.1621, 3.3869)$; (b) trajectory of red branch (left), IC: $(x_0, y_0) \approx (0.0977, 0.9019)$, (c) velocity harmonic amplitudes, (d) velocity harmonic phases; (e) pendulum rotation pattern (black), (f) pendulum rotation pattern (red). Parameters ($\alpha = 1.5, \delta = 0.75, \Omega = 1$). $X = e^{0.1t} \cos x$, $Y = e^{0.1t} \sin x$.

(d)

(e)

(f)

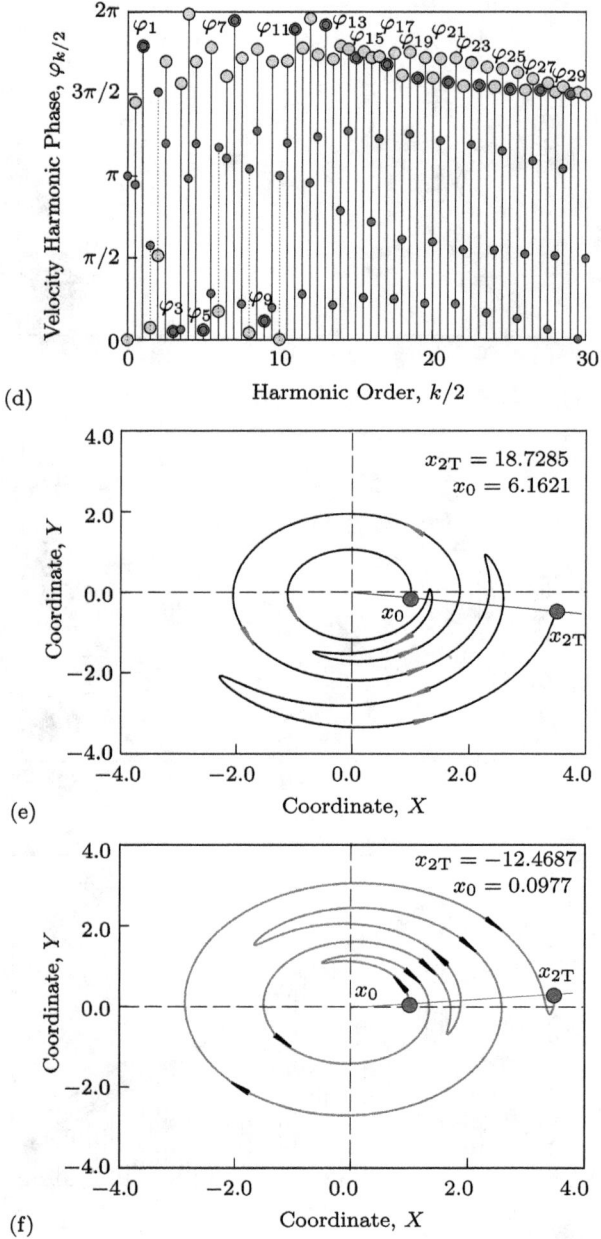

Fig. 6.28 Continued.

ity harmonic amplitudes are $A_{k/2} \in (10^{-13}, 10^{-2})$ for $k = 19, 20, \cdots, 80$. In Fig.6.29(d), velocity harmonic phases are presented for $\varphi_{k/2} \in [0, 2\pi)$. $\varphi_{k/2}^R = \mathrm{mod}\,((\varphi_{k/2}^B + k(1+2r)/2 + 1)\pi, 2\pi)$ for $r = 1$ with $t_0 = rT$ ($r \in \{0, 1\}$) and $k = 1, 2, \cdots, 80$. To illustrate displacement complexity of the travelable period-2 motions, the trajectories of fictitious coordinates in (X, Y)-space are presented in Figs.6.29(e) and (f) for the black and red branches of the travelable period-2 motions. The parameters in fictitious exponential function are $l_0 = 1$, $a = 0.1$. We have $x_0 = 1.2414$ and $x_{2T} = -5.0418$ for the black branch. The travelable period-2 motion with one negative cycle (-2π), including two positive rotations, three negative rotations plus four librations, is on the *negative* direction of $x(t)$. The negative travelable period-2 motion is of $(R_+ : R_- : L) = (2 : 3 : 4)$-kind. For the red branch, we have $x_0 = 0.1376$ and $x_T = 6.4208$ for the red branch. The travelable period-2 motion with one positive cycle (2π), including three positive rotations, two negative rotations plus four librations, is on the *positive* direction of $x(t)$. The positive travelable period-2 motion is of $(R_+ : R_- : L) = (3 : 2 : 4)$-kind.

On the bifurcation tree of travelable period-2 motion to chaos, after the period-doubling bifurcation of the travelable asymmetric period-2 motion, the travelable asymmetric period-4 motions will be obtained. Consider excitation amplitude of $Q_0 = 7.34$ for a pair of two travelable asymmetric period-4 motions. The two trajectories, velocity harmonic amplitude and phase of the travelable asymmetric period-4 motions are presented in Fig.6.30. The initial conditions are $(x_0, y_0) \approx (1.2883, 2.1487)$ (black) and $(x_0, y_0) \approx (4.8056, 4.5306)$ (red) from the analytical prediction. The trajectories of the two asymmetric period-4 motion in phase space are illustrated in Figs.6.30(a) and (b). The trajectories of period-4 motions are like the doubled period-2 motions. The velocity harmonic amplitudes of the two travelable asymmetric period-4 motions are presented in Fig.6.30(c). The two velocity constants for the travelable asymmetric period-4 motions are $-a_0^{(4)B} = a_0^{(4)R} = 0.5$ which is determined by angular velocity $\dot{x} = y$ and $y = \Omega/2 = 0.5$. The velocity harmonic amplitudes for two asymmetric period-4 motions are also of the same. For the two travelable asymmetric period-4 motions, the main velocity harmonic amplitudes are $A_{1/4} \approx 0.0205$, $A_{1/2} \approx 0.4829$, $A_{3/4} \approx 0.0285$, $A_1 \approx 5.8727$, $A_{5/4} \approx 0.0223$, $A_{3/2} \approx 0.2398$, $A_{7/4} \approx 0.0103$, $A_2 \approx 0.3567$, $A_{9/4} \approx 4.1114\mathrm{e}-3$, $A_{5/2} \approx 0.0564$, $A_{11/4} \approx 3.0961\mathrm{e}-3$, $A_3 \approx 0.0759$, $A_{13/4} \approx 3.3763\mathrm{e}-3$, $A_{7/2} \approx 0.1306$, $A_{15/4} \approx 7.3856\mathrm{e}-3$, $A_4 \approx 0.0946$, $A_{17/4} \approx 8.6219\mathrm{e}-3$, $A_{9/2} \approx 0.2366$, $A_{19/4} \approx 0.0108$, $A_5 \approx 0.1123$, $A_{21/4} \approx 0.0123$, $A_{11/2} \approx 0.0423$, $A_{23/4} \approx 6.9401\mathrm{e}-3$, $A_6 \approx 0.0865$, $A_{25/4} \approx 1.5894\mathrm{e}-3$, $A_{13/2} \approx 0.0876$, $A_{27/4} \approx 2.5433\mathrm{e}-3$, $A_7 \approx 0.0301$, $A_{29/4} \approx 3.3848\mathrm{e}-3$, $A_{15/2} \approx 5.2033\mathrm{e}-3$, $A_{31/4} \approx 1.8303\mathrm{e}-3$, $A_8 \approx 0.0173$, $A_{33/4} \approx 2.6111\mathrm{e}-4$, $A_{17/2} \approx 0.0122$, $A_{35/4} \approx 1.5399\mathrm{e}-4$, and $A_9 \approx 6.0248\mathrm{e}-3$. Expect for $A_{(4l-3)/4}$ and $A_{(4l-1)/4}$, the velocity harmonic amplitudes $A_{(4l-4)/4}$ and $A_{(4l-2)/4}$ for the period-4 motion are close to the velocity harmonic amplitudes $A_{2(l-1)/2}$ and $A_{(2l-1)/2}$. The other velocity harmonic amplitudes are $A_{k/4} \in (10^{-13}, 10^{-2})$ for $k = 37, 38, \cdots, 120$.

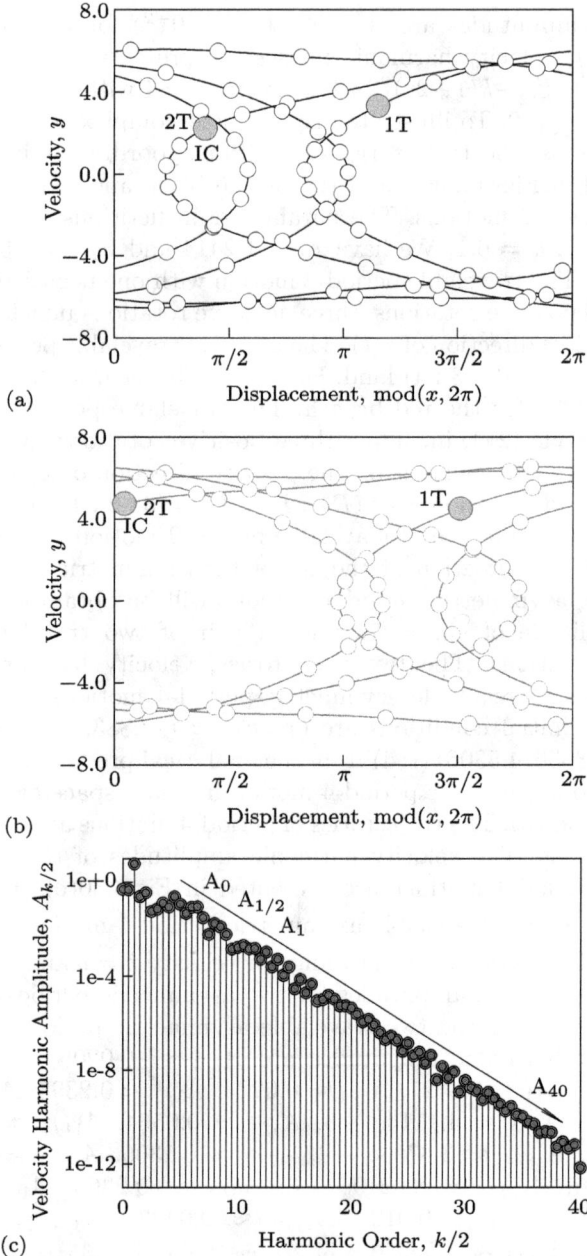

Fig. 6.29 Travelable asymmetric period-2 motions ($Q_0 = 7.37$): (a) trajectory of black branch (right), IC: $(x_0, y_0) \approx (1.2414, 2.2053)$; (b) trajectory of red branch (left), IC: $(x_0, y_0) \approx (0.1376, 4.7541)$, (c) velocity harmonic amplitudes, (d) velocity harmonic phases; (e) pendulum rotation pattern (black), (f) pendulum rotation pattern (red). Parameters ($\alpha = 1.5, \delta = 0.75, \Omega = 1$). $X = e^{0.1t} \cos x$, $Y = e^{0.1t} \sin x$.

(d)

(e)

(f)

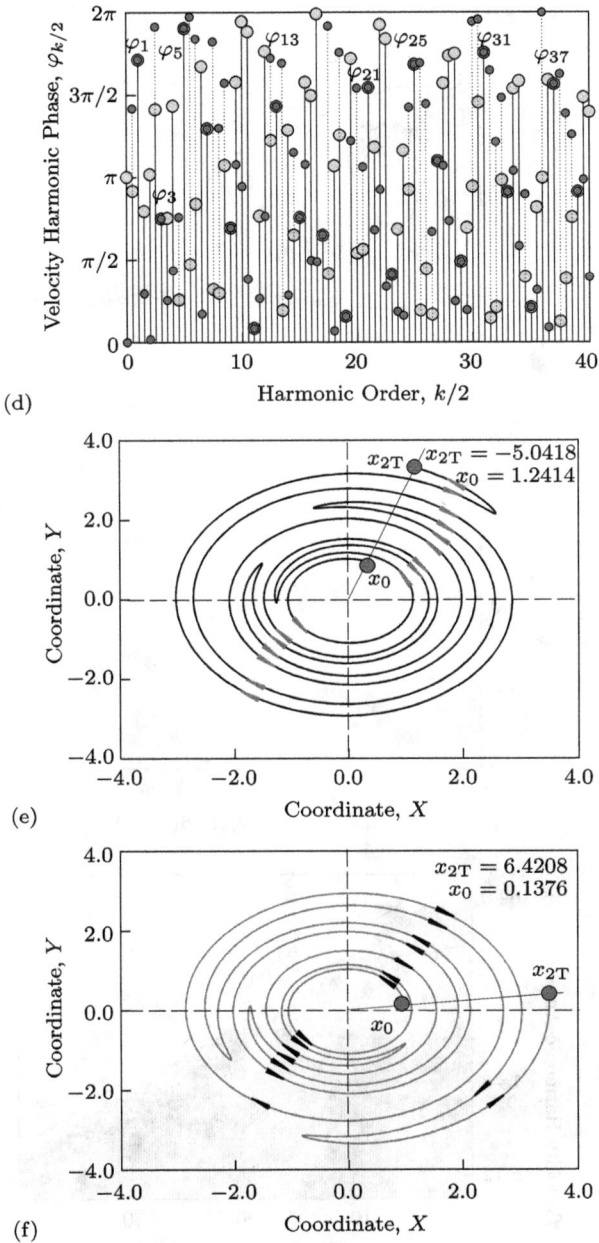

Fig. 6.29 Continued.

(a)

(b)

(c)

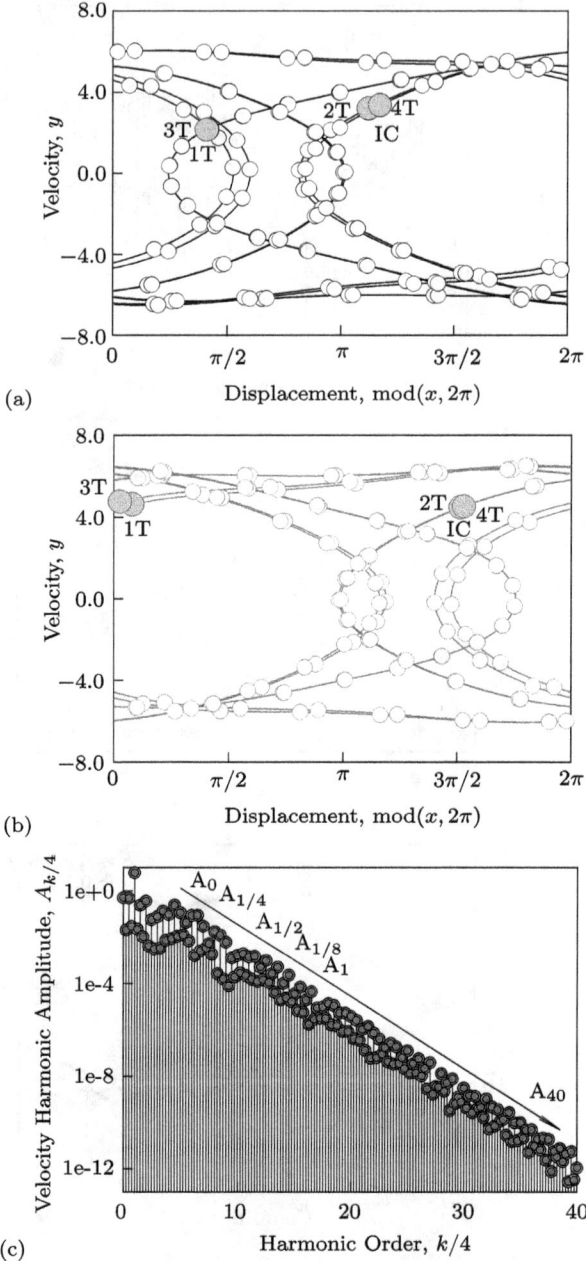

Fig. 6.30 Travelable asymmetric period-4 motions ($Q_0 = 7.34$): (a) trajectory of black branch (right), IC: $(x_0, y_0) \approx (1.2883, 2.1487)$; (b) trajectory of red branch (left), IC: $(x_0, y_0) \approx (4.8056, 4.5306)$, (c) velocity harmonic amplitudes, (d) velocity harmonic phases; (e) pendulum rotation pattern (black), (f) pendulum rotation pattern (red). Parameters ($\alpha = 1.5, \delta = 0.75, \Omega = 1$). $X = e^{0.05t} \cos x$, $Y = e^{0.05t} \sin x$.

(d)

(e)

(f)

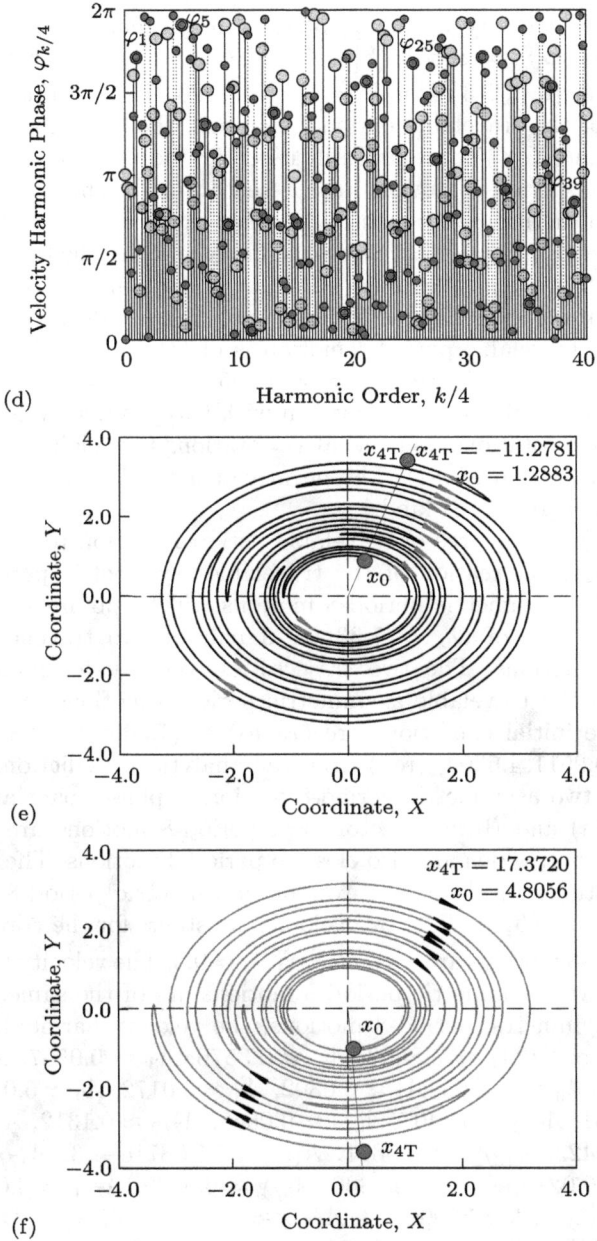

Fig. 6.30 Continued.

In Fig.6.30(d), velocity harmonic phases are presented and distributed in $\varphi_{k/4} \in [0, 2\pi)$. $\varphi_{k/4}^R = \mathrm{mod}(\varphi_{k/4}^B + k(1 + 2r)/4 + 1)\pi, 2\pi)$ with $r = 0$ with $t_0 = rT$ ($r \in \{0, 1, 2, 3\}$) for $k = 1, 2, \cdots, 120$. To illustrate displacement complexity of the travelable period-4 motions, the trajectories of fictitious coordinates in (X, Y)-space are still used for the black and red branch of the travelable period-1 motions, as presented in Figs.6.30(e) and (f). The parameters in fictitious exponential function are $l_0 = 1$, $a = 0.05$. We have $x_0 = 1.2883$ and $x_{4T} = -11.2781$ for the black branch. The travelable period-4 motion with two negative cycle (-4π), including four positive rotations, six negative rotations plus eight librations is on the *negative* direction of $x(t)$. The negative travelable period-4 motion is of $(R_+ : R_- : L) = (4 : 6 : 8)$-kind. For the red branch, we have $x_0 = 4.8056$ and $x_{4T} = 17.3720$ for the red branch. The travelable period-4 motion with two positive cycle (4π), including six positive rotations, four negative rotations plus eight librations, is on the *positive* direction of $x(t)$. The negative travelable period-1 motion is of $(R_+ : R_- : L) = (6 : 4 : 8)$-kind.

On the bifurcation tree of travelable period-2 motion to chaos, after the period-doubling bifurcation of the travelable asymmetric period-4 motion, the travelable asymmetric period-8 motions will be obtained. Consider excitation amplitude of $Q_0 = 7.33$ for a pair of two travelable asymmetric period-8 motions. The two trajectories, velocity harmonic amplitude and phase of the travelable asymmetric period-8 motions are presented in Fig.6.31. The initial conditions are $(x_0, y_0) \approx (1.3069, 2.1298)$ (black) and $(x_0, y_0) \approx (0.2511, 4.6304)$ (red) from the analytical prediction. The trajectories of the two asymmetric period-8 motion in phase space are illustrated in Figs.6.31(a) and (b). Trajectories of period-8 motions are the doubled period-4 motions, which are also close to period-4 motions. The velocity harmonic amplitudes of the two travelable asymmetric period-8 motions are presented in Fig.6.31(c). The two velocity constants for the travelable asymmetric period-8 motions are $-a_0^{(8)B} = a_0^{(8)R} = 0.5$. The velocity harmonic amplitudes for two asymmetric period-8 motions are of the same. For the two travelable asymmetric period-8 motions, the velocity harmonic amplitudes are four types: (i) $A_1 \approx 5.8666$, $A_2 \approx 0.3576$, $A_3 \approx 0.0767$, $A_4 \approx 0.0956$, $A_5 \approx 0.1126$, $A_6 \approx 0.0865$, $A_7 \approx 0.0302$, $A_8 \approx 0.0172$, $A_9 \approx 6.0406\mathrm{e} - 3$; (ii) $A_{1/2} \approx 0.4851$, $A_{3/2} \approx 0.2407$, $A_{5/2} \approx 0.0568$, $A_{7/2} \approx 0.1312$, $A_{9/2} \approx 0.2359$, $A_{11/2} \approx 0.0422$, $A_{13/2} \approx 0.0871$, $A_{15/2} \approx 5.0871\mathrm{e} - 3$, $A_{17/2} \approx 0.0121$; (iii) $A_{1/4} \approx 0.0219$, $A_{3/4} \approx 0.0306$, $A_{5/4} \approx 0.0239$, $A_{7/4} \approx 0.0110$, $A_{9/4} \approx 4.4229\mathrm{e} - 3$, $A_{11/4} \approx 3.3290\mathrm{e} - 3$, $A_{13/4} \approx 3.6368\mathrm{e} - 3$, $A_{15/4} \approx 7.9437\mathrm{e} - 3$, $A_{17/4} \approx 9.2623\mathrm{e} - 3$, $A_{19/4} \approx 0.0116$, $A_{21/4} \approx 0.0132$, $A_{23/4} \approx 7.4045\mathrm{e} - 3$, $7.4045\mathrm{e} - 3$, $A_{k/4} < 10^{-2}$ ($k = 4l - 1, 4l - 3$ and $l = 7, 8 \cdots$); (iv) $A_{k/8} < 10^{-2}$ ($k = 8l - 7, 8l - 5, 8l - 3, 8l - 1$ and $l = 1, 2 \cdots$). The other velocity harmonic amplitudes are $A_{k/8} \in (10^{-13}, 10^{-2})$ for $k = 73, 74, \cdots, 320$. In Fig.6.31(d), velocity harmonic phases are presented and distributed in $\varphi_{k/8} \in [0, 2\pi)$. $\varphi_{k/8}^R = \mathrm{mod}(\varphi_{k/8}^B + k(1 + 2r)/8 + 1)\pi, 2\pi)$ for $r = 1$ with $t_0 = rT$ ($r \in \{0, 1, \cdots, 7\}$) for $k = 1, 2, \cdots, 320$. To illustrate displacement

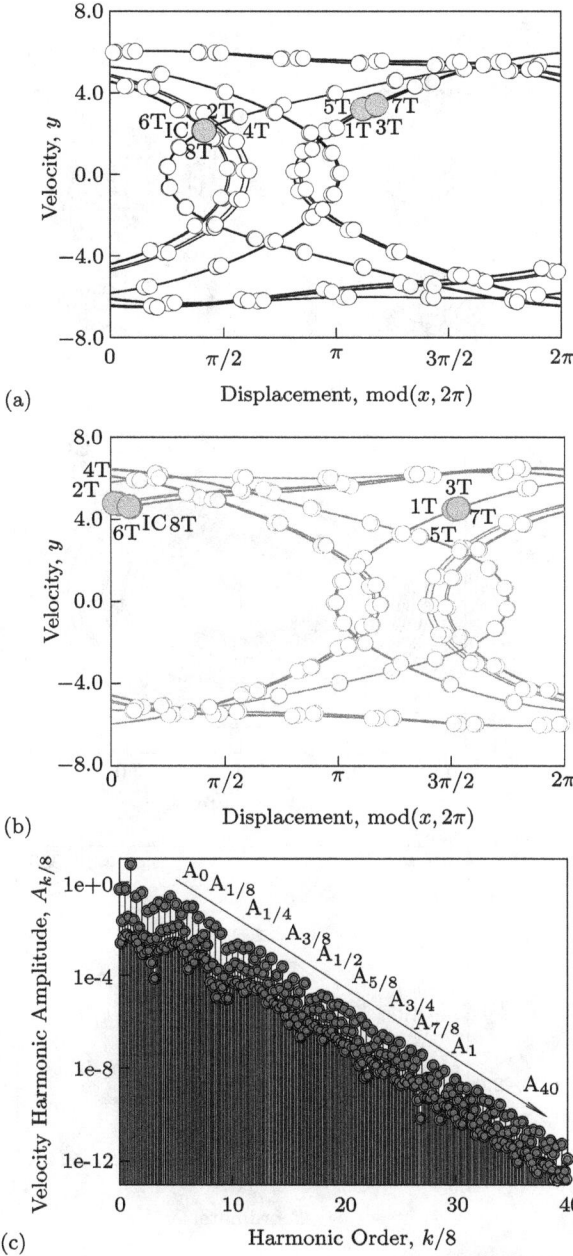

Fig. 6.31 Travelable asymmetric period-8 motions ($Q_0 = 7.33$): (a) trajectory of black branch (right), IC: $(x_0, y_0) \approx (1.3069, 2.1298)$; (b) trajectory of red branch (left), IC: $(x_0, y_0) \approx (0.2511, 4.6304)$, (c) velocity harmonic amplitudes, (d) velocity harmonic phases; (e) pendulum rotation pattern (black), (f) pendulum rotation pattern (red). Parameters ($\alpha = 1.5, \delta = 0.75, \Omega = 1$). $X = e^{0.025t} \cos x$, $Y = e^{0.025t} \sin x$.

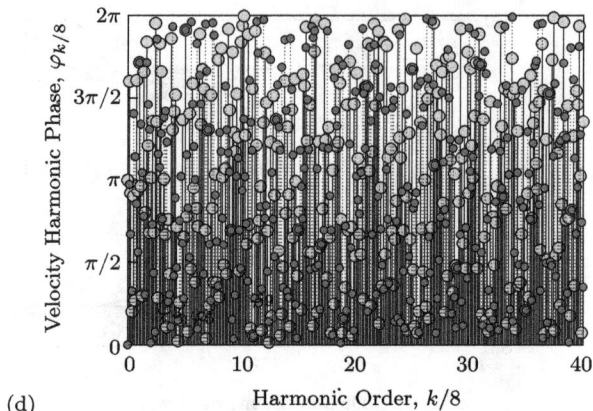

(d)

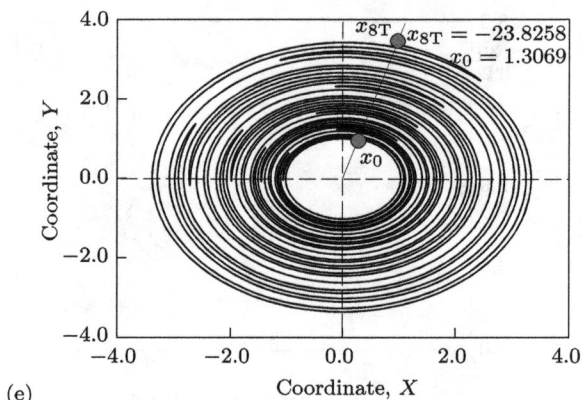

(e)

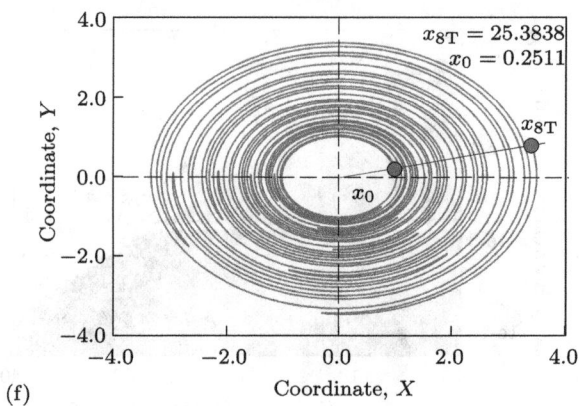

(f)

Fig. 6.31 Continued.

complexity of the travelable period-8 motions, the trajectories of fictitious coordinates in (X, Y)-space are still used for the black and red branch of the travelable period-8 motions, as presented in Figs.6.31(e) and (f). The parameters in fictitious exponential function are $l_0 = 1$, $a = 0.05$. We have $x_0 = 1.3069$ and $x_{4T} = -23.8258$ for the black branch. The travelable period-4 motion with two negative cycle (-8π), including 8 positive rotations, 12 negative rotations plus 16 librations is on the *negative* direction of $x(t)$. The negative travelable period-8 motion is of $(R_+ : R_- : L) = (8 : 12 : 16)$-kind. For the red branch, we have $x_0 = -0.2511$ and $x_{8T} = 25.3838$ for the red branch. The travelable period-4 motion with two positive cycle (8π), including 12 positive rotations, 8 negative rotations plus 16 librations, is on the *positive* direction of $x(t)$. The negative travelable period-1 motion is of $(R_+ : R_- : L) = (12 : 8 : 16)$-kind.

References

Guo, Y. and Luo, A.C.J., 2016, Routes of periodic motions to chaos in a periodically forced pendulum, *International Journal of Dynamics and Control*, DOI: 10.1007/s40435-016-0429-7.

Luo, A.C.J., 2015a, Periodic flows to chaos based on discrete implicit mappings of continuous nonlinear systems, *International Journal of Bifurcation and Chaos*, **25(3)**, Article No. 1550044 (62 pages).

Luo, A.C.J., 2015b, *Discretization and Implicit Mapping Dynamics*, HEP/Springer, Beijing/Dordrecht.

Luo, A.C.J. and Guo, Y., 2016, Periodic motions to Chao in pendulum, *International Journal of Bifurcation and Chaos*, **20(9)**, Article No. 1650159 (64pages).

Subject Index

www.ingramcontent.com/pod-product-compliance
Lightning Source LLC
Chambersburg PA
CBHW050638190326
41458CB00008B/2329